Nanofluid in Heat Exchangers for Mechanical Systems

Numerical Simulation

Nanofluid in Heat Exchangers for Mechanical Systems

Numerical Simulation

Zhixiong Li

MJU-BNUT Department-Joint Research Center on Renewable Energy and Sustainable Marine Platforms, Engineering Research Center of Fujian University for Marine Intelligent Ship Equipment, Minjiang University, Fuzhou, China.

School of Mechanical, Materials, Mechatronic and Biomedical Engineering, University of Wollongong, Wollongong, NSW, Australia

Ahmad Shafee

Institute of Research and Development, Duy Tan University, Da Nang, Vietnam

Iskander Tlili

Department of Mechanical and Industrial Engineering, College of Engineering, Majmaah University, Al-Majmaah, Saudi Arabia

M. Jafaryar

Renewable Energy Systems and Nanofluid Applications in Heat Transfer Laboratory, Babol Noshirvani University of Technology, Babol, Iran

ELSEVIER

Elsevier
Radarweg 29, PO Box 211, 1000 AE Amsterdam, Netherlands
The Boulevard, Langford Lane, Kidlington, Oxford OX5 1GB, United Kingdom
50 Hampshire Street, 5th Floor, Cambridge, MA 02139, United States

Notices

Library of Congress Cataloging-in-Publication Data
A catalog record for this book is available from the Library of Congress

British Library Cataloguing-in-Publication Data
A catalogue record for this book is available from the British Library

ISBN: 978-0-12-821923-2

For information on all Elsevier publications visit our website at
https://www.elsevier.com/books-and-journals

Publisher: Matthew Deans
Acquisitions Editor: Simon Holt
Editorial Project Manager: Fernanda A. Oliveira
Production Project Manager: Kiruthika Govindaraju
Cover Designer: Alan Studholme

Typeset by TNQ Technologies

Contents

About the Authors

Zhixiong Li

Dr. Zhixiong Li received his PhD in Transportation Engineering from Wuhan University of Technology, China, in 2013. Currently he is the co-director of MJU-BNUT Department, Joint Research Center on Renewable Energy and Sustainable Marine Platforms at the Engineering Research Center of Fujian University for Marine Intelligent Ship Equipment, Minjiang University, China; and he is also associated with the School of Mechanical, Materials, Mechatronic and Biomedical Engineering, University of Wollongong, Australia. His research interests include modeling of complex dynamic systems using artificial intelligence, intelligent manufacturing, and dynamic system optimization. This book is partially supported by NSFC (51979261) and Australia ARC DECRA (No. DE190100931).

Ahmad Shafee Hajizadeh

Ahmad Shafee Hajizadeh is a lecturer at the Public Authority of Applied Education and Training, College of Technological Studies, Applied Science department in Kuwait since 2010. Earlier he was at Kuwait University. His research interests are ordinary differential equations, special functions, nanofluid, CFD simulation, mesoscopic modeling, nonlinear science, magnetohydrodynamics, ferrohydrodynamics, electrohydrodynamics, and heat exchangers. He has written many papers. Also, he has a part-time position in Vietnam.

Iskander Tlili

Dr Iskander Tlili, Associate Professor at Majmaah University KSA, completed his MSc and PhD in Thermal Energy at the Laboratory Studies of Thermal and Energy Systems, National Engineering School of Monastir, Tunisia. He has more than 18 years of teaching and research experience in thermofluid, thermal power, renewable energy, and desalination. He participated in the implementation of several energy audits. Dr. Iskander has also published many papers in highly reputed journals. He has been conferred with internal and external grants from different international sponsors and led several units and committees at both college and university level.

Mehrdad Jafaryar

Mehrdad Jafaryar work at Renewable Energy Systems and Nanofluid Applications in Heat Transfer Laboratory, Babol, Iran. His research interests are CFD, experimental investigation, analytical solution, nanofluid, renewable energy, and aerodynamics. He has written several papers in various filed of mechanical engineering. Also, he has a book entitled as "*Nanofluid for Convective Heat Transfer in Various Geometries*" in LAMBERT publisher.

Preface

In this book, we provide readers various applications of heat exchanger, and finite volume method was applied for simulation purpose. To improve the performance of the system, turbulators have been used and nanofluid was considered as carrier fluid. This book is suitable for senior undergraduate students, postgraduate students, engineers, and scientists. In the first and second chapters, basic idea of heat exchanger and applications of nanofluid are examined. In Chapters 3–5, behaviors of nanofluid inside heat exchangers equipped with various types of twisted tapes are analyzed. In addition, entropy generation and exergy loss of nanofluid are investigated in these systems. In Chapter 6, melting process of phase change material enhanced with nanoparticles is simulated in two-dimensional enclosures. Air conditioning system with use of PCM is simulated in Chapter 7. Heat recovery system is presented in Chapter 8. Chapter 9 deals with ferrohydrodynamic effect on nanofluid flow in heat exchangers. Chapter 10 gives the reader applications of twisted tapes in solar flat plate system.

Zhixiong Li, Ahmad Shafee, Iskander Tlili, M. Jafaryar

CHAPTER

Fundamentals of heat exchangers

1

1.1 Introduction

1.1.1 Importance of heat exchangers

Heat exchangers have different applications ranging from conversion, recovery of thermal energy in different industrial, domestic, and commercial uses. Some public examples include cooling in thermal processing of chemical, condensation in power, agricultural products, pharmaceutical, steam generation, sensible heating, cogeneration plants, waste heat recovery, and fluid heating in manufacturing. Enhancement in heat exchanger's performance can make more economical design of heat exchanger which can aid to make energy, material, and cost savings related to a heat exchange process.

The importance of increasing the thermal performance of heat exchangers has caused development and use of many techniques termed as heat transfer enhancement. These methods augment convective heat transfer by reducing the thermal resistance in a heat exchanger. Utilization of augmentation techniques leads to increase in heat transfer coefficient but at the cost of increase in pressure drop. To reach high heat transfer rate while taking care of the augment pumping power, various techniques have been presented in recent decade. Recently, swirl flow devices have widely been used for increasing the convective heat transfer in various industries. This application is because of their low cost and easy setting up. The main aim of this chapter is to introduce the different ways to improve heat transfer performance. An extensive literature review of various turbulators (coiled tubes, extended surfaces (fin, louvered strip, winglet), rough surfaces (corrugated tube, rib), and swirl flow devices such as twisted tape, conical ring, snail entry turbulator, vortex rings, coiled wire) has been carried out.

1.1.2 Important definitions

In this part, a few significant terms usually used in heat transfer enhancement work are introduced. Thermal performance factor is commonly used to estimate the performance of different inserts such as wire coil, twisted tape, etc. It is a function of the heat transfer coefficient and the friction factor. The thermal performance factor of an insert device is good if this device can reach significant increase of

Nanofluid in Heat Exchangers for Mechanical Systems. https://doi.org/10.1016/B978-0-12-821923-2.00001-4

heat transfer coefficient with minimum increase of friction factor. This parameter is expressed as

$$\eta = \frac{Nu/Nu_0}{(f/f_0)^{\frac{1}{3}}} \tag{1.1}$$

where Nu, f, Nu_0, and f_0 are the Nusselt numbers and friction factors for a tube configuration with and without inserts, respectively.

The Nusselt number is a measure of the convective heat transfer occurring at the surface and is defined as hd/k, where h is the convective heat transfer coefficient, d is the diameter of the tube, and k is the thermal conductivity.

The friction factor is a measurement of pumping power. The friction factor for the tube with tabulators can be calculated from

$$f = \frac{\Delta P}{\left(\rho\, u^2/2\right)\left(L/d_H\right)} \tag{1.2}$$

where ΔP is the pressure drop across the test section, ρ is the density of fluid, d_H is the hydraulic diameter of the tube, u is the velocity of fluid, and L is the length of the tube.

The Prandtl number is introduced as the ratio of the diffusivity of momentum to the diffusivity of heat (υ/α). Pitch is defined as the distance between two points that are on the same plane, measured parallel to the axis of tape. The twist ratio is defined as the ratio of pitch to inside diameter of the tube.

1.2 Classification of enhancement techniques

Usually, heat transfer enhancement techniques are classified in three broad categories: active method, passive method, and compound method. The compound method uses complex design, so it has limited applications.

1.2.1 Active method

In these methods, some external power input needs in order to reach augment in the rate of heat transfer. Because of the need of equipment, this method has limited application in many practical applications. In comparison to the passive techniques, these techniques have not shown much potential as it is difficult to provide external power input in many cases. Various active techniques are as follows:

- **Mechanical aids:** These devices stir the fluid by mechanical means or by rotating the surface. Examples of the mechanical aids include rotating tube exchangers and scrapped surface heat and mass exchangers.
- **Surface vibration:** They have been used primarily in single-phase flows. A low or high frequency is applied to facilitate the surface vibrations which results in higher convective heat transfer coefficients.

- **Fluid vibration:** Instead of applying vibrations to the surface, pulsations are created in the fluid itself. This kind of vibration enhancement technique is employed for single phase flows.
- **Electrostatic fields:** Electrostatic field like electric or magnetic field or a combination of the two from DC or AC sources is applied in heat exchanger systems which induces greater bulk mixing, force convection, or electromagnetic pumping to enhance heat transfer. This technique is applicable in heat transfer process involving dielectric fluids.
- **Injection:** In this technique, same or other fluid is injected into the main bulk fluid through a porous heat transfer interface or upstream of the heat transfer section. This technique is used for single-phase heat transfer process.
- **Suction:** This method is used for both two-phase heat transfer and single-phase heat transfer process. Two-phase nucleate boiling involves the vapor removal through a porous heated surface, whereas in single-phase flows fluid is withdrawn through the porous heated surface.

1.2.2 Passive method

These methods need no external power and usually utilize geometrical or surface modifications to the flow tube by additional devices or joining inserts. These methods increase the rate of heat transfer by changing the flow treatment which also causes the pressure drop to increase.

1.2.3 Comparison between the active and passive thermal management

The grouping of higher powered components and more densely populated circuit boards are creating systems which also generate more heat than ever before. When components operate at excessive temperatures or are exposed to high localized temperatures, their functionality and operational life can be severely compromised. Prolonged operation in these conditions will ultimately lead to component degradation and failure. To ensure suitable performance and reliability, the excess heat must be transported away from critical components and system hot spots, and then dissipated into the ambient environment. This cooling process can be accomplished by applying the three fundamental methods of heat transfer from which all thermal solutions derive; conduction, convection, and radiation.

Conduction is the process by which heat flows from an area of higher temperature to one of lower temperature within a single medium—solid, liquid, or gas—in an attempt to equalize thermal differences. It is also used to describe heat transfer between media that are in direct physical contact such as two touching solids. Convection is the transfer of heat between a solid surface and a fluid (gas or liquid) due to the bulk motion of the fluid. In the case of natural convection, fluid absorbs heat from a nearby heated surface. It becomes warmer than the surrounding fluid and rises. Cooler fluid replaces the warmer fluid and a natural flow occurs. Radiation heat transfer results when thermal energy in the form of electromagnetic waves is emitted from a body due to its temperature. Emission and absorption is determined by the temperature and surface condition of the radiating source, and the temperature of the surrounding environment. Conduction and convection are the primary modes of heat transfer in heat exchanger with radiation typically removing only a small portion of the overall thermal load. While there are many different thermal devices and design strategies which can be used, they generally fall into two categories—passive or active. Comparison between the passive and active methods is shown in Table 1.1.

Passive thermal management refers to technologies that rely solely on the thermodynamics of conduction, convection, and radiation to complete the heat transfer process. These technologies are the most commonly used, the least expensive, and the easiest to implement. Passive techniques generally use surface or geometrical modifications to the flow by incorporating inserts or additional devices. They promote higher heat transfer coefficients by disturbing or altering the existing flow behavior (except for extended surfaces) which also leads to increase in the pressure drop. In case of extended surfaces, effective heat transfer area on the side of the extended surface is increased. Passive techniques hold the advantage over the active techniques as they do not require any direct input of external power. These techniques do not require any direct input of external power; rather they use it from the system itself which ultimately leads to an increase in fluid pressure drop.

Active thermal management refers to technologies that must introduce energy—typically from an external device—to augment the heat transfer process. A key benefit is that they increase the rate of fluid flow during convection which dramatically increases the rate of heat removal. These techniques are more complex from the use and design point of view as the method requires some

Table 1.1 Comparison between the passive and active methods.

Method	Approach	Description
Passive	Surface interruptions	Slits or offset fins interrupt the boundary layer, restarting it, creating secondary flows, and/or generating flow unsteadiness.
	Surface roughness	Accelerates transition from laminar flow to turbulent; also increases turbulent flow heat transfer.
	Surface protuberances	Ridges or three-dimensional shapes (cube, pyramid, etc.) generate secondary or unsteady flows.
Active	Forced flow unsteadiness	Surface vibration or sound waves thins or restarts boundary layer and/or induces secondary flows.
	Boundary layer injection	Enhancement primarily for multiphase flows.
	Boundary layer suction	Removal of boundary layer restarts boundary layer downstream.

external power input to cause the desired flow modification and improvement in the rate of heat transfer. It finds limited application because of the need of external power in many practical applications. In comparison to the passive techniques, these techniques have not shown much potential as it is difficult to provide external power input in many cases. In these cases, external power is used to facilitate the desired flow modification and the concomitant improvement in the rate of heat transfer.

1.2.4 The most popular passive methods

1.2.4.1 Coiled tubes

Coiled tubes are appropriate for more compact heat exchangers. These methods produce secondary flows which support higher heat transfer coefficient. Salimpour [1] performed an experimental investigation in order to study the heat transfer characteristics of temperature-dependent-property engine oil inside shell and coiled tube heat exchangers. They found that the coil-side heat transfer coefficients of the coiled tubes with larger pitches are less than those of the ones with smaller pitches; and the effect of pitch on Nusselt number is more discernible in high temperatures.

The forced convection from outside surfaces of helical coiled tubes was studied by Moawed [2]. He found that the average Nusselt number increases with the increase of D/d_0. Also, Nusselt number increases with the increase of P/d_0. Aly [3] has carried out a numerical study to investigate the heat transfer and pressure drop characteristics of water-based Al_2O_3 nanofluid flowing inside coiled tube-in-tube heat exchangers (Fig. 1.1). He showed that the friction factor increases with increase of curvature ratio, and there is no pressure drop penalty with increasing the nanoparticles volume concentration up to 2%.

1.2.4.2 Extended surfaces (Fin, Louvered strip, Winglet)

The aim of introducing these methods is to create swirl turbulence and secondary flow. The influence of baffle turbulators on heat transfer augmentation in a rectangular channel has been investigated by Sriromreun et al. [4] (Figs. 1.2 and 1.3). Their experimental results showed the significant effect of the presence of the Z-baffle on the heat transfer rate and friction loss over the smooth channel with

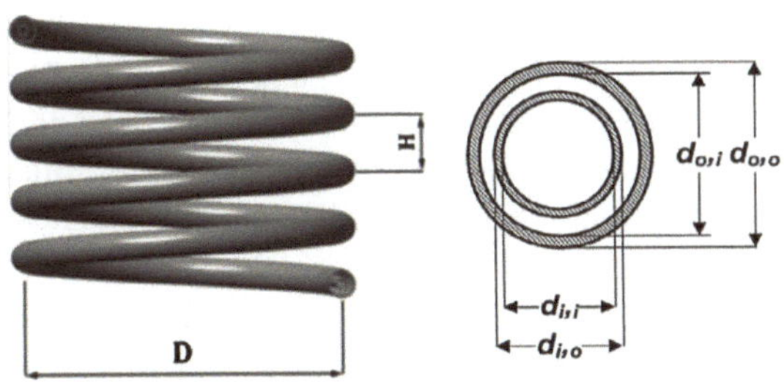

FIGURE 1.1

Schematic diagram and cross section of coiled tube-in-tube heat exchanger [3].

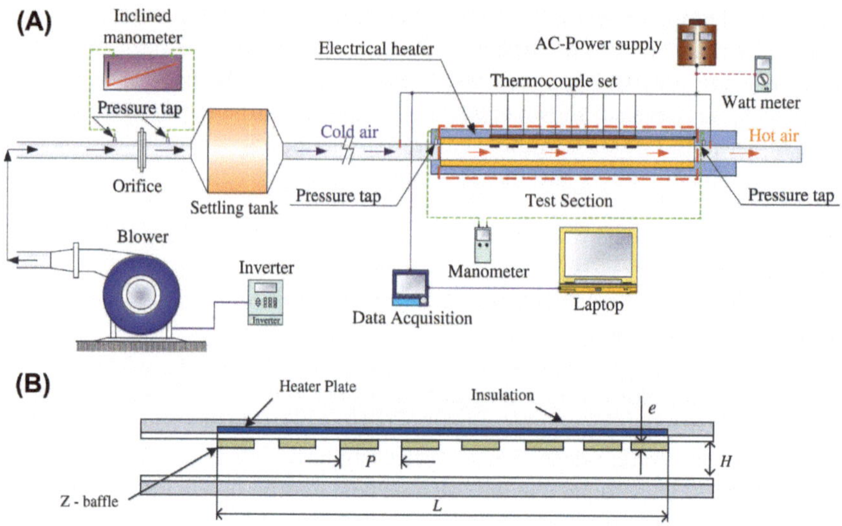

FIGURE 1.2

Schematic diagrams of (A) experimental apparatus and (B) test section [4].

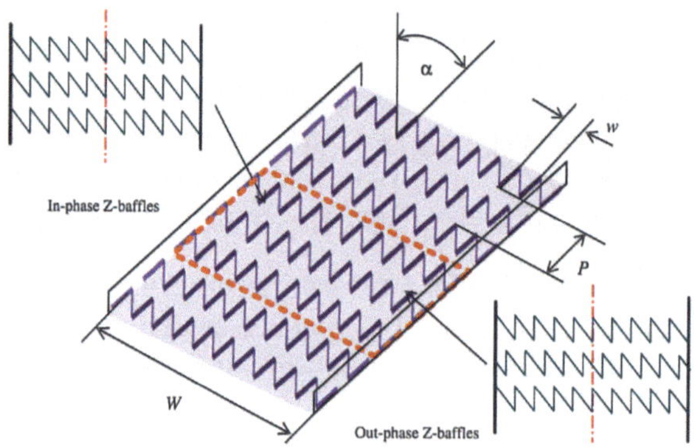

FIGURE 1.3

Test section with in-phase and out-phase Z-baffle arrangements [4].

no baffle. The Nusselt number, friction factor, and thermal performance enhancement factor for the in-phase 45 degrees Z-baffles are found to be considerably higher than those for the out-phase 45 degrees Z-baffle at a similar operating condition. A numerical investigation on periodic laminar flow and heat transfer behaviors in a three-dimensional isothermal wall square duct was presented by Kwankaomeng and Promvonge [5] (Fig. 1.4). They concluded that thermal enhancement factor for the angled baffle is much higher than unity and its maximum value is about 3.1 at BR = 0.3 and PR = 1.5, indicating higher thermal performance over the smooth duct.

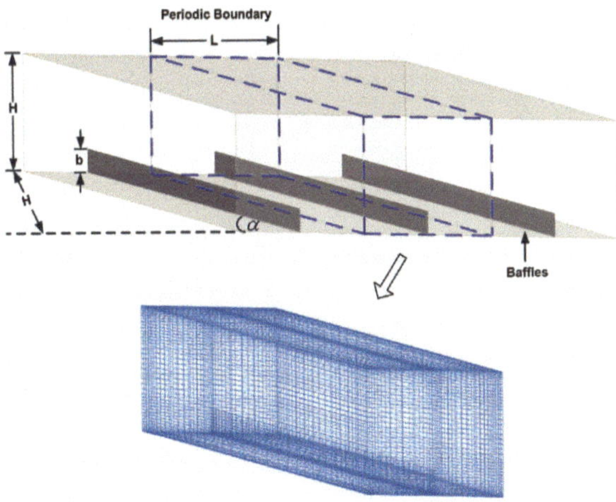

FIGURE 1.4

Duct geometry and computational domain of periodic flow [5].

Promvonge et al. [6] studied the laminar periodic flow and heat transfer in square channel with 45 in-line baffles on two opposite walls. They showed that thermal enhancement factors for both the 45 baffle arrays are found to be almost the same and much higher than unity and theirs maximum values are about 2.6 indicating higher thermal performance over the smooth channel. For the 45 baffles, the heat transfer enhancement is around 100%–200% higher than that for the 90 baffle, whereas the friction loss can be reduced at about 10%–150% (Fig. 1.5). Agyenim et al. [7] used fins to enhance rate of heat transfer in a medium temperature thermal energy storage heat exchanger (Fig. 1.6). Their experimental data have been shown to vindicate most models that ignore the thermal conductivity in the axial direction.

Mohammadi and Sabzpooshani [8] studied single-pass solar air heater with fins and baffles attached over the absorber plate (Fig. 1.7). They showed that attaching fins and baffles leads to noticeable increase in the outlet air temperature, useful energy gain, and efficiency in comparison with the simple air heater. With increasing the mass flow rate, the rate of enhancement of outlet temperature reduces sharply. Effect of porous baffles and flow pulsation on a double pipe heat exchanger performance was studied by Targui and Kahalerras [9] (Fig. 1.8). Their results revealed that the addition of an oscillating component to the mean flow affects the flow structure and enhances the heat transfer in comparison to the steady nonpulsating flow. The highest heat exchanger performance is obtained when only the flow of the hot fluid is pulsating.

Vermahmoudi et al. [10] studied the overall heat transfer coefficient of water-based iron oxide nanofluid in a compact air-cooled heat exchanger. They indicated that the overall heat transfer coefficient and the heat transfer rate of nanofluid have been improved with the enhancement in the nanofluid and air flow Reynolds

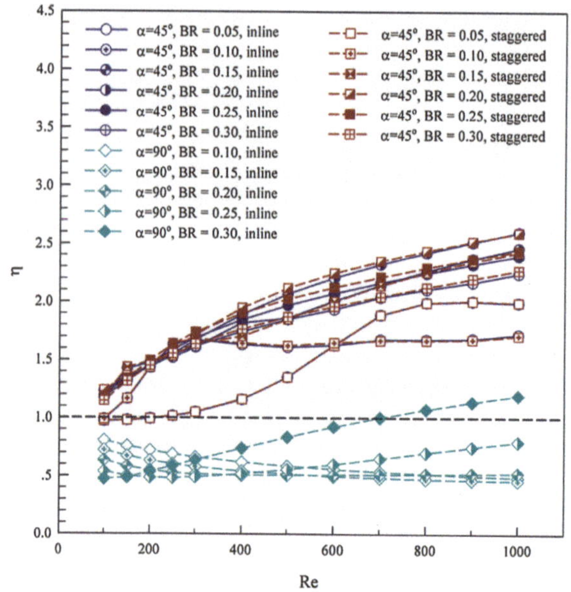

FIGURE 1.5

Comparison of thermal enhancement factor for baffles at various BRs [6].

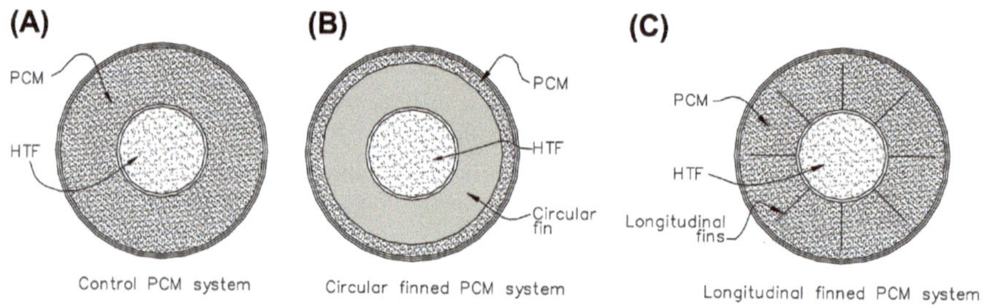

FIGURE 1.6

Cross-sectional views of the control and finned experimental concentric tube phase change material (PCM) systems [7].

numbers. The effect of using louvered strip inserts placed in a circular double pipe heat exchanger on the thermal and flow fields utilizing various types of nanofluid was studied by Mohammed et al. [11] (Fig. 1.9). They found that the Nusselt number is augmented by around four times for the louvered strip insert than that of the smooth tube. The forward louvered strip arrangement can promote the heat transfer by approximately 350%–400% at the highest slant angle of $\alpha = 30$ degrees and pitch of 30 mm, while the backward arrangement could improve the heat transfer by approximately 367%–411%.

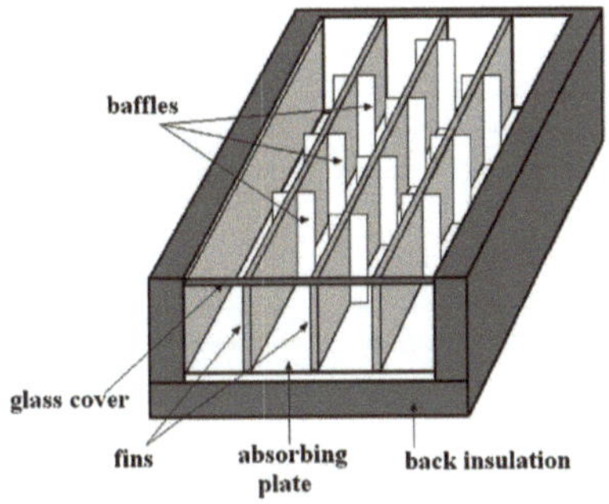

FIGURE 1.7

Schematic diagram of a single-pass upward-type solar air heater with fins and baffles attached over the absorber plate [8].

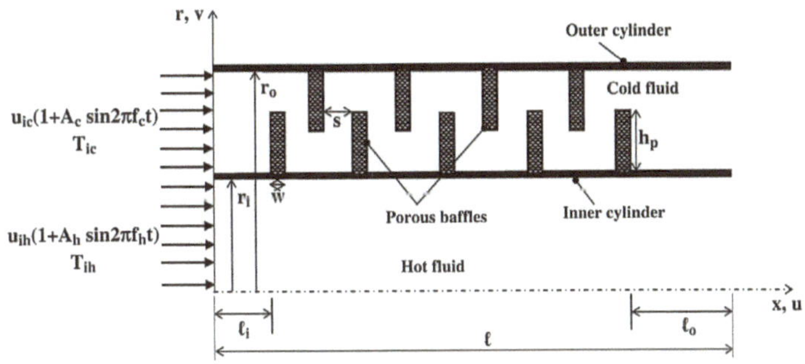

FIGURE 1.8

Physical domain [9].

Eiamsa-ard et al. [12] investigated the louvered strips inserted in a concentric tube heat exchanger. They confirmed that the use of louvered strips leads to a higher heat transfer rate over the plain tube. The use of the louvered strip with backward arrangement leads to better overall enhancement ratio than that with forward arrangement around 9%–24% (see Fig. 1.10). Kotcioglu et al. [13] investigated the second law of a cross-flow heat exchanger duct having winglets in the presence of a balance between the entropy generations. They concluded that increasing the cross-flow fluid velocity enhances the heat transfer rate and reduces the heat transfer irreversibility. Torii et al. [14] investigated two types of arrangement (staggered and in-line) of delta winglet-type vortex generators in a fin-tube heat exchanger (Fig. 1.11). The heat transfer was augmented by 10%–30% in case of staggered arrangement, and yet the pressure loss was reduced by 55%–34%. But, in case of in-line tube arrangement, these were found to be 10%–20% augmentation and 15%–8% reduction in pressure.

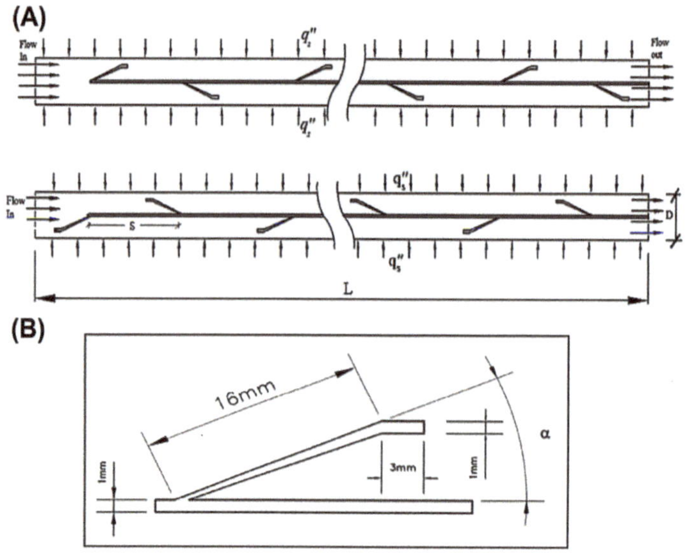

FIGURE 1.9

Schematic diagram of (A) a circular tube fitted with forward and backward arrangements of louvered strip inserts, and (B) geometry details of louvered strip insert [11].

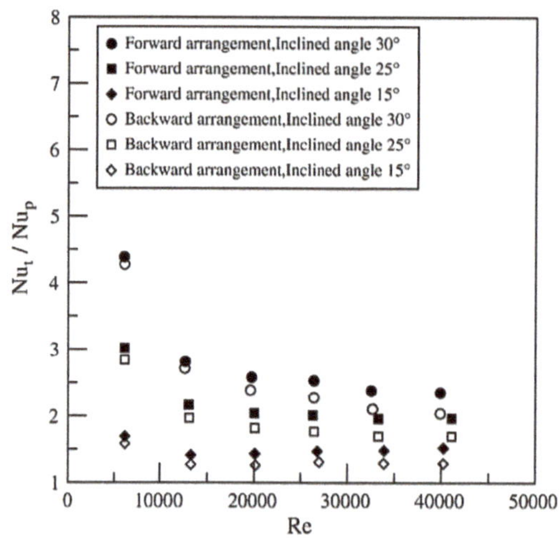

FIGURE 1.10

Relationship between Nusselt number ratio and Reynolds number [12].

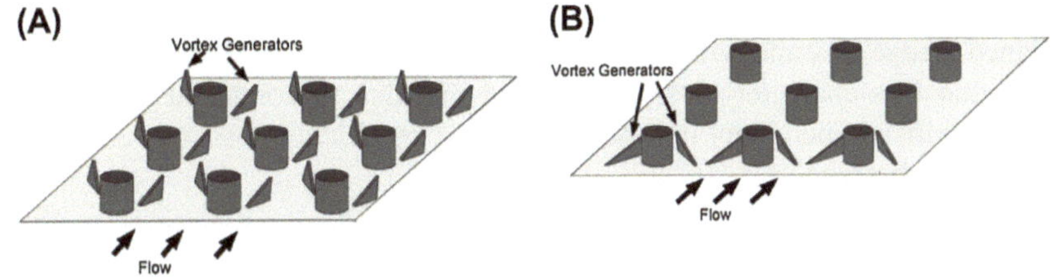

FIGURE 1.11

Configuration of winglet-type vortex generator on the fin surface-tube bank: (A) "common flow down" configuration; (B) "common flow up" configuration [14].

Zhou and Ye [15] investigated the thermohydraulic performance of curved trapezoidal winglet as shown in Fig. 1.12. They observed that curved trapezoidal winglet delta winglet had the best thermohydraulic performance in fully turbulent flow region. However, delta winglet had the best in laminar and transitional flow region. The convective heat transfer and friction behaviors of turbulent tube flow through a straight tape with double-sided delta wings (T-W) have been studied experimentally by Eiamsa-ard and Promvonge [16] (Fig. 1.13). Their experimental result revealed that for using the T-W, the increases in the mean Nusselt number and friction factor are, respectively, up to 165% and 14.8 times of the plain tube and the maximum

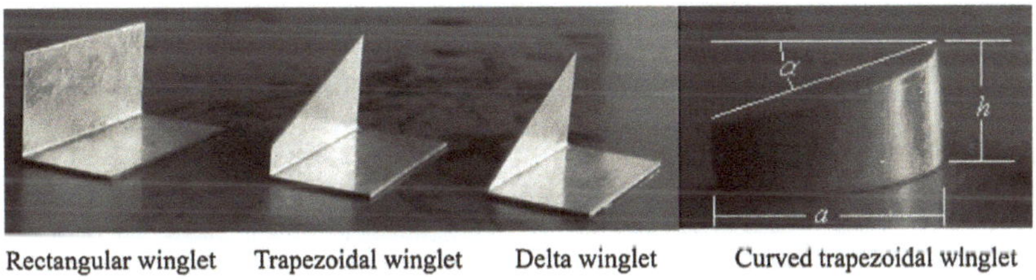

FIGURE 1.12

Pictorial diagram of vortex generators [15].

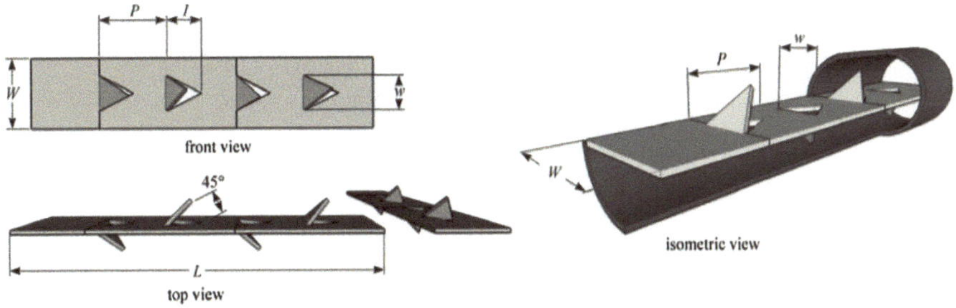

FIGURE 1.13

Geometry of tape [16].

thermal performance factor is 1.19. Chompookham et al. [17] studied the effect of combined wedge ribs and winglet-type vortex generators (WVGs) on heat transfer and friction loss characteristics. They found that due to combined ribs and the WVGs, significant increment in heat transfer rate and friction loss over the smooth channel was observed.

1.2.4.3 Rough surfaces (Corrugated tube, Rib)

Multiobjective optimal designs of high-efficiency corrugated tube heat exchangers applied in nuclear plants were presented by Han et al. [18] (Fig. 1.14). They revealed that augmenting the heat transfer performance caused by various design parameters in the optimum situation would lead to the increase of the resistance. The optimum values of $f_c/f_s = 1.22$ and $\eta = 1.42$ are obtained in the condition of specific (fixed) $Nu_c/Nu_s = 1.2$. Garcia et al. [19] analyzed hydrothermal behavior of three types of enhancement technique based on artificial roughness: corrugated tubes, dimpled tubes, and wire coils. Their results showed that the shape of the artificial roughness exerts a greater influence on the pressure drop characteristics than on the heat transfer augmentation.

Rainieri and Pagliarini [20] studied the thermal performances of corrugated wall tubes employed in a broad variety of industrial applications in order to intensify the convective heat transfer (Fig. 1.15). They showed that for not too low Reynolds

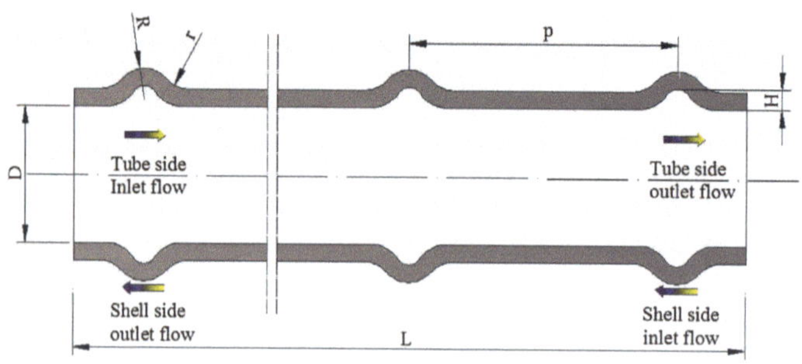

FIGURE 1.14

Structural parameters of the corrugated tube [18].

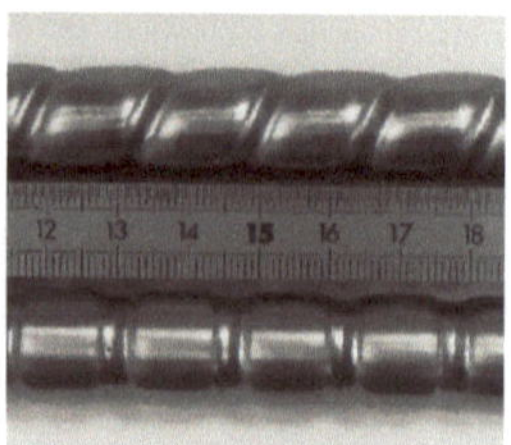

FIGURE 1.15

Helical and transverse corrugated tubes tested [20].

numbers, the helical corrugation induces significant swirl components to which, however, an as much significant heat transfer enhancement is not associated. Vicente et al. [21] studied the effect of corrugated tubes on heat transfer and friction characteristics. They showed that corrugated tubes present higher-pressure drop and heat transfer than the smooth tube under the same flow conditions. Increases from 20% to 300% in friction factor coefficient and up to 250% in Nusselt number were observed.

Buchlin [22] investigated experimentally perforated ribs in a channel flow at Reynolds numbers from 30,000 to 60,000. He tested five types of perforated ribs made in Plexiglas and found that the optimum design of ribs combines a rib pitch ratio of 5 with an open area ratio of 0.53. Naphon et al. [23] experimentally studied the heat transfer and pressure drop characteristics in horizontal double pipes with helical ribs (Fig. 1.16). Their results showed that the helical ribs have a significant effect on the heat transfer and pressure drop augmentations. The pressure drop across the tube with helical rib is produced by drag forces exerted on the flow field by the helical rib, flow blockage due to area reduction, turbulence augmentation, and rotational flow produced by the helical rib.

The heat transfer performances in discrete double inclined ribs tube (DDIR-tube) were numerically and experimentally investigated by Meng et al. [24] (Fig. 1.17). Their experimental results showed that the DDIR-tube has better comprehensive heat transfer performance than the current heat transfer enhancement tubes. Kathait and Patil [25] investigated the effect of (modified tube surface) discrete corrugated rib–roughened tube on heat transfer and frictional losses under varied fluid flow rates (Fig. 1.18). They found that heat transfer and friction losses are increased by

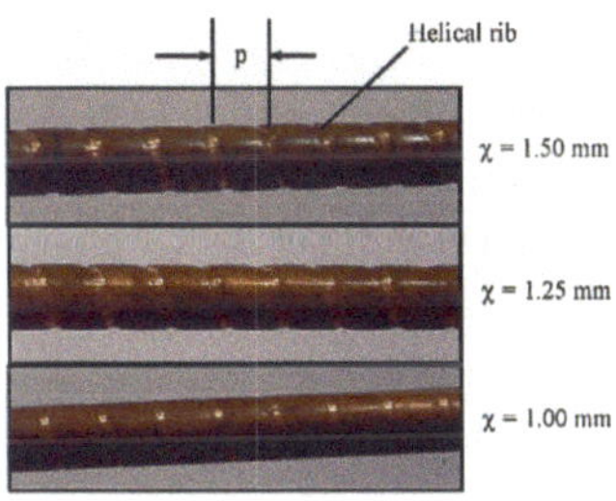

FIGURE 1.16

Photograph of the tube with helical rib used in the present study [23].

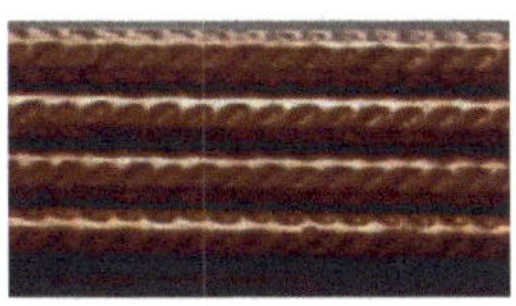

FIGURE 1.17

Photo of the novel enhanced heat transfer tube [24].

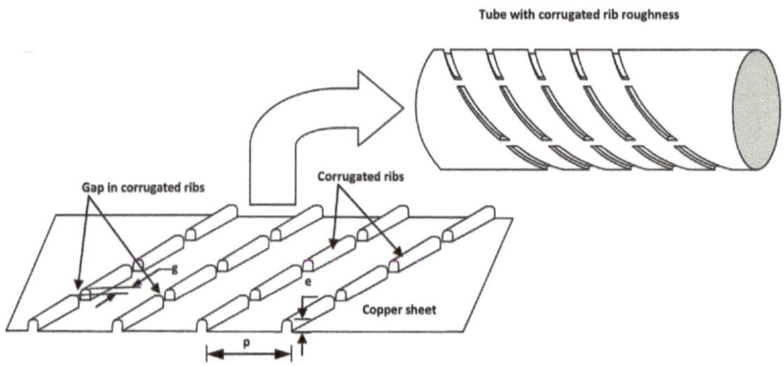

FIGURE 1.18

Typical geometry of circular tube with discrete corrugated ribs [25].

increasing the number of gaps in the corrugated roughness pattern irrespective of the change in Reynolds number. Li et al. [26] experimentally and numerically studied the turbulent heat transfer and flow resistance in an enhanced heat transfer in discrete double incline ribs (DDIR) tube (Fig. 1.19). Their results showed that the heat transfer in the DDIR tube is enhanced from 100% to 120% contrasted with a plain tube and the pressure drop is increased from 170% to 250%.

Effects of combined ribs and WVGs on forced convection heat transfer and friction loss behaviors for turbulent airflow through a constant heat flux channel were experimentally investigated by Promvonge et al. [27] (Fig. 1.20). They concluded that the values of Nusselt number and friction factor for utilizing both the rib and the WVGs are found to be considerably higher than those for using the rib or the WVGs alone. The larger the attack angle value leads to higher heat transfer and friction loss than the lower one. Promvonge et al. [28] conducted a numerical work to examine turbulent periodic flow and heat transfer characteristics in a three-dimensional square duct with in-line 60 degrees V-shaped discrete thin ribs placed on two opposite heated walls (Fig. 1.21). They showed that the rise of BR (= e/D) values leads to the increase in heat transfer and friction loss. The maximum thermal performance is around 1.8 for the rib with BR = 0.0725 where the heat transfer rate is about 4.0 times above the smooth duct at lower Reynolds number (Fig. 1.22).

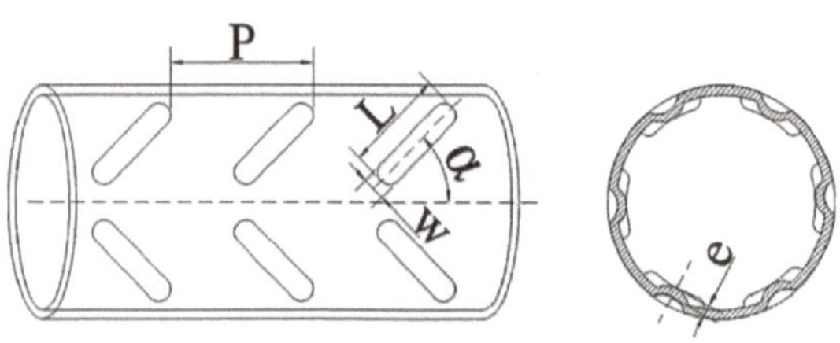

FIGURE 1.19

Structure of the DDIR tube [26].

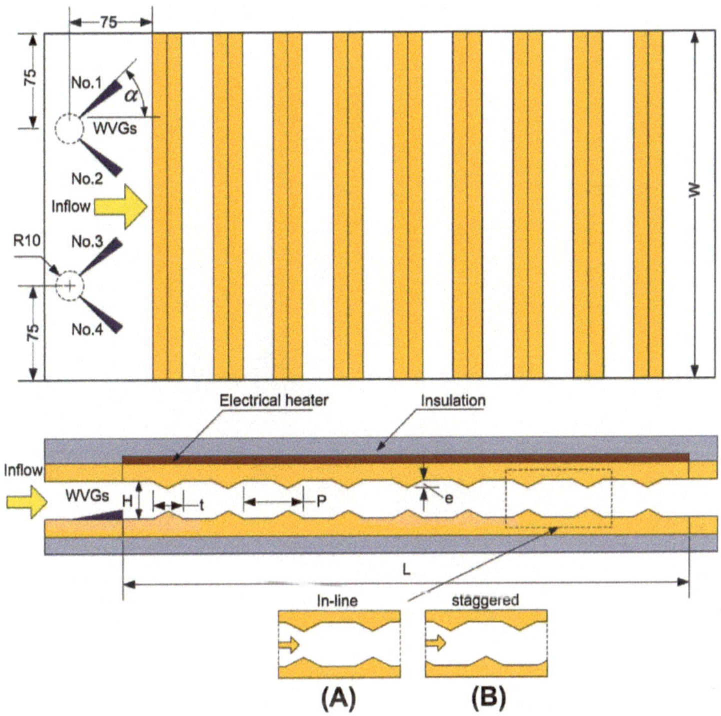

FIGURE 1.20

Test section with winglet-type vortex generators (WVGs): (A) in-line and (B) staggered rib arrangements [27].

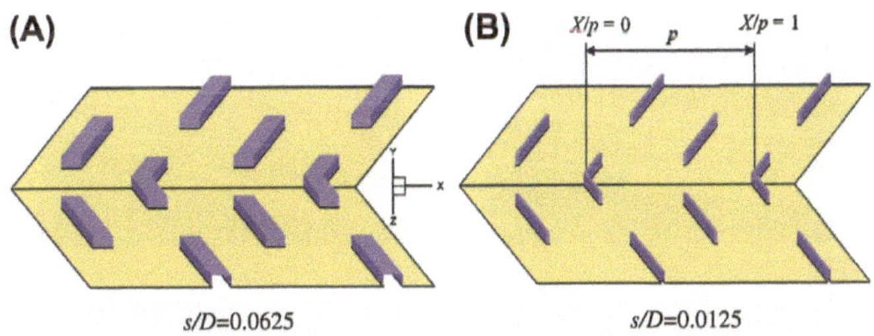

FIGURE 1.21

Arrangement of V-discrete (A) square and (B) thin ribs on lower wall [28].

Nuntadusit et al. [29] presented the heat transfer and flow characteristics in a channel with different types of transverse perforated ribs (Fig. 1.23). They showed that both inclined perforated ribs (with $\theta = 15$ degrees and 30 degrees) significantly improve heat transfer immediately downstream from the ribs when compared to straight perforated and solid ones, resulting in superior overall heat transfer

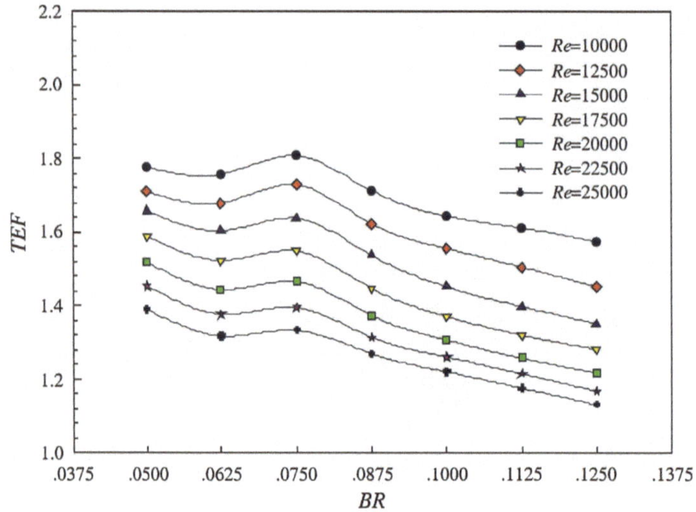

FIGURE 1.22

Effect of BRs on thermal efficiency factor (TEF) [28].

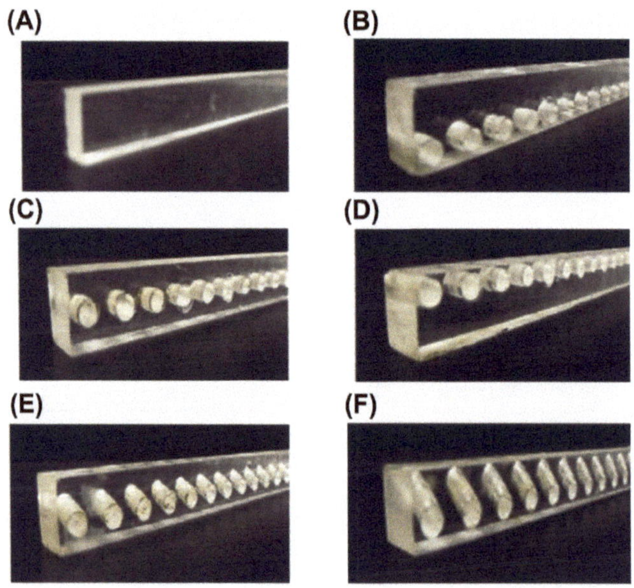

FIGURE 1.23

Photographs of solid and perforated ribs: (A) solid rib, (B) straight perforated rib ($h = 0.2H$), (C) straight perforated rib ($h = 0.5H$), (D) straight perforated rib ($h = 0.8H$), (E) inclined perforated rib ($\theta = 15$ degrees), and (F) inclined perforated rib ($\theta = 30$ degrees) [29].

performance. Kamali and Binesh [30] studied turbulent heat transfer and friction in a square duct having three different shaped ribs (square, triangular, and trapezoidal). They observed that heat transfer coefficient was strongly dependent on shape of ribs and heat transfer enhancement.

1.2.4.4 Swirl flow devices

These devices produce secondary recirculation on the axial flow for single-phase or two-phase flow heat exchanger. In this part, we investigate the previous work in which authors used twisted tape, conical ring, snail entry turbulator, vortex rings, coiled wire, and other kinds of turbulator.

1.2.4.4.1 Twisted tape

Twisted tapes are the metallic strips twisted with some appropriate techniques at preferred shape and dimension, inserted in the flow. The twisted tape inserts are widely used in heat exchangers for heat transfer augmentation because twisted tape inserts increase heat transfer rates with less friction factor penalty on pumping power. The most common twisted tape can be classified into the following eight categories.

1.2.4.4.1.1 Typical twisted tape These tapes have length equal to the length of exchanger tube. An experimental investigation for a solar water heater with twisted tape inserts having twist pitch to tube was studied by Kumar and Prasad [31]. They found that twisted tapes generate turbulency superimposed with swirlness inside the flow tube and consequently result in enhanced heat transfer. Decreasing the values of twist ratio leads to increasing values of heat transfer rate and the pressure drop. Esmaeilzadeh et al. [32] investigated heat transfer and friction factor characteristics of γ-Al_2O_3-water nanofluid through circular tube with twisted tape inserts with various thicknesses at constant heat flux (Fig. 1.24). They showed that the use of twisted tapes increases friction factor due to larger contact surface and reduction of fluid-free flow area which causes high-speed swirl flow. Finally, the convective heat transfer enhancement outweighs the effect of friction factor increase, leading enhanced thermal performance. An experimental study of fully developed turbulent flow in a dimpled tube in conjunction with a twisted tape has been performed

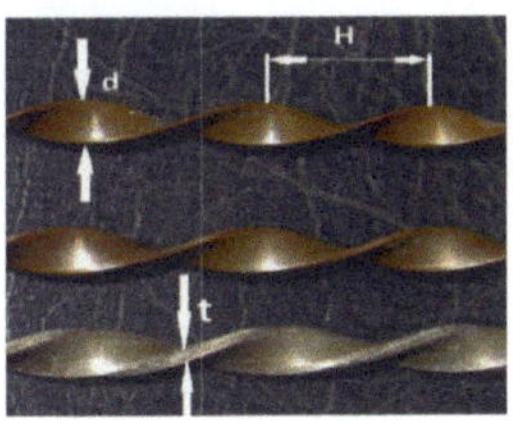

FIGURE 1.24

Schematic layout of inserts [32].

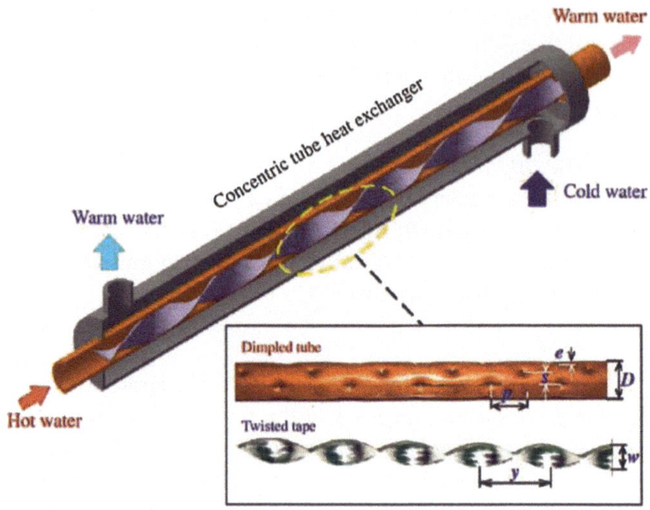

FIGURE 1.25

Dimpled tube and twisted tape [33].

by Thianpong et al. [33] (Fig. 1.25). They found that the heat transfer and friction factor increase with decreasing pitch ratio and twist ratio.

1.2.4.4.1.2 Helically twisted tapes Eiamsa-ard et al. [34] studied the heat transfer enhancement attributed to helically twisted tapes (HTTs) (Fig. 1.26). They observed

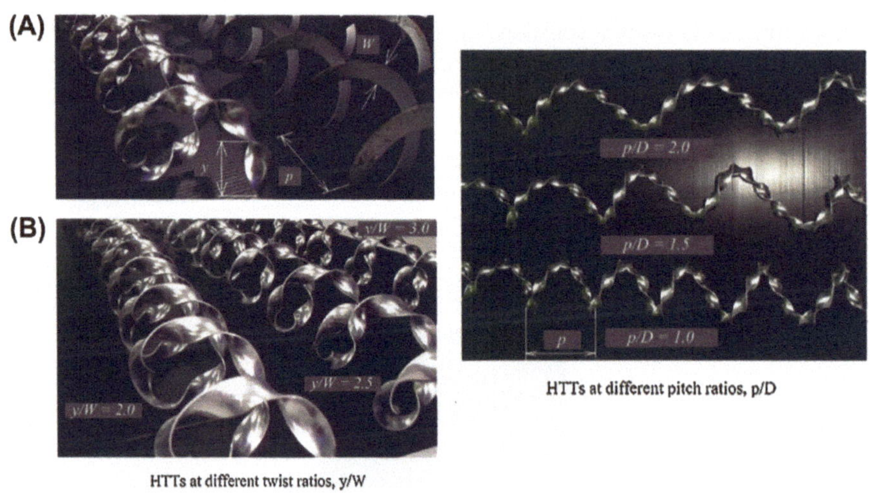

FIGURE 1.26

Photographs of helically twisted tapes (HTTs) and conventional helical tapes (CHTs): (A) geometries of HTT and CHT and (B) geometric details of HTTs [34].

FIGURE 1.27

Twisted tape geometry with twist ratio 3 (helical, left–right) [35].

that heat transfer rate and friction factor increase as the tape twist ratio and helical pitch ratio decrease, while the thermal performance shows opposite trend. Experimental investigation of heat transfer, friction factor, and thermal performance of solar water heater system fitted with helical and left–right twist has been performed by Jaisankar et al. [35] (Fig. 1.27). They showed that the helical twisted tape induces swirl flow inside the riser tubes unidirectional over the length. But, in left–right system the swirl flow is bidirectional which increases the heat transfer and pressure drop when compared to the helical system.

1.2.4.4.1.3 Varying length, alternate-axes and pitches twisted tape Sarada et al. [36] studied the enhancement of heat transfer using varying width twisted tape inserts. They showed that the enhancement of heat transfer with twisted tape inserts as compared to plain tube varied from 36% to 48% for full width and 33%–39% for reduced width inserts. This enhancement is mainly due to the centrifugal forces resulting from the spiral motion of the fluid. The thermohydraulic characteristics of the circular tubes equipped with alternate clockwise and counterclockwise twisted tapes (TA) for the Reynolds number ranging from 830 to 1990 were reported by Wongcharee and Eiamsa-ard [37] (Fig. 1.28). They found that thermal performance increases with increase of Reynolds number and decrease of pitch ratio. An experimental study on Nusselt number, friction factor, and enhancement efficiency characteristics in a round tube with short-length twisted tape insert under uniform wall heat flux boundary conditions was investigated by Eiamsa-ard and Thianpong [38] (Fig. 1.29). The heat transfer enhancement efficiency (η) is higher than unity only for the full-length tape insert at low Reynolds number.

1.2.4.4.1.4 Multiple twisted tapes Bhuiya et al. [39] presented a study that explored the effects of the double counter twisted tapes on heat transfer and fluid friction characteristics in a heat exchanger tube. Their experimental results demonstrated that the Nusselt number, friction factor, and thermal enhancement efficiency were increased with decreasing twist ratio. Their results also revealed that the heat transfer rate in the tube fitted with double counter twisted tape was significantly increased with corresponding increase in pressure drop. The maximum thermal enhancement efficiency of 1.34 was achieved by the use of double counter twisted tapes at constant blower power. Turbulent convective heat transfer characteristics in a helical ribbed tube fitted with twin twisted tapes have been investigated experimentally by Promvonge et al. [40] (Fig. 1.30). Their experimental results revealed that the co-swirling inserted tube performs much better than the ribbed/smooth tube alone at a similar operating condition. The co-swirl tube at $Y \approx 8$ yields the

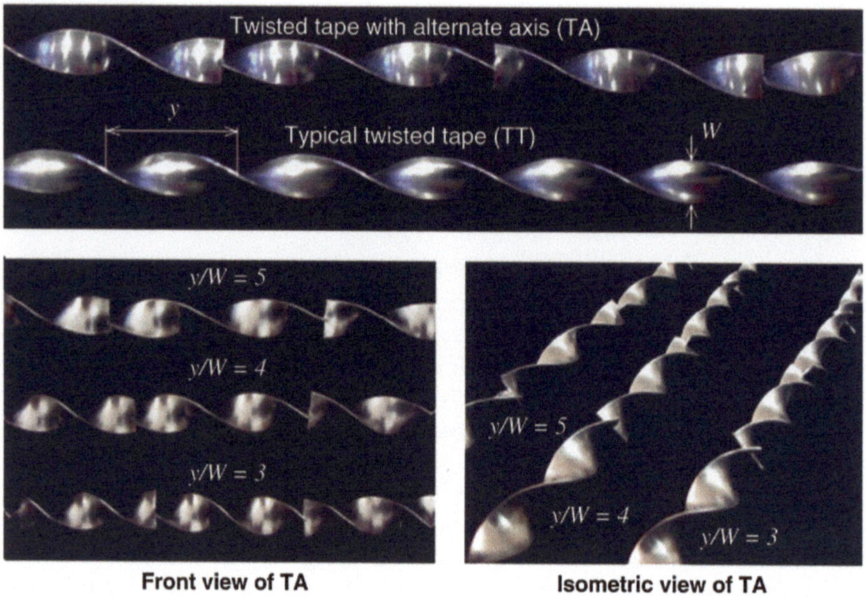

FIGURE 1.28

The twisted tapes used in the present work [37].

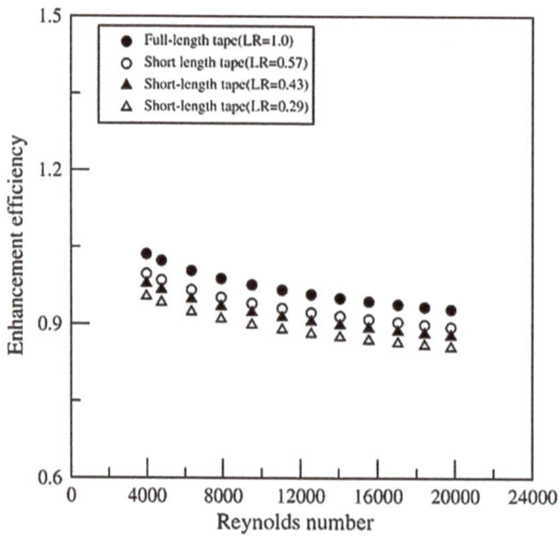

FIGURE 1.29

Heat transfer enhancement efficiency versus Reynolds number [38].

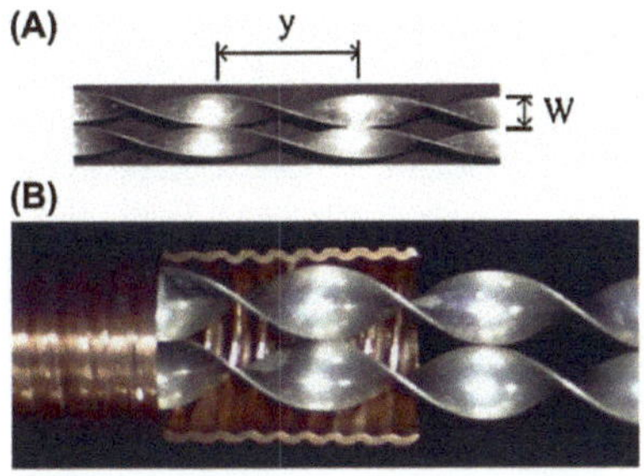

FIGURE 1.30

Details of (A) double twisted tapes and (B) helical ribbed tube with double twisted tape insert [40].

highest thermal performance at lower Reynolds number (Re). The influences of triple twisted tapes on heat transfer rate, friction factor, and thermal enhancement efficiency were experimentally investigated by Bhuiya et al. [41]. They showed that the thermal enhancement efficiency for all the cases was more than one, which indicated that the effect of heat transfer enhancement due to the enhancing tool was more dominant than the effect of the rising friction factor and vice versa.

1.2.4.4.1.5 Twisted tape with rod and varying spacer Jaisankar et al. [42] studied heat transfer and friction factor characteristics of thermosyphon solar water heater system fitted with spacer at the trailing edge of twisted tapes. They showed that the decrease in heat transfer augmentation for twist fitted with rod and spacer as compared to full length twist is nearly 17% and 29%, respectively. Also they showed that use of twist fitted with rod instead of full-length helical twist has greater advantage in friction factor with less impact on heat enhancement. Effects of insertion of a helical screw tape with or without core rod in a concentric double tube heat exchanger on heat transfer and flow friction characteristics were experimentally investigated by Eiamsa-ard and Promvonge [43]. They found that the enhancement efficiency for the helical screw-tape without core-rod is seen to be higher than that with core-rod at about two times. Ibrahim [44] used helical screw tape inserts in order to augmentation of laminar flow and heat transfer in flat tubes. Their results indicated that the helical screw element of different twist ratio and helical screw inserts with different spacer length have an effect on the results of heat transfer coefficient, friction factor. The averaged Nusselt number increases with the increase in Reynolds number and with the decrease in twist ratio and spacer length. Influence of helical tapes inserted in a tube on heat transfer enhancement was studied experimentally by Eiamsa-ard and Promvonge [45] (Fig. 1.31). They showed that the maximum mean Nusselt number may be increased by 160% for the full-length helical tape with centered rod, 150% for the full-length helical tape without rod, and 145% for the regularly spaced helical tape, $s = 0.5$, in comparison with the plain tube (Fig. 1.32).

1.2.4.4.1.6 Twisted tape with attached fins and baffles The effects of the twisted tapes consisting of center wings and alternate axes (WT-A) on thermohydraulic

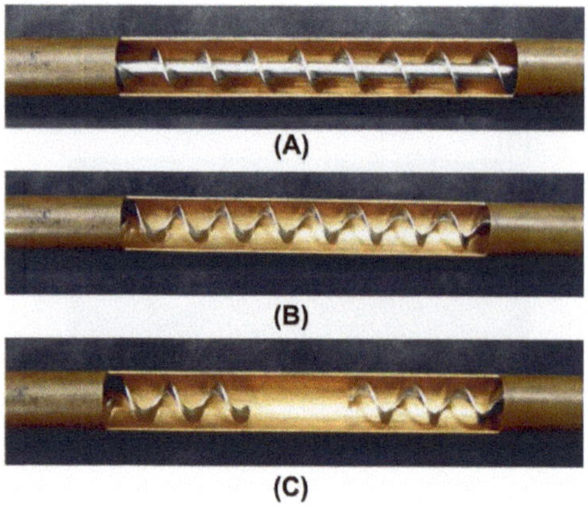

FIGURE 1.31

The inner tube fitted with various helical geometries insert: (A) Full-length helical tape with a rod; (B) Full-length helical tape without a rod; (C) Regularly spaced helical tape without a rod [45].

properties in a round tube were investigated by Eiamsa-ard et al. [46] (Fig. 1.33). They showed that the combined actions of the wing and alternate axis in the WTA resulted in a better fluid mixing and thus heat transfer enhancement compared to those induced by wing alone (WT) or alternate axis alone (T-A). Their results also revealed that Nusselt number increased with the increasing angle of attack. Ghadirijafarbeigloo et al. [47] studied the effect of a new perforated louvered twisted tape on the heat transfer coefficient and friction factor for an absorber tube of a solar parabolic trough collector. They observed high Nusselt number and friction factor for louvered twisted tape with respect to plain tube.

1.2.4.4.1.7 Twisted tapes with slots, holes, cuts Thianpong et al. [48] studied the effect of perforated twisted tapes with parallel wings on heat transfer enhancement in a heat exchanger tube (Fig. 1.34). Their results showed that heat transfer and friction factors were significantly influenced by the presence of wings and holes on perforated twisted tapes (PTTs). Both heat transfer and friction increased with the increase of wing depth ratio and the decrease of perforation hole diameter ratio. Due to the dominant effect of increased heat transfer over that of increased friction factor, the thermal performance factor was found to be increased as wing depth ratio increased and hole diameter ratio decreased. The enhancement of the heat transfer in a tube fitted with delta-winglet twisted tapes which act as swirl generator was experimentally investigated by Eiamsa-ard et al. [49]. Influences of the oblique delta-winglet twisted tape (O-DWT) and straight delta-winglet twisted tape (S-DWT) arrangements were described. They found that the O-DWT is a more effective turbulator giving higher heat transfer coefficient than the S-DWT. Shabanian et al. [50] reported the experimental and computational fluid dynamics (CFD) modeling studies on heat transfer, friction factor, and thermal performance

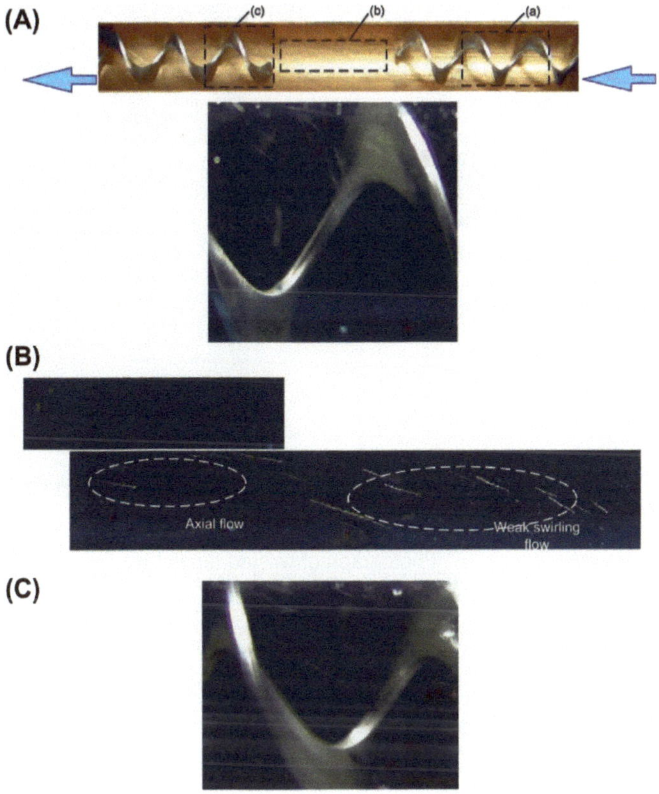

FIGURE 1.32

Photograph of the flow in a circular tube fitted the helical tape without a rod: (A) Swirling flow created by upstream helical tape; (B) Swirling flow decaying in the free space; (C) Swirling flow recreated by downstream helical tape [45].

of an air cooled heat exchanger equipped with three types of tube insert including butterfly, classic, and jagged twisted tape. They showed that in comparison between the jagged inserts and butterfly inserts, it was found that there were higher performances by the butterfly inserts in terms of thermal performance factor. In the studied range of Reynolds number, the thermal performance factor of butterfly insert varied between 1.28 and 1.62, while it was between 1 and 1.23 for the jagged inserts (Fig. 1.35).

Murugesan et al. [51] studied the effect of V-cut twisted tape insert on heat transfer, friction factor, and thermal performance factor characteristics in a circular tube. They found that the V-cut twisted tape offered a higher heat transfer rate, friction factor, and also thermal performance factor compared to the plain twisted tape. In addition, the influence of the depth ratio was more dominant than that of the width ratio for all the Reynolds number. The effects of twisted tapes with alternate axes

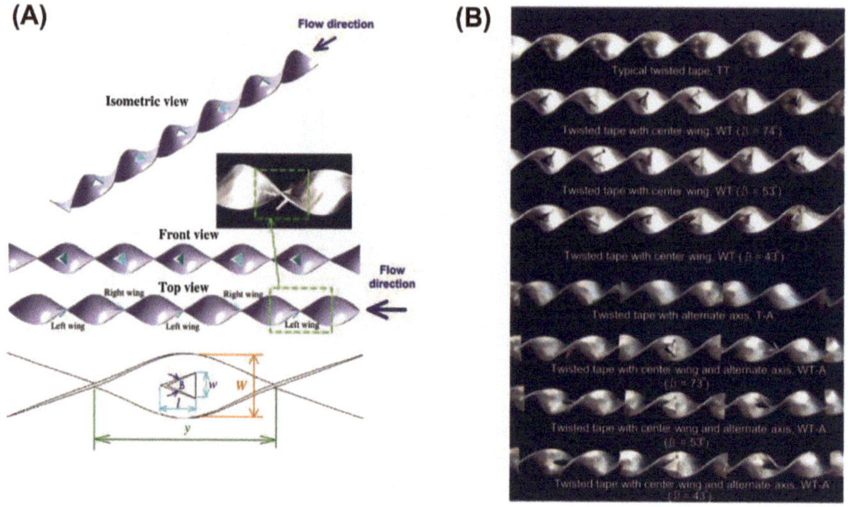

FIGURE 1.33

(A) Geometry of twisted tapes. (B) Photograph of twisted tapes [46].

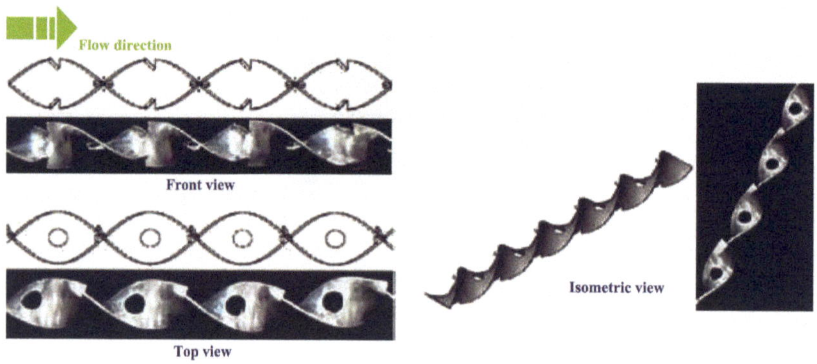

FIGURE 1.34

Photograph and sketch of PTT with parallel wings [48].

and wings on heat transfer, flow friction, and thermal performance characteristics in a round tube were investigated by Wongcharee and Eiamsa-ard [52] (Fig. 1.36). They concluded that the maximum thermal performance factor of 1.42 is found with the use of twisted tape with trapezoidal wings (T-Tra) at wing-chord ratio (d/W) of 0.3 and Reynolds number of 5500, where heat transfer rate and friction factor, respectively, increase to 2.84 and 8.02 times of those in the plain tube.

The effects of a peripherally-cut twisted tape with alternate axis (PTA) on the heat transfer, flow friction, and thermal performance factor characteristics were experimentally studied by Seemawute et al. [53] (Fig. 1.37). They proved that at

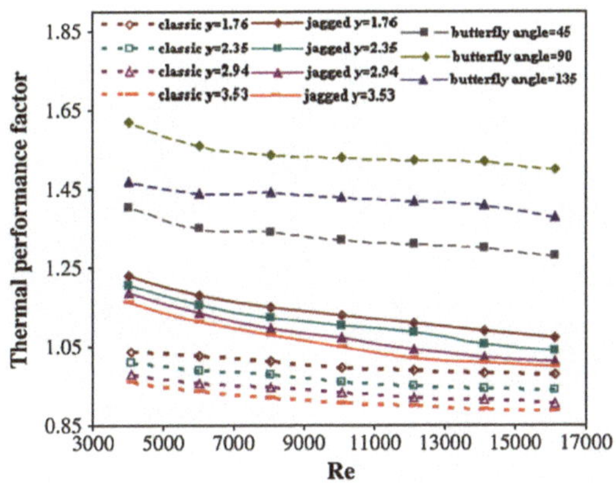

FIGURE 1.35

The performance ratio for various tube inserts [50].

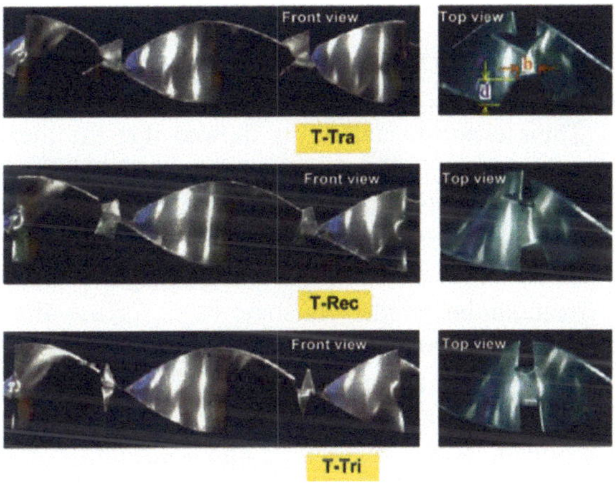

FIGURE 1.36

Twisted tapes used in the present work and their geometries [52].

constant pumping power, the thermal performance factors obtained in the present study are as high as 1.25 for PT-A, 1.11 for PT, and 1.02 for TT. Over the range studied, the maximum thermal performance factor for each twisted tape insert is achieved at the lowest values of Reynolds number (Re = 5000). Effects of peripherally-cut twisted tape insert on heat transfer, friction loss, and thermal performance factor characteristics in a round tube were investigated by Eiamsa-ard et al. [54]. They showed that the peripherally-cut twisted tape offered higher heat transfer rate, friction factor, and also thermal performance factor compared to the typical twisted tape. An additional turbulence of fluid in the vicinity of the tube wall and vorticity behind the cuts generated by the modified twisted tape are referred as the reason for a better heat transfer enhancement.

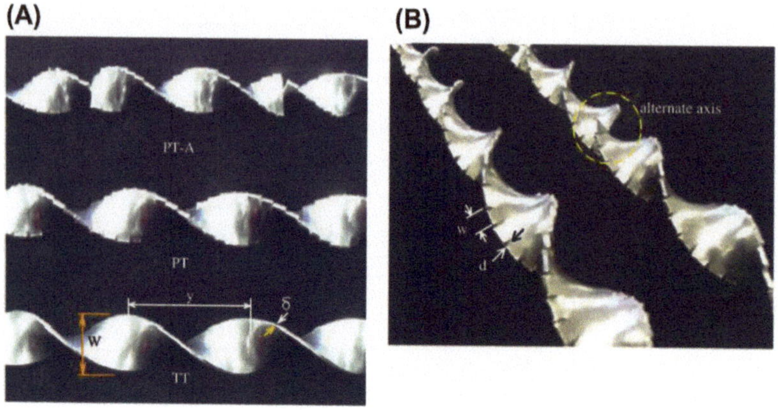

FIGURE 1.37

Photographs and geometries of the various twisted tapes (TT, PT, and PT-A) that are used in the present work: (A) front view and (B) isometric view [53].

Eiamsa-ard and Promvonge [55] investigated the effect of twisted tape with serrated-edge insert on heat transfer and pressure loss behaviors in a constant heat-fluxed tube. Their experimental result showed that the mean heat transfer rate of the tube fitted with the STT increases at about 72.2% in comparison with the plain tube. The heat transfer rate increases with increase in depth ratio and decrease in serration width ratio.

1.2.4.4.1.8 Twisted tape with wire nail Some insulating materials are provided to tapes so that fin effect can be avoided. In some cases surface dimpled material is used for tape fabrication. Double pipe heat exchanger equipped with twisted tape consisting wire nails (WN-TT) and plain twisted tapes (P-TT) was investigated by Murugesan et al. [56] (Fig. 1.38). They found that Nusselt number, friction factor, and thermal enhancement factor in the tube equipped with WN-TT appreciably higher than those in the tube fitted with P-TT and plain tube.

1.2.4.4.2 Conical ring

To increase convection heat transfer in a uniform heat flux tube by a passive method, several conical rings used as turbulators were mounted over the test tube by

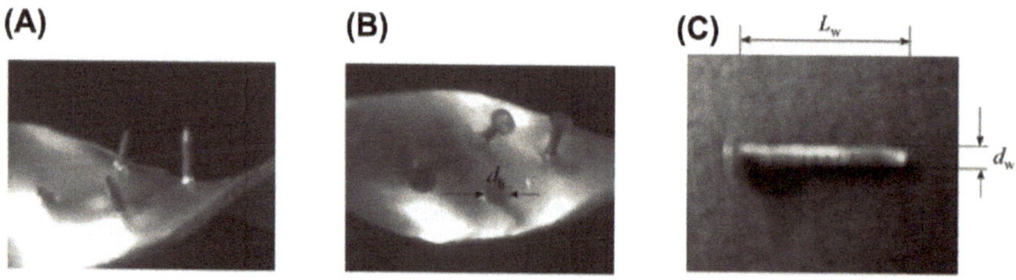

FIGURE 1.38

Geometries of twisted tapes [56]. (A and B) Details of WN-TT; (C) wire nail.

Promvonge [57]. They reported an augmentation of up to 197%, 333%, and 237% in Nusselt number is obtained in the turbulent flow for the CR, DR, and CDR arrays, respectively, although the effect of using the conical ring causes a substantial increase in friction factor. Their results were correlated in the form of Nusselt number as a function of Reynolds number, Prandtl number, and diameter ratio.

For the converging conical ring array (CR):

$$Nu = 0.09155\mathrm{Re}^{0.655}\,\mathrm{Pr}^{0.4}\left(\frac{d}{D}\right)^{-1.31} \tag{1.3}$$

$$f = 1.12\mathrm{Re}^{-0.258}\left(\frac{d}{D}\right)^{-4.4} \tag{1.4}$$

For the diverging conical ring array (DR):

$$Nu = 0.863\mathrm{Re}^{0.459}\mathrm{Pr}^{0.4}\left(\frac{d}{D}\right)^{-1.32} \tag{1.5}$$

$$f = 39.46\,\mathrm{Re}^{-0.591}\left(\frac{d}{W}\right)^{0.195}\left(\frac{w}{W}\right)^{-0.205} \tag{1.6}$$

For the converging–diverging conical ring array (CDR):

$$Nu = 0.244\mathrm{Re}^{0.625}\mathrm{Pr}^{0.4}\left(\frac{d}{W}\right)^{0.168}\left(\frac{w}{W}\right)^{-0.112} \tag{1.7}$$

$$f = 1.038\mathrm{Re}^{-0.23}\left(\frac{d}{D}\right)^{-4.58} \tag{1.8}$$

Heat transfer and turbulent flow in a circular tube fitted with conical-nozzle turbulators were studied by Promvonge and Eiamsa-ard [58]. They found that the insertion of nozzle turbulators has significant effect on the enhancement of heat transfer which the turbulator is used as a reverse/turbulence flow generator. Heat transfer, friction factor, and enhancement efficiency characteristics in a circular tube fitted with conical ring turbulators and a twisted-tape swirl generator have been investigated experimentally by Promvonge and Eiamsa-ard [59] (Fig. 1.39). They found that for all the devices used, the enhancement efficiency tends to decrease with the rise of Reynolds number and to be nearly uniform for Reynolds number over 16,000. In addition, correlations for Nusselt number, friction factor, and performance evaluation criteria to assess the real benefits in using the turbulator and swirl generator of the enhanced tube were determined. Kongkaitpaiboon et al. [60] used perforated conical ring (PCR) as turbulator devices for enhancing the heat transfer rate in a heat exchanger system (Fig. 1.40). They found that the heat transfer rate and friction factor of PCRs increase with decreasing pitch ratio (PR) and decreasing number of perforated hole (N). However, the thermal performance factor increases with increasing number of perforated whole and decreasing pitch ratio.

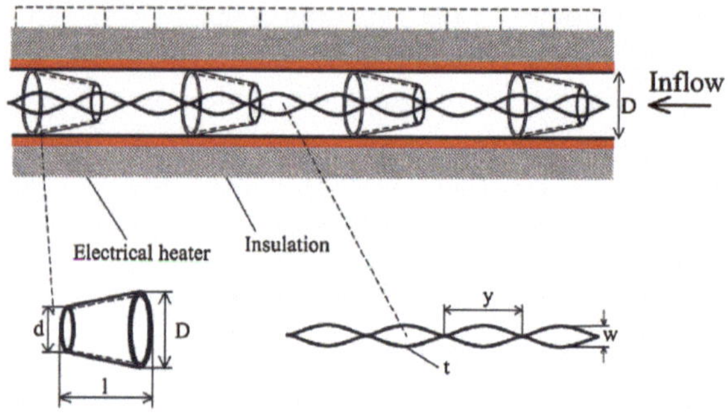

FIGURE 1.39

Test tube fitted with conical ring turbulator and twisted tape inserts [59].

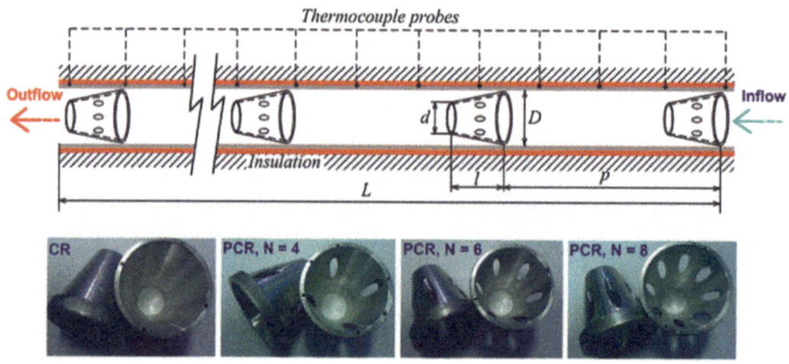

FIGURE 1.40

The arrangement of CR/PCR in a round tube [60]. CR, converging conical ring; PCR, perforated conical ring.

1.2.4.4.3 Snail entry turbulator

Promvonge and Eiamsa-ard [61] presented the effect of a free-spacing snail entry together with conical-nozzle turbulators on turbulent heat transfer and friction characteristics in a uniform heat-flux tube (Fig. 1.41). They noted that the enhancement efficiency shows a rapid decrease with the rise of Reynolds number for all pitch ratios. At the same Reynolds number, the small pitch ratio value provides higher enhancement efficiency than the larger one. The enhancement efficiency is peak at the lowest Reynolds number and pitch ratio values. The C-nozzle with free-spacing snail entry at the lowest pitch ratio (PR = 2.0) yields the maximum enhancement efficiency of about 0.93, better than that without free-space entry around 9%. Influences of V-nozzle turbulator inserts in conjunction with a snail entry on heat transfer and friction loss characteristics in a circular tube were experimentally investigated by Promvonge and Eiamsa-ard [62]. They found that the values of Nusselt number and friction factor for utilizing both the V-nozzle and the snail entry were found to be considerably higher than that for using the

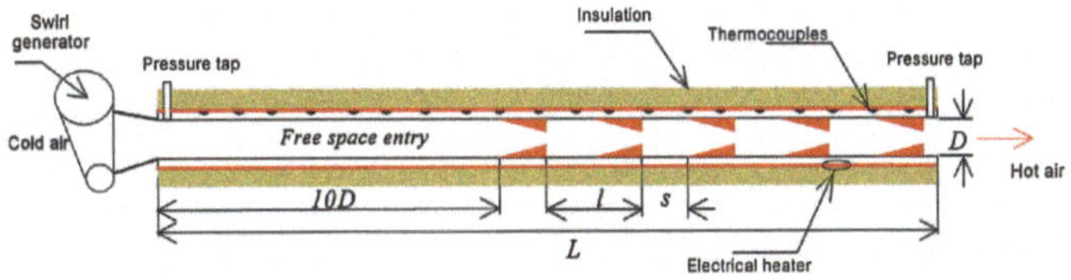

FIGURE 1.41

Test tube fitted with C-nozzle turbulators and snail with free-spacing entry [61].

V-nozzle alone or the plain tube. Durmus et al. [63] used snail entrance in order to increase heat transfer in concentric double-pipe heat exchangers. They concluded that the swirl flow effect of the snail caused some increase in pressure drop while this effect was unimportant compared with the improvement in heat transfer capacity.

1.2.4.4.4 Vortex rings

The influence of inclined vortex rings (VR) on heat transfer augmentation in a uniform heat-fluxed tube has been investigated by Promvonge et al. [64] (Fig. 1.42). Their experimental results showed that the larger blockage ratio value provides higher heat transfer and pressure loss than the smaller one but the pitch ratio gives an opposite trend. Kongkaitpaiboon et al. [65] presented the effect of the circular ring turbulator (CRT) on the heat transfer and fluid friction characteristics in a heat exchanger tube (Fig. 1.43). They found that heat transfer rates in the tube fitted with CRTs are augmented around 57%–195% compared to that in the plain tube, depending upon operating conditions. Heat transfer, friction factor, and thermal performance characteristics in a tube equipped with twisted rings (TRs) are experimentally investigated by Thianpong et al. [66] (Fig. 1.44). Their experimental results revealed that most TRs yield lower Nusselt numbers and friction factor than CRs, except at the largest width ratio (W/D = 0.15) and the smallest pitch ratio (p/D = 1.0). In addition, Nusselt number and friction factor increase as

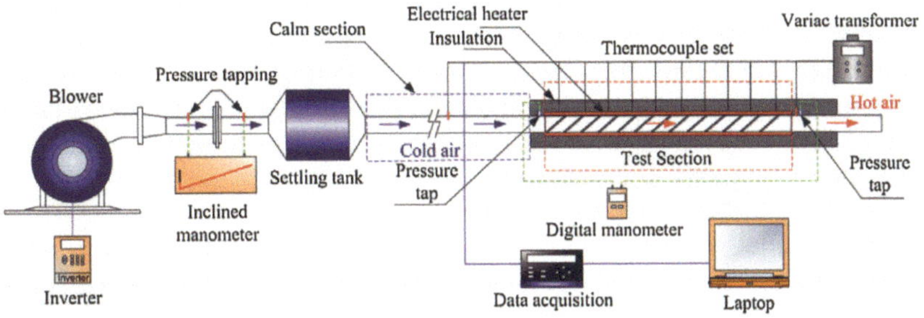

FIGURE 1.42

Schematic diagram of experimental apparatus [64].

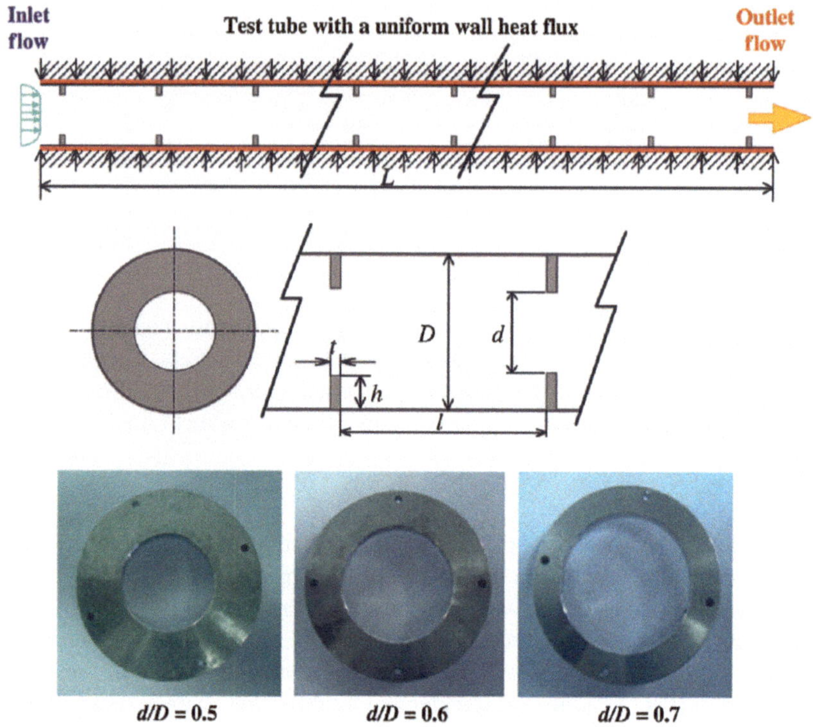

FIGURE 1.43

Test section with circular ring turbulators (CRT) [65].

width ratio increases and pitch ratio decreases. However, a maximum thermal performance factor is associated by TRs with the smallest width ratio and pitch ratio.

The influences of CRT and twisted tape (TT) swirl generators on the heat transfer enhancement, pressure drop, and thermal performance factor characteristics in a round tube were reported by Eiamsa-ard et al. [67] (Fig. 1.45). They found that the increases of mean Nusselt number, friction factor, and thermal performance, in the tube equipped with combined devices, respectively, are 25.8%, 82.8%, and 6.3% over those in the tube with the CRT alone. Akansu [68] studied heat transfers and pressure drops for porous-ring turbulators in a circular pipe. He found that the recirculation regions are larger at higher blockage ratio. Blockage ratio and Nusselt number increments were not proportional.

1.2.4.4.5 Coiled wire

This insert is a swirling flow device which is widely utilized in various heat transfer applications such as chemical process plants, refrigeration systems and air conditioning, food and dairy processes, heat recovery processes. The wire coils have some gains over the other enhancement techniques such as low cost, easy to insert

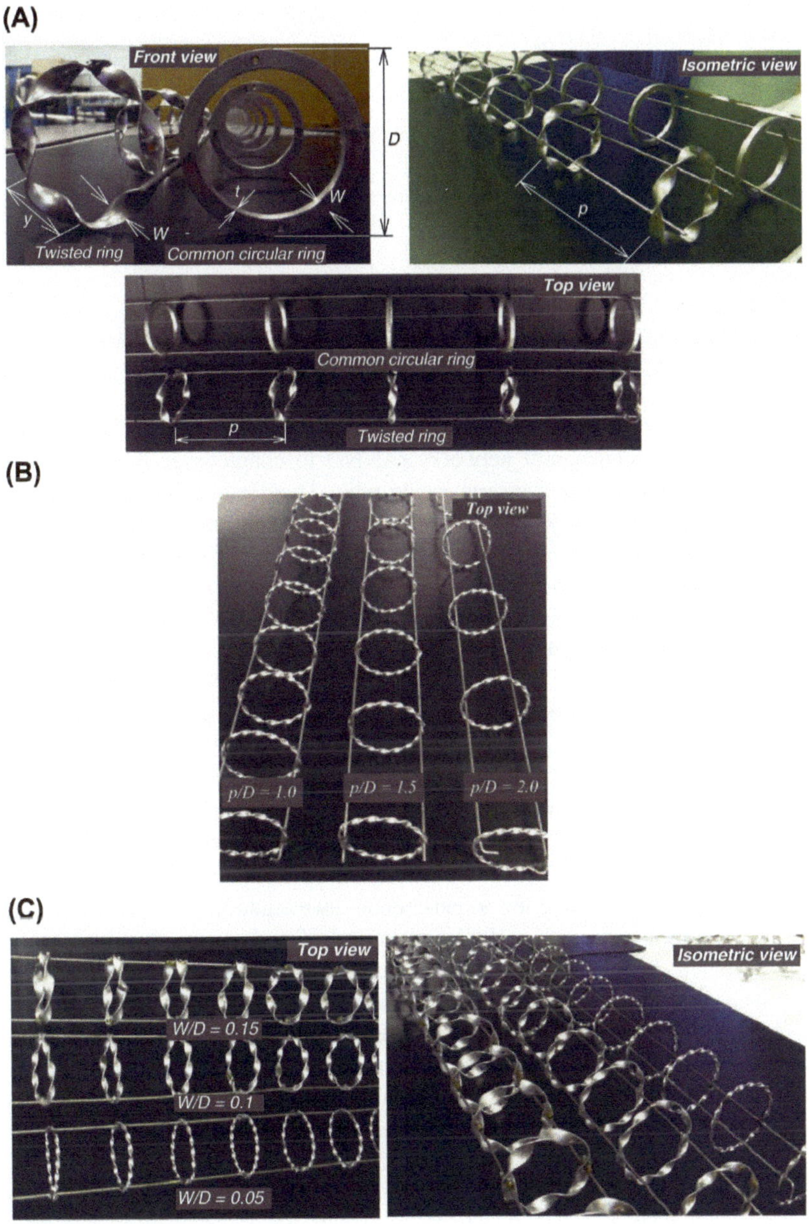

FIGURE 1.44

Photographs of twisted rings (TRs) at different arrangements: (A) geometric details of TR, (B) TR at different pitch lengths (p/D), and (C) TR at different widths lengths (W/D) [66].

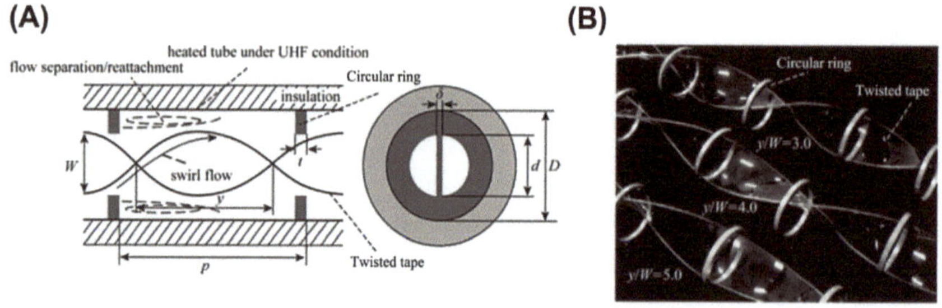

FIGURE 1.45

Diagram and photograph of circular rings combined with a twisted tape [67]. (A) Diagram of tube with enhancement devices; (B) Photograph of enhancement devices.

and remove from the tube, simple manufacturing. Jafari Nasr et al. [69] presented an application of artificial neural networks (ANNs) to characterize thermohydraulic behavior of helical wire coil inserts inside tube. They found that using the ANN in order to predict the performance of thermal systems in engineering applications is recommended. Helical wire coils fitted inside a round tube have been experimentally studied by Garcia et al. [70]. They found that wire coil inserts offer their best performance within the transition region where they show a considerable benefit over other enhancement techniques.

Convective heat transfer for an annulus, having its inner cylinder wrapped with a helical wire, has been numerically studied by Sreenivasulu and Prasad [71]. They found that thermal hydraulic performance ratio for the wrapped wire annulus is considerably higher than unity. Comini et al. [72] investigated the forced convection heat transfer from banks of helical coiled resistance wires in a cross flow of air. They found that no significant interactions were detected between upstream and downstream coils, possibly because of the high percentage of interspaces between wires. Gunes et al. [73] presented the experimental investigation of heat transfer and pressure drop in a tube with coiled wire inserts placed separately from the tube wall in turbulent flow regime (Fig. 1.46). They showed that the use of coiled wire inserts leads to a considerable increase in heat transfer and pressure drop over the smooth tube. The Nusselt number and friction factor increase with decreasing pitch ratio (P/D) and distance (s) for coiled wire inserts.

Gunes et al. [74] investigated the heat transfer and pressure drop in a coiled wire inserted tube in turbulent flow regime. Their experimental results revealed that the best operating regime of all coiled wire inserts is detected at low Reynolds number, leading to more compact heat exchanger. The heat transfer enhancement by using coiled wire inserts during forced convection condensation of R-22 inside a horizontal tube was experimentally investigated by Agrawal et al. [75]. They found that the use of helically coiled wires were found to increase the condensing heat transfer coefficients by as much as 100% above the plain tube values on a nominal area basis. Garcia et al. [19] used three types of enhancement techniques based on artificial roughness: corrugated tubes, dimpled tubes, and wire coils

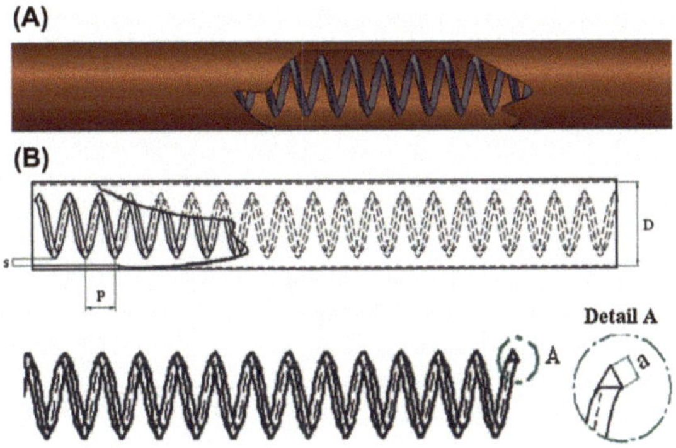

FIGURE 1.46

(A) The coiled wire inserts placed separately from the tube wall, (B) the details of coiled wire [73].

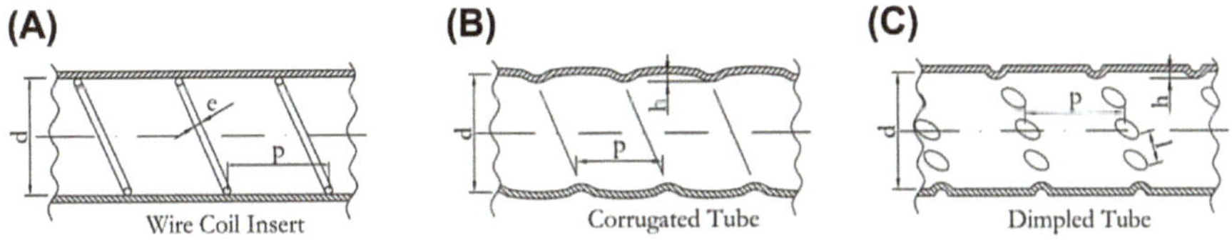

FIGURE 1.47

Types of surface roughness that this chapter contemplates. (A) Wire coils: helical pitch *p*, wire diameter *e*; (B) Corrugated tubes: corrugation pitch *p*, corrugation height h; (C) Dimpled tubes: corrugation pitch *p*, length between dimples *l*, dimple height *h* [19].

(Fig. 1.47). Their study concluded that for Reynolds numbers lower than 200, the use of smooth tubes is recommended. For Reynolds numbers between 200 and 2000, the employment of wire coils is more advantageous, while for Reynolds numbers higher than 2000, the use of corrugated and dimpled tubes is favored over the wire coils because of the lower pressure drop encountered for similar heat transfer coefficient levels (Fig. 1.48).

Heat transfer coefficient and friction factor of TiO_2 nanofluid flowing in a double pipe heat exchanger with and without helical coil inserts were studied experimentally by Reddy and Rao [76]. They showed that heat transfer coefficient and friction factor further get enhanced by 13.85% and 10.69%, respectively, for 0.02% nanofluid when compared with base fluid flowing in a tube with helical coil insert of P/d = 2.5 (see Fig. 1.49). Heat transfer in counter-flow heat exchangers with helical turbulators was studied by Zamankhan [77]. His results demonstrated that the model could be used as a useful tool for optimization of heat exchanger performance in the presence of a turbulator. Comparisons with experimental data showed reasonable agreement with large eddy simulation results. Akpinar [78] studied the effects of

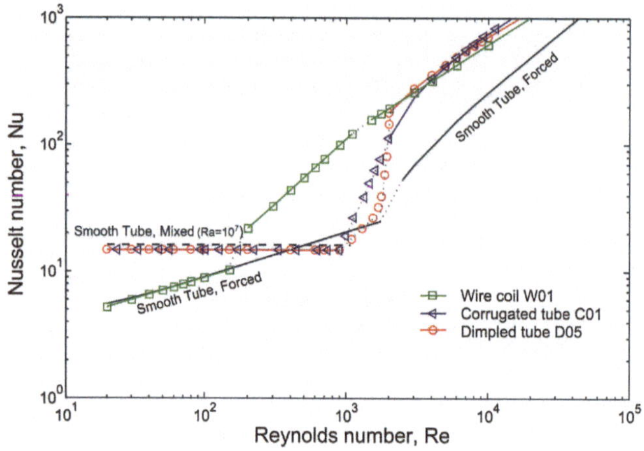

FIGURE 1.48

Nusselt number versus Reynolds number. Experimental correlations for the wire coil, the corrugated tube, and the dimpled tube [19].

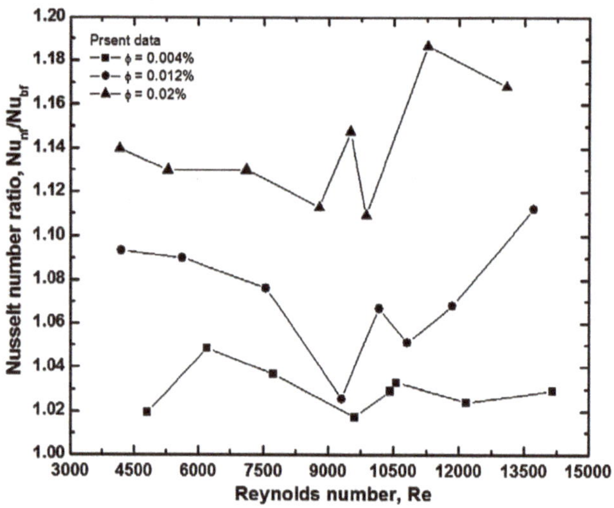

FIGURE 1.49

Nusselt number ratio as a function of Reynolds number for TiO_2 nanofluids with different volume concentrations [76].

mounting helical (spring shaped) wires on heat transfer, friction factor, and dimensionless exergy loss in a double pipe heat exchanger. He observed that heat transfer rates increased with decreasing pitch and with increasing helical number of the helical wires used in the experiments. The heat transfer rates in this heat exchanger increased up to 2.64 times with the help of the helical wires. Esparza and Rojas [79] used helical wire coils fitted inside a round pipe in order to improve the overall performance of heat exchangers. They confirmed that the increase of pitch ratio decreases the friction factor.

Promvonge [80] studied the effects of wires with square cross section forming a coil used as a turbulator on the heat transfer and turbulent flow friction characteristics in a uniform heat flux (Fig. 1.50). Their experimental results revealed that the use of coiled square wire turbulators leads to a considerable increase in heat transfer and friction loss over those of a smooth wall tube (Fig. 1.51). Thermal augmentation in circular tube with twisted tape and wire coil turbulators has been studied by Promvonge [81] (Fig. 1.52). Their results indicated that the

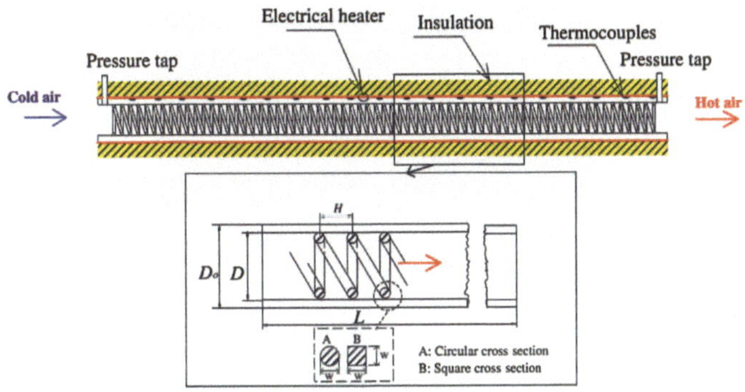

FIGURE 1.50

Test tube with coiled wire insert [80].

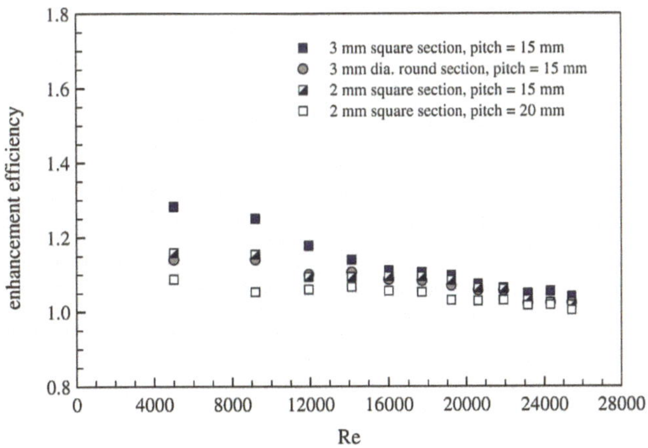

FIGURE 1.51

Variation of enhancement efficiency with Reynolds number [80].

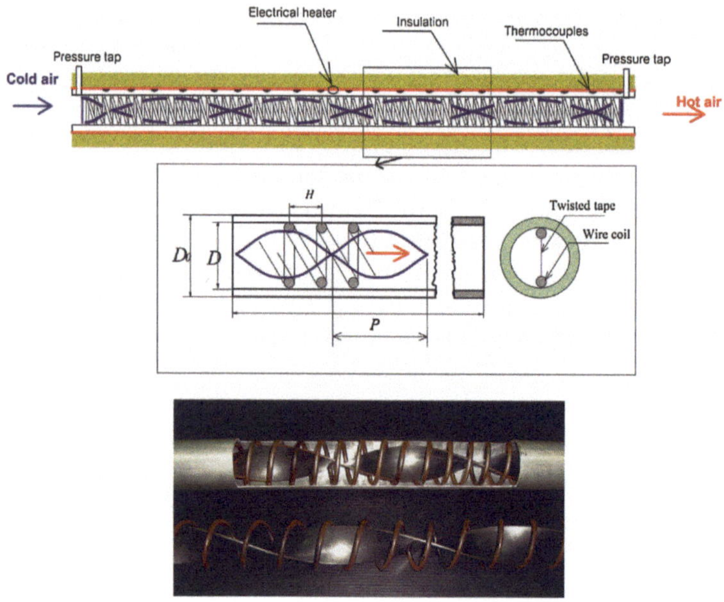

FIGURE 1.52

Test tube with wire coil and twisted tape inserts [81].

presence of wire coils together with twisted tapes leads to a double increase in heat transfer over the use of wire coil or twisted tape alone. Heat transfer, friction factor, and thermal performance behaviors in a tube equipped with the combined devices between the twisted tape (TT) and constant/periodically varying wire coil pitch ratio were experimentally investigated by Eiamsa-ard et al. [82]. They reported that the highest thermal performance factor of around 1.25 is found by using DI coil (decreasing coil pitch ratio arrangement). Heat transfer and friction factor analysis of CuO/water nanofluid flowing through a tube under turbulent flow conditions and with twisted tape (TT) and wire coil (WC) inserts were presented by Naik et al. [83]. They found that the thermal performance factor of 0.3% nanofluid in tube with twisted tape and wire coil inserts are 1.24 and 1.36 compared against water data, respectively. Promvonge [84] studied the influences of the insertion of coiled wires in conjunction with a snail-type swirl generator mounted at the tube entrance on heat transfer and turbulent flow friction characteristics in a uniform heat-flux circular tube. They found that the coiled square wire should be applied instead of the round one to obtain higher heat transfer and performance at about 10%−15%, leading to more compact heat exchanger.

1.2.4.4.6 Other kinds of turbulator

Naphon and Suchana [85] studied heat transfer enhancement and pressure drop of the horizontal concentric tube with twisted wires brush inserts (Fig. 1.53). They

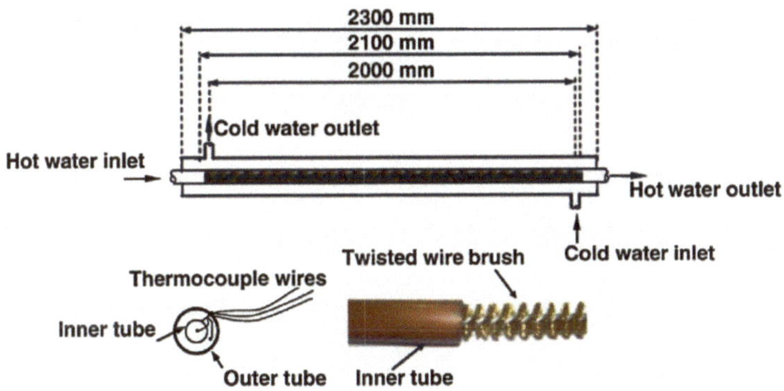

FIGURE 1.53

Schematic diagram of the test section [85].

found that twisted wire brush inserts have a significant effect on the enhancement of heat transfer; however, the pressure drops also increase too. Eiamsa-ard and Promvonge [86] studied the heat transfer and friction factor characteristics of the fully developed turbulent airflow through a uniform heat flux tube fitted with diamond-shaped turbulators in tandem arrangements (Fig. 1.54). They showed that the increase in the heat transfer rate with increasing the cone angle and decreasing the tail length ratio ($TR = l_t/l_h$) is due to the higher turbulence intensity imparted to the flow between the turbulators and the heating wall. Karakaya and Durmus [87] used conical-shaped spring as turbulators (Fig. 1.55). They concluded that the exergy loss decreases with increasing conical angle. This is due to the turbulators being placed into the flow medium, which cause increasing heat transfer rates and

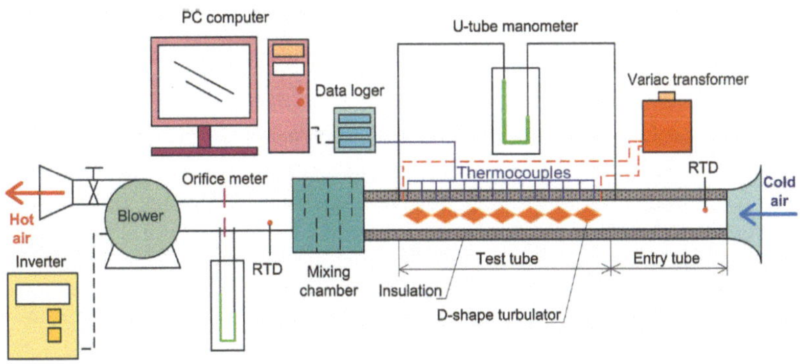

FIGURE 1.54

Schematic diagram of the experiment setup: (1) inverter, (2) blower, (3) rotameter, (4) insulated test tube, (5) electrical heating coil, (6) variac transformer, (7) U-tube manometer, (8) a set of thermocouple, and (9) data logger [86].

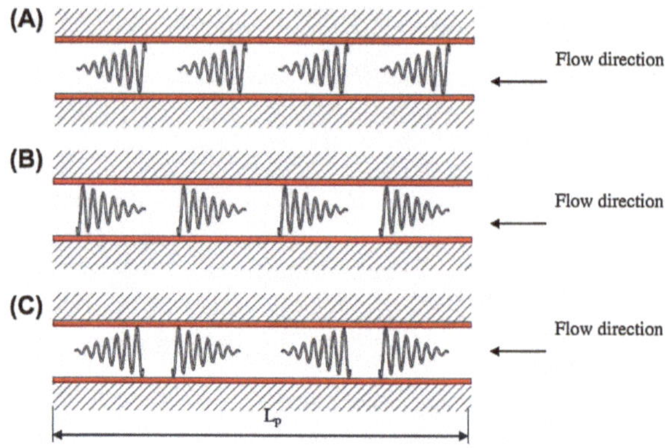

FIGURE 1.55

Test tube fitted with various CST arrays: (A) diverging conical ring array, (B) converging conical ring array, and (C) converging–diverging conical ring array [87].

more effective operation of the exchanger. Besides, there are several other passive methods such as microchannel, surface tension devices, additives for liquids [88–168], additives for gas, etc.

Table 1.2 shows the summary of important investigations of different swirl flow devices in laminar and turbulent flow. A twisted tape insert mixes the bulk flow well and therefore performs better in a laminar flow than any other insert because in a laminar flow the thermal resistance is not limited to a thin region. Twisted tape inserts perform better in laminar flow than turbulent flow. Compared with wire coil, twisted tape is not effective in turbulent flow because it blocks the flow and therefore the pressure drop is large. Hence, the thermohydraulic performance of a twisted tape is not good compared with wire coil in turbulent flow. Therefore, it may be concluded that, for compact heat exchanger design, wire coil is a good choice in turbulent flow. However, the other several passive techniques such as ribs, conical nozzle, and conical ring are generally more efficient in the turbulent flow than in the laminar flow. If the Prandtl number is high, the twisted tape will not provide good thermohydraulic performance compared with other inserts such as wire coil. In this case, the performance of the wire coil is good because, for a high Prandtl number, the thickness of the thermal boundary layer is small compared with the hydrodynamic boundary layer and the wire coil breaks this boundary layer easily.

1.3 Conclusion

In the present chapter, an effort has been made to report useful passive methods which are used in order to enhance the heat transfer performance. Insertion of swirl flow devices enhances the convective heat transfer by making swirl into the bulk

Table 1.2 Summary of important investigations of passive methods in laminar and turbulent flow.

Authors	Type of swirl flow devices	Working fluid	Configuration	Conditions	Remark
Chompookham et al. [17]	Extended surfaces	Air	Combined wedge ribs and winglet-type vortex generators (WVGs)	$5000 < Re < 22000$, Uniform heat-flux tube	• The combination of staggered wedge rib and the WVGs has efficiently performed and should be applied to obtain higher thermal performance at about 17%–20% of a single use of turbulators
Eiamsa et al. [23]	Extended surfaces	Water	Louvered strip	$6000 < Re < 42000$, forward or backward arrangements	• Louvered strip insertions can be used efficiently to augment heat transfer rate because of the turbulence intensity
Naphon et al. [12]	Rough surface	Water	Horizontal double pipes with helical ribs	$15000 < Re < 60000$	• Helical ribs have a significant effect on the heat transfer and pressure drop augmentations
Li et al. [26]	Rough surface	Water	Discrete double inclined ribs	$15000 < Re < 60000$	• Visualization of the flow field shows that in addition to the front and rear vortices around the ribs, main vortices and induced vortices are also generated by the ribs in the DDIR tube

Continued

Table 1.2 Summary of important investigations of passive methods in laminar and turbulent flow.—*cont'd*

Authors	Type of swirl flow devices	Working fluid	Configuration	Conditions	Remark
Promvonge et al. [27]	Rough surface	Air	Combined wedge ribs and winglet-type vortex generators (WVGs)	$5000 < Re < 22000$, attack angle (30°, 45°, 60°)	• The nusselt number augmentation tends to decrease slightly with the rise in Reynolds number • The best operating regime for using these compound turbulators is found at the lower attack angle and/or Reynolds number values
Liao and Xin [110]	Twisted tape	Water, ethylene glycol, turbine oil	Segmented twisted tape and three-dimensional extended surfaces	$80 < Re < 50000$ $5.5 < Pr < 590$	• In a tube with three-dimensional extended surfaces and twisted tape Increases average stanton number up to 5.8 times compared with empty smooth tube
Noothong et al. [111]	Twisted tape	Water	Typical twisted tape (TT)	Twist ratios $y = 5$ and 7,$2000 < Re < 12000$	• Secondary fluid motion is generated by the tape twist, and the resulting twist mixing improves the convection heat transfer • The enhancement efficiency and nusselt number increases with decreasing the twist ratio and friction factor also increases with decreasing the twist ratio

Sarada et al. [36]	Twisted tape	Air	Typical twisted tape (TT)	Widths of the twisted tapes range from 10 to 20 mm, $6000 < Re < 13500$	• This enhancement is mainly due to the centrifugal forces resulting from the spiral motion of the fluid • The enhancement of heat transfer with twisted tape inserts as compared to plain tube
Wongcharee and Eiamsaard [37]	Twisted tape	Water	Alternate clockwise and counterclockwise twisted tapes (TA)	$830 < Re < 1990$, Twist ratios $y = 3, 4$ and 5	• The friction factor associated by TA is higher than that induced by TT, and friction factor increases with decreasing twist ratio
Eiamsa-ard et al. [112]	Twisted tape	Air	Dual twisted tapes	$400 < Re < 19000$, Twist ratios $y = 3, 4$ and 5	• The smaller space ratio of the dual twisted tapes in tandem is more attractive in heat transfer application due to higher enhancement efficiency than the single one
Eiamsa-ard et al. [49]	Twisted tape	Water	Oblique delta-winglet twisted tape (O-DWT) and straight delta-winglet twisted tape (S-DWT). Uniform wall Heat flux tube	$3000 < Re < 27000$, twist ratios $y = 5$ and 7, depth of wing cut ratios ($DR = d/w = 0.11, 0.21$ and 0.32)	• Nusselt number and friction factor increase with decreasing of twist ratio and increasing depth of wing cut ratio (DR) for all Reynolds numbers studied • DWT performs better heat transfer enhancement

Continued

Table 1.2 Summary of important investigations of passive methods in laminar and turbulent flow.—*cont'd*

Authors	Type of swirl flow devices	Working fluid	Configuration	Conditions	Remark
					than that of typical twisted tape • DWT can be replaced any of the TT efficiently to reduce the size of the heat exchanger
Jaisankar et al. [113]	Twisted tape	Water	Helical twisted tape	Twist ratios $y = 3, 4$ and 5	• Thermal performance of twisted tape collector with minimum twist ratio ($y = 3$) is better than the other twist ratios
Saha [114]	Twisted tape	Air	Twisted-tape inserts with and without oblique teeth	$10000 < Re < 100000$	• Full-length and short-length twisted tapes with oblique teeth in combination with axial corrugations show only marginal improvements over the twisted tapes without oblique teeth
Thianpong et al. [33]	Twisted tape	Water	Dimpled tube fitted wwith a twisted tape sswirl generator	$12000 < Re < 44000$, twist ratios $y = 3, 5$ and 7	• A dimpled tube in common with a twisted tape has significant effects on the heat transfer enhancement and friction factor. The heat transfer and friction factor are increased with decreasing both of pitch ratio (PR) and twist ratio (y/w)

Sivashanmugam and Nagarajan [115]	Twisted tape	Water	Right–left helical screw inserts of equal length	$200 < Re < 3000$, twist ratios $y = 2.93 - 4.89$	• The heat transfer coefficient enhancement for right –left helical screw inserts is higher than that for straight helical twist for a given twist ratio
Gararcia et al. [70]	Coiled wire	Water propyleneGlycol mixtures	$1.17 \leq p/d \leq 2.68$ (Helical pitch), $0.07 \leq e/d \leq 0.1$ (wire diameter)	$80 < Re < 90000$, $2.8 < Pr < 150$	• In laminar flow, results show that wire coils behave mmainly as a smooth tube • In turbulent flow, wire coils cause a high pressure drop Increase which depends mainly on pitch to wire-diameter ratio p/e
Promvonge [80]	Coiled wire	Air	Wires with square cross section	$5000 < Re < 25000$	• The nusselt number augmentation tends to decrease rapidly with the rise of Reynolds number • The coiled square wire should be applied instead of the rround one to obtain higher heat transfer and performance, leading to more compact heat exchanger
Promvonge [81]	Coiled wire	Air	Wire coils in conjunction with twisted tapes	$3000 < Re < 18000$	• The best operating regime for both turbulators is found at lower Reynolds number values for the lowest

Continued

Table 1.2 Summary of important investigations of passive methods in laminar and turbulent flow.—*cont'd*

Authors	Type of swirl flow devices	Working fluid	Configuration	Conditions	Remark
					values of the coil spring pitch and twist ratio • The combined wire coil and twisted tape turbulators are compared with a smooth tube at a constant pumping power. Heat transfer performance is doubled at low Reynolds number
Akhavan-Behabadi et al. [116]	Coiled wire	Oil	Seven coiled wires having pitches of *12–69* mm	$10 < Re < 1500$	• Wire coil inserts with lower wire diameters have better performance, especially at low Reynolds numbers. Also, the increase in the coil pitch made a moderate decrease in performance parameter
Kurtbs et al. [117]	Conical injector	Air	Conical injector–type swirl generator (CITSG)	$9400 < Re < 35000$, CITSGs' angle ($\alpha = 30°, 45°, 60°$)	• The heat transfer ratio decreases with increase in Re • The effect of α on heat transfer ratio is negligible for higher Re

Promvonge and Eiamsa-ard [59]	Conical ring	Air	Conical-ring turbulators and a twisted tape swirl generator	$6000 < Re < 26000$	• For all the devices used, the enhancement efficiency tends to decrease with the rise of Reynolds number and to be nearly uniform for Reynolds number over 16,000 • The smaller twist ratio, the larger the heat transfer and friction factor for all Reynolds numbers
Kongkaitpaiboon et al. [60]	Conical ring	Air	Perforated conical ring (PCR)	$4000 < Re < 20000$, pitch ratios ($PR = p/D = 4, 6, 12$), perforated holes ($N = 4, 6, 8$)	• The heat transfer rate and friction factor of PCRs increase with decreasing pitch ratio (PR) and decreasing number of perforated hole (N). However, the thermal performance factor increases with increasing number of perforated hole and decreasing pitch ratio

flow and disrupting the boundary layer at the tube surface due to repeated changes in the surface geometry. That is to say such devices induce turbulence and superimposed vortex motion (swirl flow) which induces a thinner boundary layer and consequently results in a better heat transfer coefficient and higher Nusselt number due to the changes in the twisted tape geometry. However, the pressure drop inside the tube will be increased by introducing the twisted tape to insert. Based on the literature review, following conclusions are drawn.

- The heat transfer coefficients of the coiled tubes with larger pitches are less than those of the ones with smaller pitches, and the effect of pitch on Nusselt number is more visible in high temperatures.
- Delta winglets generate the vortexes which increase the heat transfer without much increase in friction factor in solar air heater or heat exchangers. Curved trapezoidal delta winglet had the best thermohydraulic performance in fully turbulent flow region.
- Full-length twisted tape increases the pressure drop comparing to an empty tube. Most of the research studies need to reduce the extra pressure drop by using short-length twisted tape. Twisted tape in turbulent flow insert is not very effective.
- The helical ribs have a significant effect on the heat transfer and pressure drop augmentations. The pressure drop across the tube with helical rib is produced by drag forces, flow blockage, turbulence augmentation, and rotational flow produced by the helical rib.
- Thermal performance factor increases with increasing number of perforated whole in conical ring.
- Wire coil gives better overall performance if the pressure drop penalty is considered. The use of coiled square wire turbulators leads to a considerable increase in heat transfer and friction loss over those of a smooth wall tube.
- Twisted tape inserts perform better in laminar flow. The other several passive techniques to enhance the heat transfer in a flow, such as ribs, conical nozzle, are generally more efficient in the turbulent flow than in the laminar flow.

References

[1] M.R. Salimpour, Heat transfer characteristics of a temperature-dependent-property fluid in shell and coiled tube heat exchangers, International Communications in Heat and Mass Transfer 35 (2008) 1190–1195.

[2] M. Moawed, Experimental study of forced convection from helical coiled tubes with different parameters, Energy Conversion and Management 52 (2011) 1150–1156.

[3] W.I.A. Aly, Numerical study on turbulent heat transfer and pressure drop of nanofluid in coiled tube-in-tube heat exchangers, Energy Conversion and Management 79 (2014) 304–316.

[4] P. Sriromreun, C. Thianpong, P. Promvonge, Experimental and numerical study on heat transfer enhancement in a channel with Z-shaped baffles, International Communications in Heat and Mass Transfer 39 (2012) 945–952.

[5] S. Kwankaomeng, P. Promvonge, Numerical prediction on laminar heat transfer in square duct with 30° angled baffle on one wall, International Communications in Heat and Mass Transfer 37 (2010) 857–866.

[6] P. Promvonge, S. Sripattanapipat, S. Kwankaomeng, Laminar periodic flow and heat transfer in square channel with 45 inline baffles on two opposite walls, International Journal of Thermal Sciences 49 (2010) 963–975.

[7] F. Agyenim, P. Eames, M. Smyth, A comparison of heat transfer enhancement in a medium temperature thermal energy storage heat exchanger using fins, Solar Energy 83 (2009) 1509–1520.

[8] K. Mohammadi, M. Sabzpooshani, Comprehensive performance evaluation and parametric studies of single pass solar air heater with fins and baffles attached over the absorber plate, Energy 57 (2013) 741–750.

[9] N. Targui, H. Kahalerras, Analysis of a double pipe heat exchanger performance by use of porous baffles and pulsating flow, Energy Conversion and Management 76 (2013) 43–54.

[10] Y. Vermahmoudi, S.M. Peyghambarzadeh, S.H. Hashemabadi, M. Naraki, Experimental investigation on heat transfer performance of Fe_2O_3/water nanofluid in an air-finned heat exchanger, European Journal of Mechanics – B: Fluids 44 (2014) 32–41.

[11] H.A. Mohammed, A. Husam, M.A. Wahid, Heat transfer enhancement of nanofluids in a double pipe heat exchanger with louvered strip inserts, International Communications in Heat and Mass Transfer 40 (2013) 36–46.

[12] S. Eiamsa-ard, S. Pethkool, C. Thianpong, P. Promvonge, Turbulent flow heat transfer and pressure loss in a double pipe heat exchanger with louvered strip inserts, International Communications in Heat and Mass Transfer 35 (2008) 120–129.

[13] I. Kotcioglu, S. Caliskan, A. Cansiz, S. Baskaya, Second law analysis and heat transfer in a cross-flow heat exchanger with a new winglet-type vortex generator, Energy 35 (2010) 3686–3695.

[14] K. Torii, K.M. Kwak, K. Nishino, Heat transfer enhancement accompanying pressure-loss reduction with winglet-type vortex generators for fin-tube heat exchangers, International Journal of Heat and Mass Transfer 45 (2002) 3795–3801.

[15] G. Zhou, Q. Ye, Experimental investigations of thermal and flow characteristics of curved trapezoidal winglet type vortex generators, Applied Thermal Engineering 37 (2012) 241–248.

[16] S. Eiamsa-ard, P. Promvonge, Influence of double-sided delta-wing tape insert with alternate-axes on flow and heat transfer characteristics in a heat exchanger tube, Fluid flow and transport phenomena, Chinese Journal of Chemical Engineering 19 (3) (2011) 410–423.

[17] T. Chompookham, C. Thianpong, S. Kwankaomeng, P. Promvonge, Heat transfer augmentation in a wedge-ribbed channel using winglet vortex generators, International Communications in Heat and Mass Transfer 37 (2010) 163–169.

[18] H. Han, B. Li, W. Shao, Multi-objective optimization of outward convex corrugated tubes using response surface methodology, Applied Thermal Engineering 70 (2014) 250–262.

[19] A. Garcia, J.P. Solano, P.G. Vicente, A. Viedma, The influence of artificial roughness shape on heat transfer enhancement: corrugated tubes, dimpled tubes and wire coils, Applied Thermal Engineering 35 (2012) 196–201.

[20] S. Rainieri, G. Pagliarini, Convective heat transfer to temperature dependent property fluids in the entry region of corrugated tubes, International Journal of Heat and Mass Transfer 45 (2002) 4525–4536.

[21] P.G. Vicente, A. Garcia, A. Viedma, Experimental investigation on heat transfer and frictional characteristics of spirally corrugated tubes in turbulent flow at different Prandtl numbers, International Journal of Heat and Mass Transfer 47 (2004) 671–681.

[22] J.M. Buchlin, Convective heat transfer in a channel with perforated ribs, International Journal of Thermal Sciences 41 (2002) 332–340.

[23] P. Naphon, M. Nuchjapo, J. Kurujareon, Tube side heat transfer coefficient and friction factor characteristics of horizontal tubes with helical rib, Energy Conversion and Management 47 (2006) 3031–3044.

[24] J.A. Meng, X.G. Liang, Z.X. Li, Field synergy optimization and enhanced heat transfer by multi-longitudinal vortexes flow in tube, International Journal of Heat and Mass Transfer 48 (2005) 3331–3337.

[25] P.S. Kathait, A.K. Patil, Thermo-hydraulic performance of a heat exchanger tube with discrete corrugations, Applied Thermal Engineering 66 (2014) 162–170.

[26] X. Li, J. Meng, Z. Guo, Turbulent flow and heat transfer in discrete double inclined ribs tube, International Journal of Heat and Mass Transfer 52 (2009) 962–970.

[27] P. Promvonge, T. Chompookham, S. Kwankaomeng, C. Thianpong, Enhanced heat transfer in a triangular ribbed channel with longitudinal vortex generators, Energy Conversion and Management 51 (2010) 1242–1249.

[28] P. Promvonge, W. Changcharoen, S. Kwankaomeng, C. Thianpong, Numerical heat transfer study of turbulent square-duct flow through inline V-shaped discrete ribs, International Communications in Heat and Mass Transfer 38 (2011) 1392–1399.

[29] C. Nuntadusit, M. Wae-hayee, A. Bunyajitradulya, S. Eiamsa-ard, Thermal visualization on surface with transverse perforated ribs, International Communications in Heat and Mass Transfer 39 (2012) 634–639.

[30] R. Kamali, R. Binesh, The importance of rib shape effects on the local heat transfer and flow friction characteristics of square ducts with ribbed internal surfaces, International Communications in Heat and Mass Transfer 35 (2008) 1032–1040.

[31] A. Kumar, B.N. Prasad, Investigation of twisted tape inserted solar water heaters-heat transfer, friction factor and thermal performance results, Renewable Energy 19 (2000) 379–398.

[32] E. Esmaeilzadeh, H. Almohammadi, A. Nokhosteen, A. Motezaker, A.N. Omrani, Study on heat transfer and friction factor characteristics of γ-Al_2O_3/water through circular tube with twisted tape inserts with different thicknesses, International Journal of Thermal Sciences 82 (2014) 72–83.

[33] C. Thianpong, P. Eiamsa-ard, K. Wongcharee, S. Eiamsa-ard, Compound heat transfer enhancement of a dimpled tube with a twisted tape swirl generator, International Communications in Heat and Mass Transfer 36 (2009) 698–704.

[34] S. Eiamsa-ard, K. Yongsiri, K. Nanana, C. Thianpong, Heat transfer augmentation by helically twisted tapes as swirl and turbulence promoters, Chemical Engineering and Processing 60 (2012) 42–48.

[35] S. Jaisankar, T.K. Radhakrishnan, K.N. Sheeba, Experimental studies on heat transfer and thermal performance characteristics of thermosyphon solar water heating system with helical and left−right twisted tapes, Energy Conversion and Management 52 (2011) 2048−2055.

[36] S. Naga Sarada, A.V. Sita Rama Raju, K. Kalyani Radha, L. Shyam Sunder, Enhancement of heat transfer using varying width twisted tape inserts, International Journal of Engineering, Science and Technology 2 (2010) 107−118.

[37] K. Wongcharee, S. Eiamsa-ard, Friction and heat transfer characteristics of laminar swirl flow through the round tubes inserted with alternate clockwise and counter-clockwise twisted-tapes, International Communications in Heat and Mass Transfer 38 (2011) 348−352.

[38] S. Eiamsa-ard, C. Thianpong, P. Eiamsa-ard, P. Promvonge, Convective heat transfer in a circular tube with short-length twisted tape insert, International Communications in Heat and Mass Transfer 36 (2009) 365−371.

[39] M.M.K. Bhuiya, A.S.M. Sayem, M. Islam, M.S.U. Chowdhury, M. Shahabuddin, Performance assessment in a heat exchanger tube fitted with double counter twisted tape inserts, International Communications in Heat and Mass Transfer 50 (2014) 25−33.

[40] P. Promvonge, S. Pethkool, M. Pimsarn, C. Thianpong, Heat transfer augmentation in a helical-ribbed tube with double twisted tape inserts, International Communications in Heat and Mass Transfer 39 (2012) 953−959.

[41] M.M.K. Bhuiya, M.S.U. Chowdhury, M. Shahabuddin, M. Saha, L.A. Memon, Thermal characteristics in a heat exchanger tube fitted with triple twisted tape inserts, International Communications in Heat and Mass Transfer 48 (2013) 124−132.

[42] S. Jaisankar, T.K. Radhakrishnan, K.N. Sheeba, Experimental studies on heat transfer and friction factor characteristics of thermosyphon solar water heater system fitted with spacer at the trailing edge of twisted tapes, Applied Thermal Engineering 29 (2009) 1224−1231.

[43] S. Eiamsa-ard, P. Promvonge, Heat transfer characteristics in a tube fitted with helical screw-tape with/without core-rod inserts, International Communications in Heat and Mass Transfer 34 (2007) 176−185.

[44] E.Z. Ibrahim, Augmentation of laminar flow and heat transfer in flat tubes by means of helical screw-tape inserts, Energy Conversion and Management 52 (2011) 250−257.

[45] S. Eiamsa-ard, P. Promvonge, Enhancement of heat transfer in a tube with regularly-spaced helical tape swirl generators, Solar Energy 78 (2005) 483−494.

[46] S. Eiamsa-ard, K. Wongcharee, P. Eiamsa-ard, C. Thianpong, Thermohydraulic investigation of turbulent flow through a round tube equipped with twisted tapes consisting of centre wings and alternate-axes, Experimental Thermal and Fluid Science 34 (2010) 1151−1161.

[47] S. Ghadirijafarbeigloo, A.H. Zamzamian, M. Yaghoubi, 3-D numerical simulation of heat transfer and turbulent flow in a receiver tube of solar parabolic trough concentrator with louvered twisted-tape inserts, Energy Procedia 49 (2014) 373−380.

[48] C. Thianpong, P. Eiamsa-ard, P. Promvonge, S. Eiamsa-ard, Effect of perforated twisted-tapes with parallel wings on heat transfer enhancement in a heat exchanger tube, Energy Procedia 14 (2012) 1117−1123.

[49] S. Eiamsa-ard, K. Wongcharee, P. Eiamsa-ard, C. Thianpong, Heat transfer enhancement in a tube using delta-winglet twisted tape inserts, Applied Thermal Engineering 30 (2010) 310−318.

[50] S.R. Shabanian, M. Rahimi, M. Shahhosseini, A.A. Alsairafi, CFD and experimental studies on heat transfer enhancement in an air cooler equipped with different tube inserts, International Communications in Heat and Mass Transfer 38 (2011) 383–390.

[51] P. Murugesan, K. Mayilsamy, S. Suresh, P.S.S. Srinivasan, Heat transfer and pressure drop characteristics in a circular tube fitted with and without V-cut twisted tape insert, International Communications in Heat and Mass Transfer 38 (2011) 329–334.

[52] K. Wongcharee, S. Eiamsa-ard, Heat transfer enhancement by twisted tapes with alternate-axes and triangular, rectangular and trapezoidal wings, Chemical Engineering and Processing 50 (2011) 211–219.

[53] P. Seemawute, S. Eiamsa-ard, Thermohydraulics of turbulent flow through a round tube by a peripherally-cut twisted tape with an alternate axis, International Communications in Heat and Mass Transfer 37 (2010) 652–659.

[54] S. Eiamsa-ard, P. Seemawute, K. Wongcharee, Influences of peripherally-cut twisted tape insert on heat transfer and thermal performance characteristics in laminar and turbulent tube flows, Experimental Thermal and Fluid Science 34 (2010) 711–719.

[55] S. Eiamsa-ard, P. Promvonge, Thermal characteristics in round tube fitted with serrated twisted tape, Applied Thermal Engineering 30 (2010) 1673–1682.

[56] P. Murugesan, K. Mayilsamy, S. Suresh, Heat transfer and friction factor studies in a circular tube fitted with twisted tape consisting of wire-nails, Chinese Journal of Chemical Engineering 18 (6) (2010) 1038–1042.

[57] P. Promvonge, Heat transfer behaviors in round tube with conical ring inserts, Energy Conversion and Management 49 (2008) 8–15.

[58] P. Promvonge, S. Eiamsa-ard, Heat transfer and turbulent flow friction in a circular tube fitted with conical-nozzle turbulators, International Communications in Heat and Mass Transfer 34 (2007) 72–82.

[59] P. Promvonge, S. Eiamsa-ard, Heat transfer behaviors in a tube with combined conical-ring and twisted-tape insert, International Communications in Heat and Mass Transfer 34 (2007) 849–859.

[60] V. Kongkaitpaiboon, K. Nanan, S. Eiamsa-ard, Experimental investigation of heat transfer and turbulent flow friction in a tube fitted with perforated conical-rings, International Communications in Heat and Mass Transfer 37 (2010) 560–567.

[61] P. Promvonge, S. Eiamsa-ard, Heat transfer in a circular tube fitted with free-spacing snail entry and conical-nozzle turbulators, International Communications in Heat and Mass Transfer 34 (2007) 838–848.

[62] P. Promvonge, S. Eiamsa-ard, Heat transfer augmentation in a circular tube using V-nozzle turbulator inserts and snail entry, Experimental Thermal and Fluid Science 32 (2007) 332–340.

[63] A. Durmus, A. Durmus, M. Esen, Investigation of heat transfer and pressure drop in a concentric heat exchanger with snail entrance, Applied Thermal Engineering 22 (2002) 321–332.

[64] P. Promvonge, N. Koolnapadol, M. Pimsarn, C. Thianpong, Thermal performance enhancement in a heat exchanger tube fitted with inclined vortex rings, Applied Thermal Engineering 62 (2014) 285–292.

[65] V. Kongkaitpaiboon, K. Nanan, S. Eiamsa-ard, Experimental investigation of convective heat transfer and pressure loss in a round tube fitted with circular-ring turbulators, International Communications in Heat and Mass Transfer 37 (2010) 568–574.

[66] C. Thianpong, K. Yongsiri, K. Nanan, S. Eiamsa-ard, Thermal performance evaluation of heat exchangers fitted with twisted-ring turbulators, International Communications in Heat and Mass Transfer 39 (2012) 861–868.

[67] S. Eiamsa-ard, V. Kongkaitpaiboon, K. Nanan, Thermohydraulics of turbulent flow through heat exchanger tubes fitted with circular-rings and twisted tapes, Fluid dynamics and transport phenomena, Chinese Journal of Chemical Engineering 21 (6) (2013) 585–593.

[68] S.O. Akansu, Heat transfers and pressure drops for porous-ring turbulators in a circular pipe, Applied Energy 83 (2006) 280–298.

[69] M.R. Jafari Nasr, A. Habibi Khalaj, S.H. Mozaffari, Modeling of heat transfer enhancement by wire coil inserts using artificial neural network analysis, Applied Thermal Engineering 30 (2010) 143–151.

[70] A. Garcia, P.G. Vicente, A. Viedma, Experimental study of heat transfer enhancement with wire coil inserts in laminar-transition-turbulent regimes at different Prandtl numbers, International Journal of Heat and Mass Transfer 48 (2005) 4640–4651.

[71] T. Sreenivasulu, B.V.S.S.S. Prasad, Flow and heat transfer characteristics in an annulus wrapped with a helical wire, International Journal of Thermal Sciences 48 (2009) 1377–1391.

[72] G. Comini, S. Savino, E. Bari, A. Bison, Forced convection heat transfer from banks of helical coiled resistance wires, International Journal of Thermal Sciences 47 (2008) 442–449.

[73] S. Gunes, V. Ozceyhan, O. Buyukalaca, The experimental investigation of heat transfer and pressure drop in a tube with coiled wire inserts placed separately from the tube wall, Applied Thermal Engineering 30 (2010) 1719–1725.

[74] S. Gunes, V. Ozceyhan, O. Buyukalaca, Heat transfer enhancement in a tube with equilateral triangle cross sectioned coiled wire inserts, Experimental Thermal and Fluid Science 34 (2010) 684–691.

[75] K.N. Agrawal, M.A. Anil Kumar, A. Behabadi, H.K. Varma, Heat transfer augmentation by coiled wire inserts during forced convection condensation of R-22 inside horizontal tubes, International Journal of Multiphase Flow 24 (1998) 635–650.

[76] B. Rezaeianjouybari, M. Sheikholeslami, A. Shafee, H. Babazadeh, A novel bayesian optimization for flow condensation enhancement using nanorefrigerant: a combined analytical and experimental study, Chemical Engineering Science 215 (2020) 115465, https://doi.org/10.1016/j.ces.2019.115465.

[77] P. Zamankhan, Heat transfer in counter flow heat exchangers with helical turbulators, Communications in Nonlinear Science and Numerical Simulation 15 (2010) 2894–2907.

[78] E.K. Akpinar, Evaluation of heat transfer and exergy loss in a concentric double pipe exchanger equipped with helical wires, Energy Conversion and Management 47 (2006) 3473–3486.

[79] D. Muñoz-Esparza, E. Sanmiguel-Rojas, Numerical simulations of the laminar flow in pipes with wire coil inserts, Computers and Fluids 44 (2011) 169–177.

[80] P. Promvonge, Thermal performance in circular tube fitted with coiled square wires, Energy Conversion and Management 49 (2008) 980–987.

[81] P. Promvonge, Thermal augmentation in circular tube with twisted tape and wire coil turbulators, Energy Conversion and Management 49 (2008) 2949–2955.

[82] S. Eiamsa-ard, P. Nivesrangsan, S. Chokphoemphun, P. Promvonge, Influence of combined non-uniform wire coil and twisted tape inserts on thermal performance characteristics, International Communications in Heat and Mass Transfer 37 (2010) 850–856.

[83] M.T. Naik, S.S. Fahad, L. Syam Sundar, M.K. Singh, Comparative study on thermal performance of twisted tape and wire coil inserts in turbulent flow using CuO/water nanofluid, Experimental Thermal and Fluid Science 57 (2014) 65–76.

[84] P. Promvonge, Thermal enhancement in a round tube with snail entry and coiled-wire inserts, International Communications in Heat and Mass Transfer 35 (2008) 623–629.

[85] P. Naphon, T. Suchana, Heat transfer enhancement and pressure drop of the horizontal concentric tube with twisted wires brush inserts, International Communications in Heat and Mass Transfer 38 (2011) 236–241.

[86] S. Eiamsa-ard, P. Promvonge, Thermal characterization of turbulent tube flows over diamond-shaped elements in tandem, International Journal of Thermal Sciences 49 (2010) 1051–1062.

[87] H. Karakaya, A. Durmus, Heat transfer and exergy loss in conical spring turbulators, International Journal of Heat and Mass Transfer 60 (2013) 756–762.

[88] M. Sheikholeslami, Effect of uniform suction on nanofluid flow and heat transfer over a cylinder, Journal of the Brazilian Society of Mechanical Sciences and Engineering 37 (2015) 1623–1633.

[89] M. Sheikholeslami, Lattice Boltzmann Method simulation of MHD non-Darcy nanofluid free convection, Physica B 516 (2017) 55–71.

[90] M. Sheikholeslami, Magnetic field influence on CuO -H_2O nanofluid convective flow in a permeable cavity considering various shapes for nanoparticles, International Journal of Hydrogen Energy 42 (2017) 19611–19621.

[91] M. Sheikholeslami, Numerical simulation for solidification in a LHTESS by means of Nano-enhanced PCM, Journal of the Taiwan Institute of Chemical Engineers 86 (2018) 25–41.

[92] M. Sheikholeslami, Numerical modeling of Nano enhanced PCM solidification in an enclosure with metallic fin, Journal of Molecular Liquids 259 (2018) 424–438.

[93] Y. Qin, J. Luo, Z. Chen, G. Mei, L.-E. Yan, Measuring the albedo of limited-extent targets without the aid of known-albedo masks, Solar Energy 171 (2018) 971–976.

[94] Y. Qin, A review on the development of cool pavements to mitigate urban heat island effect, Renewable and Sustainable Energy Reviews 52 (2015) 445–459.

[95] M. Sheikholeslami, Solidification of NEPCM under the effect of magnetic field in a porous thermal energy storage enclosure using CuO nanoparticles, Journal of Molecular Liquids 263 (2018) 303–315.

[96] M. Sheikholeslami, Influence of magnetic field on Al_2O_3-H_2O nanofluid forced convection heat transfer in a porous lid driven cavity with hot sphere obstacle by means of LBM, Journal of Molecular Liquids 263 (2018) 472–488.

[97] M. Sheikholeslami, New computational approach for exergy and entropy analysis of nanofluid under the impact of Lorentz force through a porous media, Computer Methods in Applied Mechanics and Engineering 344 (2019) 319–333.

[98] M. Sheikholeslami, Numerical approach for MHD Al_2O_3-water nanofluid transportation inside a permeable medium using innovative computer method, Computer Methods in Applied Mechanics and Engineering 344 (2019) 306–318.

[99] X. Ma, M. Sheikholeslami, M. Jafaryar, S. Ahmad, T. Nguyen-Thoi, Z. Li, Solidification inside a clean energy storage unit utilizing phase change material with copper oxide nanoparticles, Journal of Cleaner Production (2019), https://doi.org/10.1016/j.jclepro.2019.118888.

[100] Y. Qin, Pavement surface maximum temperature increases linearly with solar absorption and reciprocal thermal inertial, International Journal of Heat and Mass Transfer 97 (2016) 391–399.

[101] M. Sheikholeslami, O. Mahian, Enhancement of PCM solidification using inorganic nanoparticles and an external magnetic field with application in energy storage systems, Journal of Cleaner Production 215 (2019) 963–977.

[102] M. Sheikholeslami, B. Rezaeianjouybari, M. Darzi, A. Shafee, Z. Li, T.K. Nguyen, Application of nano-refrigerant for boiling heat transfer enhancement employing an experimental study, International Journal of Heat and Mass Transfer 141 (2019) 974–980.

[103] M. Sheikholeslami, M. Jafaryar, M. Hedayat, A. Shafee, Z. Li, T.K. Nguyen, M. Bakouri, Heat transfer and turbulent simulation of nanomaterial due to compound turbulator including irreversibility analysis, International Journal of Heat and Mass Transfer 137 (2019) 1290–1300.

[104] M. Sheikholeslami, Application of Darcy law for nanofluid flow in a porous cavity under the impact of Lorentz forces, Journal of Molecular Liquids 266 (2018) 495–503.

[105] M. Sheikholeslami, Finite element method for PCM solidification in existence of CuO nanoparticles, Journal of Molecular Liquids 265 (2018) 347–355.

[106] M. Sheikholeslami, M. Jafaryar, A. Shafee, Z. Li, R.-ul Haq, Heat transfer of nanoparticles employing innovative turbulator considering entropy generation, International Journal of Heat and Mass Transfer 136 (2019) 1233–1240.

[107] M. Sheikholeslami, R.-ul Haq, A. Shafee, Z. Li, Y.G. Elaraki, I. Tlili, Heat transfer simulation of heat storage unit with nanoparticles and fins through a heat exchanger, International Journal of Heat and Mass Transfer 135 (2019) 470–478.

[108] M. Sheikholeslami, R.-ul Haq, A. Shafee, Z. Li, Heat transfer behavior of Nanoparticle enhanced PCM solidification through an enclosure with V shaped fins, International Journal of Heat and Mass Transfer 130 (2019) 1322–1342.

[109] Y. Qin, H. He, X. Ou, T. Bao, Experimental study on darkening water-rich mud tailings for accelerating desiccation, Journal of Cleaner Production (2019) 118235, https://doi.org/10.1016/j.jclepro.2019.118235.

[110] Y. Qin, J.E. Hiller, D. Meng, Linearity between pavement thermophysical properties and surface temperatures, Journal of Materials in Civil Engineering (2019), https://doi.org/10.1061/(ASCE)MT.1943-5533.0002890.

[111] M. Sheikholeslami, S.A. Shehzad, Z. Li, A. Shafee, Numerical modeling for alumina nanofluid magnetohydrodynamic convective heat transfer in a permeable medium using Darcy law, International Journal of Heat and Mass Transfer 127 (2018) 614–622.

[112] M. Sheikholeslami, Z. Li, A. Shafee, Lorentz forces effect on NEPCM heat transfer during solidification in a porous energy storage system, International Journal of Heat and Mass Transfer 127 (2018) 665–674.

[113] M. Sheikholeslami, M. Jafaryar, S. Saleem, Z. Li, A. Shafee, Y. Jiang, Nanofluid heat transfer augmentation and exergy loss inside a pipe equipped with innovative turbulators, International Journal of Heat and Mass Transfer 126 (2018) 156–163.

[114] Y. Qin, M. Zhang, J.E. Hiller, Theoretical and experimental studies on the daily accumulative heat gain from cool roofs, Energy 129 (2017) 138–147.

[115] Y. Qin, H. He, A new simplified method for measuring the albedo of limited extent targets, Solar Energy 157 (Suppl. C) (2017) 1047–1055.

[116] M. Sheikholeslami, A. Ghasemi, Z. Li, A. Shafee, S. Saleem, Influence of CuO nanoparticles on heat transfer behavior of PCM in solidification process considering radiative source term, International Journal of Heat and Mass Transfer 126 (2018) 1252–1264.

[117] M. Sheikholeslami, M. Darzi, Z. Li, Experimental investigation for entropy generation and exergy loss of nano-refrigerant condensation process, International Journal of Heat and Mass Transfer 125 (2018) 1087–1095.

[118] M. Sheikholeslami, S.A. Shehzad, Z. Li, Water based nanofluid free convection heat transfer in a three dimensional porous cavity with hot sphere obstacle in existence of Lorenz forces, International Journal of Heat and Mass Transfer 125 (2018) 375–386.

[119] Y. Qin, Y. He, J.E. Hiller, G. Mei, A new water-retaining paver block for reducing runoff and cooling pavement, Journal of Cleaner Production 199 (2018) 948–956.

[120] Y. Qin, Y. Zhao, X. Chen, L. Wang, F. Li, T. Bao, Moist curing increases the solar reflectance of concrete, Construction and Building Materials 215 (2019) 114–118.

[121] M. Sheikholeslami, M. Jafaryar, Z. Li, Nanofluid turbulent convective flow in a circular duct with helical turbulators considering CuO nanoparticles, International Journal of Heat and Mass Transfer 124 (2018) 980–989.

[122] M. Sheikholeslami, A. Ghasemi, Solidification heat transfer of nanofluid in existence of thermal radiation by means of FEM, International Journal of Heat and Mass Transfer 123 (2018) 418–431.

[123] Y. Qin, M. Zhang, G. Mei, A new simplified method for measuring the permeability characteristics of highly porous media, Journal of Hydrology 562 (2018) 725–732.

[124] Y. Qin, Urban canyon albedo and its implication on the use of reflective cool pavements, Energy and Buildings 96 (2015) 86–94.

[125] M. Sheikholeslami, S.A. Shehzad, CVFEM simulation for nanofluid migration in a porous medium using Darcy model, International Journal of Heat and Mass Transfer 122 (2018) 1264–1271.

[126] M. Sheikholeslami, M. Darzi, M.K. Sadoughi, Heat transfer improvement and pressure drop during condensation of refrigerant-based nanofluid; an experimental procedure, International Journal of Heat and Mass Transfer 122 (2018) 643–650.

[127] Y. Qin, Y. He, B. Wu, S. Ma, X. Zhang, Regulating top albedo and bottom emissivity of concrete roof tiles for reducing building heat gains, Energy and Buildings 156 (Suppl. C) (2017) 218–224.

[128] M. Sheikholeslami, S.A. Shehzad, Simulation of water based nanofluid convective flow inside a porous enclosure via Non-equilibrium model, International Journal of Heat and Mass Transfer 120 (2018) 1200–1212.

[129] M. Sheikholeslami, M. Seyednezhad, Simulation of nanofluid flow and natural convection in a porous media under the influence of electric field using CVFEM, International Journal of Heat and Mass Transfer 120 (2018) 772–781.

[130] M. Sheikholeslami, H.B. Rokni, Numerical simulation for impact of Coulomb force on nanofluid heat transfer in a porous enclosure in presence of thermal radiation, International Journal of Heat and Mass Transfer 118 (2018) 823–831.

[131] Y. Qin, J. Liang, H. Yang, Z. Deng, Gas permeability of pervious concrete and its implications on the application of pervious pavements, Measurement 78 (2016) 104–110.

[132] Y. Qin, J.E. Hiller, Understanding pavement-surface energy balance and its implications on cool pavement development, Energy and Buildings 85 (2014) 389–399.

[133] M. Sheikholeslami, S.A. Shehzad, Numerical analysis of $Fe_3O_4 - H_2O$ nanofluid flow in permeable media under the effect of external magnetic source, International Journal of Heat and Mass Transfer 118 (2018) 182–192.

[134] M. Sheikholeslami, M.K. Sadoughi, Simulation of CuO- water nanofluid heat transfer enhancement in presence of melting surface, International Journal of Heat and Mass Transfer 116 (2018) 909–919.

[135] M. Sheikholeslami, H.B. Rokni, Simulation of nanofluid heat transfer in presence of magnetic field: a review, International Journal of Heat and Mass Transfer 115 (2017) 1203–1233.

[136] M. Sheikholeslami, T. Hayat, A. Alsaedi, On simulation of nanofluid radiation and natural convection in an enclosure with elliptical cylinders, International Journal of Heat and Mass Transfer 115 (2017) 981–991.

[137] M. Sheikholeslami, S.A. Shehzad, CVFEM for influence of external magnetic source on $Fe_3O_4 - H_2O$ nanofluid behavior in a permeable cavity considering shape effect, International Journal of Heat and Mass Transfer 115 (2017) 180–191.

[138] M. Sheikholeslami, M. Seyednezhad, Nanofluid heat transfer in a permeable enclosure in presence of variable magnetic field by means of CVFEM, International Journal of Heat and Mass Transfer 114 (2017) 1169–1180.

[139] M. Sheikholeslami, H.B. Rokni, Melting heat transfer influence on nanofluid flow inside a cavity in existence of magnetic field, International Journal of Heat and Mass Transfer 114 (2017) 517–526.

[140] L. Yang, J.N. Huang, W. Ji, M. Mao, Investigations of a new combined application of nanofluids in heat recovery and air purification, Powder Technology (2020), https://doi.org/10.1016/j.powtec.2019.10.053.

[141] M. Sheikholeslami, M. Sadoughi, Mesoscopic method for MHD nanofluid flow inside a porous cavity considering various shapes of nanoparticles, International Journal of Heat and Mass Transfer 113 (2017) 106–114.

[142] M. Sheikholeslami, M.A. Sheremet, A. Shafee, I. Tlili, Simulation of nanoliquid thermogravitational convection within a porous chamber imposing magnetic and radiation impacts, Physica A: Statistical Mechanics and its Applications (2020), https://doi.org/10.1016/j.physa.2019.124058.

[143] M. Sheikholeslami, M.M. Bhatti, Forced convection of nanofluid in presence of constant magnetic field considering shape effects of nanoparticles, International Journal of Heat and Mass Transfer 111 (2017) 1039–1049.

[144] M. Sheikholeslami, M.M. Bhatti, Active method for nanofluid heat transfer enhancement by means of EHD, International Journal of Heat and Mass Transfer 109 (2017) 115–122.

[145] M. Sheikholeslami, S.A. Shehzad, Thermal radiation of ferrofluid in existence of Lorentz forces considering variable viscosity, International Journal of Heat and Mass Transfer 109 (2017) 82–92.

[146] Y. Qin, J. Liang, K. Tan, F. Li, A side by side comparison of the cooling effect of building blocks with retro-reflective and diffuse-reflective walls, Solar Energy 133 (2016) 172–179.

[147] M. Sheikholeslami, H.B. Rokni, Nanofluid two phase model analysis in existence of induced magnetic field, International Journal of Heat and Mass Transfer 107 (2017) 288–299.

[148] M. Sheikholeslami, A.N. Keshteli, H. Babazadeh, Nanoparticles favorable effects on performance of thermal storage units, Journal of Molecular Liquids 300 (2020) 112329.

[149] M. Sheikholeslami, S.A. Shehzad, Magnetohydrodynamic nanofluid convection in a porous enclosure considering heat flux boundary condition, International Journal of Heat and Mass Transfer 106 (2017) 1261–1269.

[150] M. Sheikholeslami, R. Ellahi, Three dimensional mesoscopic simulation of magnetic field effect on natural convection of nanofluid, International Journal of Heat and Mass Transfer 89 (2015) 799–808.

[151] L. Yang, K. Du, Z. Zhang, Heat transfer and flow optimization of a novel sinusoidal minitube filled with non-Newtonian SiC/EG-water nanofluids, International Journal of Mechanical Sciences 168 (2019), https://doi.org/10.1016/j.ijmecsci.2019.105310.

[152] M. Sheikholeslami, Numerical investigation of nanofluid free convection under the influence of electric field in a porous enclosure, Journal of Molecular Liquids 249 (2018) 1212–1221.

[153] M. Sheikholeslami, CuO-water nanofluid flow due to magnetic field inside a porous media considering Brownian motion, Journal of Molecular Liquids 249 (2018) 921–929.

[154] M. Sheikholeslami, Magnetic field influence on nanofluid thermal radiation in a cavity with tilted elliptic inner cylinder, Journal of Molecular Liquids 229 (2017) 137–147.

[155] M. Sheikholeslami, Influence of Lorentz forces on nanofluid flow in a porous cylinder considering Darcy model, Journal of Molecular Liquids 225 (2017) 903–912.

[156] M. Sheikholeslami, Influence of Coulomb forces on Fe_3O_4-H_2O nanofluid thermal improvement, International Journal of Hydrogen Energy 42 (2017) 821–829.

[157] M. Sheikholeslami, S.A. Shehzad, F.M. Abbasi, Z. Li, Nanofluid flow and forced convection heat transfer due to Lorentz forces in a porous lid driven cubic enclosure with hot obstacle, Computer Methods in Applied Mechanics and Engineering 338 (2018) 491–505.

[158] M. Sheikholeslami, M. Barzegar Gerdroodbary, R. Moradi, A. Shafee, Z. Li, Application of neural network for estimation of heat transfer treatment of Al_2O_3- H_2O nanofluid through a channel, Computer Methods in Applied Mechanics and Engineering 344 (2019) 1–12.

[159] M. Sheikholeslami, A. Arabkoohsar, I. Khan, A. Shafee, Z. Li, Impact of Lorentz forces on Fe_3O_4-water ferrofluid entropy and exergy treatment within a permeable semi annulus, Journal of Cleaner Production 221 (2019) 885–898.

[160] M. Sheikholeslami, K. Vajravelu, M.M. Rashidi, Forced convection heat transfer in a semi annulus under the influence of a variable magnetic field, International Journal of Heat and Mass Transfer 92 (2016) 339–348.

[161] A. Shafee, M. Sheikholeslami, M. Jafaryar, H. Babazadeh, Utilizing copper oxide nanoparticles for expedition of solidification within a storage system, Journal of Molecular Liquids 302 (2020) 112371, https://doi.org/10.1016/j.molliq.2019.112371.

[162] M. Sheikholeslami, T. Hayat, A. Alsaedi, S. Abelman, Numerical analysis of EHD nanofluid force convective heat transfer considering electric field dependent viscosity, International Journal of Heat and Mass Transfer 108 (2017) 2558–2565.

[163] M. Sheikholeslami, Influence of magnetic field on nanofluid free convection in an open porous cavity by means of Lattice Boltzmann method, Journal of Molecular Liquids 234 (2017) 364–374.

[164] M. Sheikholeslami, Magnetohydrodynamic nanofluid forced convection in a porous lid driven cubic cavity using Lattice Boltzmann method, Journal of Molecular Liquids 231 (2017) 555–565.

[165] M. Sheikholeslami, Numerical investigation for CuO-H_2O nanofluid flow in a porous channel with magnetic field using mesoscopic method, Journal of Molecular Liquids 249 (2018) 739–746.

[166] M. Sheikholeslami, H.B. Rokni, Numerical modeling of nanofluid natural convection in a semi annulus in existence of Lorentz force, Computer Methods in Applied Mechanics and Engineering 317 (2017) 419–430.

[167] F. Li, M. Sheikholeslami, R.N. Dara, M. Jafaryar, A. Shafee, T. Nguyen-Thoi, Z. Li, Numerical study for nanofluid behavior inside a storage finned enclosure involving melting process, Journal of Molecular Liquids 297 (2020) 111939.

[168] M. Sheikholeslami, A. Zeeshan, Analysis of flow and heat transfer in water based nanofluid due to magnetic field in a porous enclosure with constant heat flux using CVFEM, Computer Methods in Applied Mechanics and Engineering 320 (2017) 68–81.

CHAPTER

Nanofluid for heat exchangers

2

2.1 Introduction

Nanofluids are produced by dispersing the nanometer-scale solid particles into base liquids with low thermal conductivity such as water, ethylene glycol (EG), oils, etc. Control of heat transfer in many energy systems is crucial due to the increase in energy prices. In recent years, nanofluids technology is proposed and studied by some researchers experimentally or numerically to control heat transfer in a process. The nanofluid can be applied to engineering problems, such as heat exchangers, cooling of electronic equipment and chemical processes. There are two ways for simulation of nanofluid: single phase and two phase. In the first method, researchers assumed that nanofluids treated as the common pure fluid and conventional equations of mass, momentum, and energy are used, and the only effect of nanofluid is its thermal conductivity and viscosity which are obtained from the theoretical models or experimental data. These researchers assumed that nanoparticles are in thermal equilibrium and there are not any slip velocities between the nanoparticles and fluid molecules, thus they have a uniform mixture of nanoparticles. In the second method, researchers assumed that there are slip velocities between nanoparticles and fluid molecules. So the volume fraction of nanofluids may not be uniform anymore and there would be a variable concentration of nanoparticles in a mixture. There are several numerical and semianalytical methods which have been used by several authors in order to simulate nanofluid flow and heat transfer.

2.1.1 Definition of nanofluid

Low thermal conductivity of conventional heat transfer fluids such as water, oil, and ethylene glycol mixture is a serious limitation in improving the performance and compactness of many engineering equipment such as heat exchangers and electronic devices. To overcome this disadvantage, there is strong motivation to develop advanced heat transfer fluids with substantially higher conductivity. An innovative way of improving the thermal conductivities of fluids is to suspend small solid particles in the fluid. Various types of powders such as metallic, nonmetallic, and polymeric particles can be added into fluids to form slurries. The thermal conductivities of fluids with suspended particles are expected to be higher than that of

Nanofluid in Heat Exchangers for Mechanical Systems. https://doi.org/10.1016/B978-0-12-821923-2.00002-6

common fluids. Nanofluids are a new kind of heat transfer fluid containing a small quantity of nanosized particles (usually less than 100 nm) that are uniformly and stably suspended in a liquid. The dispersion of a small amount of solid nanoparticles in conventional fluids changes their thermal conductivity remarkably. Compared with the existing techniques for enhancing heat transfer, the nanofluids show a superior potential for increasing heat transfer rates in a variety of cases [1].

2.1.2 Model description

In the literature, convective heat transfer with nanofluids can be modeled using mainly the two-phase or single-phase approach. In the two-phase approach, the velocity between the fluid and particles might not be zero [2] due to several factors such as gravity, friction between the fluid and solid particles, Brownian forces, Brownian diffusion, sedimentation, and dispersion. In the second approach, the nanoparticles can be easily fluidized and therefore, one may assume that the motion slip between the phases, if any, would be considered negligible [3]. The latter approach is simpler and more computationally efficient.

2.1.3 Conservation equations

2.1.3.1 Single-phase model

Although nanofluids are solid–liquid mixtures, the approach conventionally used in most studies of natural convection handles the nanofluid as a single-phase (homogenous) fluid. In fact, due to the extreme size and low concentration of the suspended nanoparticles, the particles are assumed to move with same velocity as the fluid. Also, by considering the local thermal equilibrium, the solid particle–liquid mixture may then be approximately considered to behave as a conventional single-phase fluid with properties that are to be evaluated as functions of those of the constituents. The governing equations for a homogenous analysis of natural convection are continuity, momentum, and energy equations with their density, specific heat, thermal conductivity, and viscosity modified for nanofluid application. The specific governing equations for various studied enclosures are not shown here, and they can be found in different references [4]. It should be mentioned that sometime this assumption is not correct. For example, Ding and Wen [5], this assumption may not always remain true for a nanofluid. They investigated the particle migration in a nanofluid for a pipe flow and stated that at Peclet numbers exceeding 10 the particle distribution is significantly nonuniform. Nevertheless, many studies have performed the numerical simulation using single-phase assumption and reported acceptable results for the heat transfer and hydrodynamic properties of the flow.

2.1.3.2 Two-phase model

Several authors have tried to establish convective transport models for nanofluids [6]. Nanofluid is a two-phase mixture in which the solid phase consists of nanosized particles. In view of the nanoscale size of the particles, it may be questionable whether the theory of conventional two-phase flow can be applied in describing the flow characteristics of nanofluid. On the other hand, several factors such as gravity, friction between the fluid and solid particles and Brownian forces, the phenomena of Brownian diffusion, sedimentation, and dispersion may affect a nanofluid flow. Consequently, the slip velocity between the fluid and particles cannot be neglected for simulating nanofluid flows. Since the two-phase approach considers the movement between the solid and fluid molecule, it may have better prediction in nanofluid study. To fully describe and predict the flow and behavior of complex flows, different multiphase theories have been proposed and used. The large number of published articles concerning multiphase flows typically employed the mixture theory to predict the behavior of nanofluids [7]. A comprehensive survey of convective transport in nanofluids was made by Buongiorno [8], using a model in which Brownian motion and thermophoresis are accounted for. Buongiorno developed a two-component four-equation nonhomogeneous equilibrium model for mass, momentum, and heat transfer in nanofluids. The nanofluid is treated as a two-component mixture (base fluidþnanoparticles) with the following assumptions: no chemical reactions; negligible external forces; dilute mixture ($\phi = 1$); negligible viscous dissipation; negligible radiative heat transfer; nanoparticle and base fluid locally in thermal equilibrium. Invoking the above assumptions, the following equations represent the mathematical formulation of the nonhomogenous single-phase model for the governing equations as formulated by Buongiorno [8]:

2.1.3.2.1 Continuity equation

$$\nabla . v = 0 \tag{2.1}$$

where v is the velocity.

2.1.3.2.2 Nanoparticle continuity equation

$$\frac{\partial \phi}{\partial t} + v . \nabla \phi = \nabla . \left(D_B \nabla \phi + D_T \frac{\nabla T}{T} \right) \tag{2.2}$$

Here ϕ is nanoparticle volume fraction, D_B is the Brownian diffusion coefficient given by the Einstein–Stokes's equation:

$$D_B = \frac{k_B T}{3 \pi \mu d_p} \tag{2.3}$$

where μ is the viscosity of the fluid, d_p is the nanoparticle diameter, $k_B = 1.385 \times 10^{-23}$ is Boltzmann constant and D_T is the thermophoretic diffusion coefficient, which is defined as

$$D_T = \left(\frac{\mu}{\rho} \right) \left(0.26 \frac{k}{k + k_p} \right) \tag{2.4}$$

In Eq. (2.4), k and k_p are the thermal conductivity of the fluid and particle materials, respectively.

2.1.3.2.3 Momentum equation

$$v.\nabla v = -\frac{1}{\rho_{nf}}\nabla p + \nabla.\tau + g \tag{2.5}$$

where

$$\tau = -\mu_{nf}\left(\nabla v + (\nabla v)^t\right) \tag{2.6}$$

where the superscript "t" indicates the transpose of ∇v. Also p is pressure.

2.1.3.2.4 Energy equation

$$v.\nabla T = \nabla(\alpha_{nf}\nabla T) + \frac{\rho_p c_p}{\rho_{nf} c_{nf}}\left(D_B \nabla\phi.\nabla T + D_T \frac{\nabla T.\nabla T}{T}\right) \tag{2.7}$$

where ϕ and T are nanoparticle concentration and temperature of nanofluid, respectively.

This nanofluid model can be characterized as a "two-fluid" (nanoparticles + base fluid), four-equation (mass, momentum, energy), nonhomogeneous (nanoparticle/fluid slip velocity allowed) equilibrium (nanoparticle/fluid temperature differences not allowed) model. Note that the conservation equations are strongly coupled. That is, v depends on ϕ via viscosity; ϕdepends on T mostly because of thermophoresis; T depends on ϕ via thermal conductivity and also via the Brownian and thermophoretic terms in the energy equation; ϕ and T obviously depends on v because of the convection terms in the nanoparticle continuity and energy equations, respectively.

In a numerical study by Behzadmehr et al. [9], for the first time a two-phase mixture model was implemented to investigate the behavior of Cu–water nanofluid in a tube and the results were also compared with previous works using a single-phase approach. The authors claimed that the simulation done by assuming that basefluid and particles behave separately possessed results that are more precise compared to the previous computational modeling. They implemented the mixture theory for their work. It was suggested that the continuity, momentum, and energy equations be written for a mixture of fluid and a solid phase. Some assumptions were also stated for the model such as a strong coupling between two phases and the fluid being closely followed by the particles with each phase owning a different velocity leading to a term called slip velocity of nanoparticles as in Eq. (2.8):

$$V_{pf} = V_P - V_f = \frac{\rho_p d_p^2}{18\mu_f f_{drag}} \frac{(\rho_p - \rho_m)}{\rho_p} a, \quad a = g - (V_m.\nabla)V_m, f_{drag} = \begin{cases} 1 + 0.15\mathrm{Re}_p^{0.687} \\ 0.0183\mathrm{Re}_p \end{cases} \tag{2.8}$$

The conservation equations (continuity, momentum, and energy, respectively) will be written for the mixture as follows:

$$\nabla.(\rho_m V_m) = 0 \tag{2.9}$$

$$\nabla.(\rho_m V_m V_m) = -\nabla P_m + \nabla.[\tau - \tau_t] + \rho_m g + \nabla.\left(\sum_{k=1}^{n} \phi_k \rho_k V_{dr,k} V_{dr,k}\right) \tag{2.10}$$

$$\nabla.(\phi_p V_k(\rho_k h_k + p)) = \nabla.(k_{eff} \nabla T - C_p \rho_m vt) \tag{2.11}$$

where $V_{dr,p}$ is the particle draft velocity that is related to the slip velocity and is defined as:

$$V_{dr,p} = V_P - V_f = V_{pf} - \sum_{k=1}^{n} \frac{\phi_k \rho_k}{\rho_m} V_{fk} \tag{2.12}$$

2.1.4 Physical properties of the nanofluids for single-phase model

Base nanofluid properties have been published over the past few years in literature. However, only recently have some data on temperature-dependent properties been provided, even though they are only for nanofluid effective thermal conductivity and effective absolute viscosity.

2.1.4.1 Density

In the absence of experimental data for nanofluid densities, constant-value temperature independent values, based on nanoparticle volume fraction, are used:

$$\rho_{nf} = \rho_f(1 - \phi) + \rho_p \phi \tag{2.13}$$

2.1.4.2 Specific heat capacity

It has been suggested that the effective specific heat can be calculated using the following equation as reported in Refs. [10] as

$$(C_p)_{nf} = (C_p)_f(1 - \phi) + (C_p)_p \phi \tag{2.14}$$

Other authors suggest an alternative approach based on heat capacity concept [11]:

$$(\rho C_p)_{nf} = (\rho C_p)_f(1 - \phi) + (\rho C_p)_p \phi \tag{2.15}$$

These two formulations may of course lead to different results for specific heat. Due to the lack of experimental data, both formulations are considered equivalent in estimating nanofluid specific heat capacity [12].

2.1.4.3 Thermal expansion coefficient

Thermal expansion coefficient of nanofluid can be obtained as follows [1]:

$$(\rho\beta)_{nf} = (\rho\beta)_f(1 - \phi) + (\rho\beta)_p \phi \tag{2.16}$$

2.1.4.4 The electrical conductivity

The effective electrical conductivity of nanofluid was presented by Maxwell [13] as below:

$$\sigma_{nf} / \sigma_f = 1 + 3(\sigma_P / \sigma_f - 1)\phi/((\sigma_P / \sigma_f + 2) - (\sigma_P / \sigma_f - 1)\phi) \quad (2.17)$$

2.1.4.5 Dynamic viscosity

Various models have been suggested to model the viscosity of a nanofluid mixture that takes into account the percentage of nanoparticles suspended in the base fluid. The classic Brinkman model [14] seems to be a proper one which has been extensively used in the studies on numerical simulation concerning nanofluids. Eq. (2.1) shows the relation between the nanofluid viscosity, basefluid viscosity, as well as the nanoparticle concentration in this model.

$$\mu_{nf} = \mu_f/(1 - \phi)^{2.5} \quad (2.18)$$

However, in some recent computational studies, other models have been selected to be used in the numerical process, like the work done by Abu-neda and Chamkha [15] to investigate the convection of CuO–ethylene glycol–water nanofluid in an enclosure where Namburu correlation for viscosity [16] was applied:

$$\log(\mu_{nf}) = Ae^{-BT} \quad (2.19)$$

where

$$\begin{aligned} A &= 1.837\phi^2 - 29.643\phi + 165.65 \\ B &= 4 \times 10^{-6}\phi^2 - 0.001\phi + 0.0186 \end{aligned} \quad (2.20)$$

In their study, the results where compared with that of viscosity modeled by Brinkman. It was outlined that as far as a value for normalized average Nusselt number for the fluid is concerned, for various values of Rayleigh number, Brinkman model owns a prediction of higher value compared with that for Namburu model showing the notable role of viscosity model used in the calculations. The authors also state that a combination of different models might as well be implemented that will show different dependence on volume concentration as well as the geometry aspect ratio yet along with the limitation that the models include only the ones mentioned in the study. Other studies have also shown that different models might lead to different results, like that by a number of suggested relations for viscosity models used in numerical studies are also presented in Table 2.1.

2.1.4.6 Thermal conductivity

Different nanofluid models based on a combination of the different formulas for the thermal conductivity adopted in the studies of natural convection are summarized in Table 2.2. Also Table 2.3 demonstrates values of thermophysical properties for different materials used as suspended particles in nanofluids. Scientists provided several applications of new carrier fluids [33–112].

Table 2.1 Different models for viscosity of nanofluids used in simulation.

Model	Equation
Einstein model [17]	$\mu_{nf} = (2.5\phi + 1)\mu_f, \ \phi < 0.05$
Pak and Cho's correlation [10]	$\mu_{nf} = \mu_f(1 + 39.11\phi + 533.9\phi^2)$
Jang et al. model [18]	$\mu_{nf} = (2.5\phi + 1)\mu_f\left[1 + \eta(d_p/H)^{-2\varepsilon}\phi^{2/3}(\varepsilon + 1)\right]$
Koo and Kleinstreuer [19]	$\mu_{nf} = 5 \times 10^4 \beta\phi\rho_f\sqrt{\frac{k_B T}{d_p\rho_p}}f(T,\phi), \ \begin{cases} \beta = 0.0137(100\phi)^{-0.8229} \ for \ \phi < 1\% \\ \beta = 0.0011(100\phi)^{-0.7272} \ for \ \phi > 1\% \end{cases}$
Maiga model [20]	$\mu_{nf} = \mu_f(1 + 7.3\phi + 123\phi^2)$
Brownian model [21]	$\mu_{nf} = \mu_f(1 + 2.5\phi + 6.17\phi^2)$
Nguyen model [22]	$\mu_{nf} = \mu_f(1 + 0.025\phi + 0.015\phi^2)$
Masoumi et al. [23]	$\mu_{nf} = \mu_f + \rho_p V_B d_p^2 / (72C\delta)$
Gherasim et al. [24]	$\mu_{nf} = \mu_f 0.904 e^{14.8\phi}$

Table 2.2 Different models for thermal conductivity of nanofluids used in simulation.

Model	Equation
Koo and Kleinstreuer [19]	$k_{nf} = k_f\left[\frac{k_p + 2k_f - 2\phi(k_f - k_p)}{k_p + 2k_f + \phi(k_f - k_p)}\right] + 5 \times 10^4 \beta\phi\rho_f (C_p)_f \sqrt{\frac{k_B T}{d_p\rho_p}} f(T,\phi)$
Jang and Choi model [25]	$\frac{k_{nf}}{k_f} = (1 - \phi) + Bk_p\phi + 18 \times 10^6 \frac{3d_f}{d_p} k_f \mathrm{Re}_{dp}^2 \Pr \phi$
Bruggeman model [26]	$k_{nf} = 0.25k_f (3\phi - 1)k_p / k_f + [3(1 - \phi) - 1] + \sqrt{\Delta_B}$ $\Delta_B = [(3\phi - 1)k_p/k_f + (3(1 - \phi) - 1)]^2 + 8k_p / k_f$
Chon et al. model [27]	$k_{nf} = k_f\left[1 + 64.7\phi^{0.7640}(d_f/d_p)^{0.369}(k_f/k_p)^{0.7476}\Pr_T^{0.9955}\mathrm{Re}^{1.2321}\right]$
Charuyakorn et al. [28]	$\frac{k_{nf}}{k_f} = \left[\frac{k_p + 2k_f - 2\phi(k_f - k_p)}{k_p + 2k_f + \phi(k_f - k_p)}\right]\left(1 + b\phi Pe_p^m\right)$
Stationary model [29]	$k_{nf} = k_f[1 + k_p\phi d_f / (k_f(1 - \phi)d_p)]$
Yu and Choi [30]	$\frac{k_{nf}}{k_f} = \left[\frac{k_p + 2k_f - 2\phi(k_f - k_p)(1+\eta)^3}{k_p + 2k_f + \phi(k_f - k_p)(1+\eta)^3}\right]$
Patel et al. [31]	$\frac{k_{nf}}{k_f} = 1 + \frac{k_p d_f \phi}{k_f d_p (1 - \phi)}\left[1 + c\frac{2k_B T d_p}{\pi\alpha_f\mu_f d_p^2}\right]$
Mintsa et al. [32]	$k_{nf} = k_f(1.72\phi + 1.0)$

Table 2.3 The thermophysical properties of the nanofluid.

	$\rho(kg/m^3)$	$C_p(j/kgk)$	k(W/m.k)	$\beta(K^{-1})$	$\sigma(\Omega \times m)^{-1}$
Pure water	997.1	4179	0.613	21×10^{-5}	0.05
Copper (Cu)	8933	385	401	1.67×10^{-5}	5.96×10^7
Silver (Ag)	10 500	235	429	1.89×10^{-5}	3.60×10^7
Alumina (Al_2O_3)	3970	765	40	0.85×10^{-5}	1×10^{-10}
Titanium oxide (TiO_2)	4250	686.2	8.9538	0.9×10^{-5}	1×10^{-12}

References

[1] K. Khanafer, K. Vafai, M. Lightstone, Buoyancy-driven heat transfer enhancement in a two-dimensional enclosure utilizing nanofluids, International Journal of Heat and Mass Transfer 46 (2003) 3639–3653.

[2] Y. Xuan, Q. Li, Heat transfer enhancement of nanofluids, International Journal of Heat and Fluid Flow 21 (2000) 58–64.

[3] Y. Xuan, W. Roetzel, Conceptions for heat transfer correlations of nanofluids, International Journal of Heat and Mass Transfer 43 (2000) 3701–3707.

[4] E. Abu-Nada, H.F. Oztop, Effects of inclination angle on natural convection in enclosures filled with Cu–water nanofluid, International Journal of Heat and Fluid Flow 30 (2009) 669–678.

[5] Y. Ding, D. Wen, Particle migration in a flow of nanoparticle suspensions, Powder Technology 149 (2–3) (2005) 84–92.

[6] A. Akbarinia, R. Laur, Investigation the diameter of solid particles effects on a laminar nanofluid flow in a curved tube using a two phase approach, International Journal of Heat and Fluid Flow 30 (2009) 706–714.

[7] S. Mirmasoumi, A. Behzadmehr, Numerical study of laminar mixed convection of a nanofluid in a horizontal tube using two-phase mixture model, Applied Thermal Engineering 28 (2008) 717–727.

[8] J. Buongiorno, Convective transport in nanofluids, ASME. Journal of Heat Transfer 128 (2006) 240–250.

[9] A. Behzadmehr, M. Saffar-Avval, N. Galanis, Prediction of turbulent forced convection of a nanofluid in a tube with uniform heat flux using a two phase approach, International Journal of Heat and Fluid Flow 28 (2) (2007) 211–219.

[10] B.C. Pak, Y.I. Cho, Hydrodynamic and heat transfer study of dispersed fluids with submicron metallic oxide particles, Experimental Heat Transfer 11 (1998) 151–170.

[11] J.A. Eastman, S.U.S. Choi, S. Li, G. Soyez, L.J. Thompson, R.J. Di Melfi, Novel thermal properties of nanostructured materials, Materials Science Forum 312–314 (1999) 629–634.

[12] S.J. Palm, G. Roy, C.T. Nguyen, Heat transfer enhancement with the use of nanofluids in radial flow cooling systems considering temperature dependent properties, Applied Thermal Engineering 26 (2006) 2209–2218.

[13] J.C. Maxwell, A Treatise on Electricity and Magnetism, second ed., Oxford University Press, Cambridge, 1904, pp. 435–441.
[14] H. Brinkman, The viscosity of concentrated suspensions and solutions, The Journal of Chemical Physics 20 (1952) 571.
[15] E. Abu-Nada, A.J. Chamkha, Effect of nanofluid variable properties on natural convection in enclosures filled with a CuO–EG–Water nanofluid, International Journal of Thermal Sciences 49 (12) (2010) 2339–2352.
[16] P.K. Namburu, D.P. Kulkarni, D. Misra, D.K. Das, Viscosity of copper oxide nanoparticles dispersed in ethylene glycol and water mixture, Experimental Thermal and Fluid Science 32 (2) (2007) 397–402.
[17] A. Einstein, Investigation on the Theory of Brownian Motion, Dover, New York, 1956.
[18] S.P. Jang, J.H. Lee, K.S. Hwang, S.U.S. Choi, Particle concentration and tube size dependence of viscosities of Al_2O_3-water nanofluids flowing through microand minitubes, Applied Physics Letters 91 (2007) 243112.
[19] J. Koo, C. Kleinstreuer, A new thermal conductivity model for nanofluids, Journal of Nanoparticle Research 6 (2004) 577–588.
[20] S.E.B. Maïga, S.M. Palm, C.T. Nguyen, G. Roy, N. Galanis, Heat transfer enhancement by using nanofluids in forced convection flows, International Journal of Heat and Fluid Flow 26 (2005) 530–546.
[21] D. Orozco, Hydrodynamic behavior of suspension of polar particles, Encyclopedia Surface Colloid Science 4 (2005) 2375–2396.
[22] C.T. Nguyen, F. Desgranges, N. Galanis, G. Roy, T. Mare, S. Boucher, H.A. Mintsa, Viscosity data for Al_2O_3-water nanofluid-hysteresis: is heat transfer enhancement using nanofluids reliable? International Journal of Thermal Sciences 47 (2008) 103–111.
[23] N. Masoumi, N. Sohrabi, A. Behzadmehr, A new model for calculating the effective viscosity of nanofluids, Journal of Physics D Applied Physics 42 (2009) 055501.
[24] I. Gherasim, G. Roy, C.T. Nguyen, D. Vo-Ngoc, Experimental investigation of nanofluids in confined laminar radial flows, International Journal of Thermal Sciences 48 (2009) 1486–1493.
[25] S.P. Jang, S.U.S. Choi, The role of Brownian motion in the enhanced thermal conductivity of nanofluids, Applied Physics Letters 84 (2004) 4316–4318.
[26] B.X. Wang, L.R. Zhou, X.F. Peng, A fractal model for predicting the effective thermal conductivity of liquid with suspension of nanoparticles, International Journal of Heat and Mass Transfer 46 (2003) 2665–2672.
[27] C.H. Chon, K.D. Kihm, S.P. Lee, S.U.S. Choi, Empirical correlation finding the role of temperature and particle size for nanofluid (Al2O3) thermal conductivity enhancement, Applied Physics Letters 87 (2005), 153107, https://doi.org/10.1063/1.2093936.
[28] P. Charuyakorn, S. Sengupta, S.K. Roy, Forced convection heat transfer in micro encapsulated phase change material slurries, International Journal of Heat and Mass Transfer 34 (1991) 819–833.
[29] J.A. Eastman, S.R. Phillpot, U.S. Choi, P. Keblinski, Thermal transport in nanofluids, Annual Review of Materials Research 34 (2000) 219–246.
[30] W. Yu, S.U.S. Choi, The role of interfacial layers in the enhanced thermal conductivity of nanofluids: a renovated Maxwell model, Journal of Nanoparticle Research 5 (2003) 167–171.

[31] H.E. Patel, T. Sundarrajan, T. Pradeep, A. Dasgupta, N. Dasgupta, S.K. Das, A microconvection model for thermal conductivity of nanofluid, Pramana Journal of Physics 65 (2005) 863–869.

[32] H. Angue Mintsa, G. Roy, C.T. Nguyen, D. Doucet, New temperature dependent thermal conductivity data for water-based nanofluids, International Journal of Thermal Sciences 48 (2009) 363–371.

[33] M. Sheikholeslami, Effect of uniform suction on nanofluid flow and heat transfer over a cylinder, Journal of the Brazilian Society of Mechanical Sciences and Engineering 37 (2015) 1623–1633.

[34] Y. Qin, Y. He, J.E. Hiller, G. Mei, A new water-retaining paver block for reducing runoff and cooling pavement, Journal of Cleaner Production 199 (2018) 948–956.

[35] M. Sheikholeslami, Lattice Boltzmann Method simulation of MHD non-Darcy nanofluid free convection, Physica B 516 (2017) 55–71.

[36] M. Sheikholeslami, Magnetic field influence on CuO -H_2O nanofluid convective flow in a permeable cavity considering various shapes for nanoparticles, International Journal of Hydrogen Energy 42 (2017) 19611–19621.

[37] Y. Qin, Y. Zhao, X. Chen, L. Wang, F. Li, T. Bao, Moist curing increases the solar reflectance of concrete, Construction and Building Materials 215 (2019) 114–118.

[38] M. Sheikholeslami, Numerical simulation for solidification in a LHTESS by means of nano-enhanced PCM, Journal of the Taiwan Institute of Chemical Engineers 86 (2018) 25–41.

[39] M. Sheikholeslami, Numerical modeling of nano enhanced PCM solidification in an enclosure with metallic fin, Journal of Molecular Liquids 259 (2018) 424–438.

[40] Y. Qin, H. He, X. Ou, T. Bao, Experimental study on darkening water-rich mud tailings for accelerating desiccation, Journal of Cleaner Production (2019) 118235, https://doi.org/10.1016/j.jclepro.2019.118235.

[41] Y. Qin, J. Luo, Z. Chen, G. Mei, L.-E. Yan, Measuring the albedo of limited-extent targets without the aid of known-albedo masks, Solar Energy 171 (2018) 971–976.

[42] Y. Qin, A review on the development of cool pavements to mitigate urban heat island effect, Renewable and Sustainable Energy Reviews 52 (2015) 445–459.

[43] M. Sheikholeslami, Solidification of NEPCM under the effect of magnetic field in a porous thermal energy storage enclosure using CuO nanoparticles, Journal of Molecular Liquids 263 (2018) 303–315.

[44] M. Sheikholeslami, Influence of magnetic field on Al_2O_3-H_2O nanofluid forced convection heat transfer in a porous lid driven cavity with hot sphere obstacle by means of LBM, Journal of Molecular Liquids 263 (2018) 472–488.

[45] Y. Qin, J.E. Hiller, D. Meng, Linearity between pavement thermophysical properties and surface temperatures, Journal of Materials in Civil Engineering (2019), https://doi.org/10.1061/(ASCE)MT.1943-5533.0002890.

[46] M. Sheikholeslami, New computational approach for exergy and entropy analysis of nanofluid under the impact of Lorentz force through a porous media, Computer Methods in Applied Mechanics and Engineering 344 (2019) 319–333.

[47] Y. Qin, Urban canyon albedo and its implication on the use of reflective cool pavements, Energy and Buildings 96 (2015) 86–94.

[48] M. Sheikholeslami, Numerical approach for MHD Al_2O_3-water nanofluid transportation inside a permeable medium using innovative computer method, Computer Methods in Applied Mechanics and Engineering 344 (2019) 306–318.

[49] Y. Qin, M. Zhang, G. Mei, A new simplified method for measuring the permeability characteristics of highly porous media, Journal of Hydrology 562 (2018) 725–732.

[50] X. Ma, M. Sheikholeslami, M. Jafaryar, A. Shafee, T. Nguyen-Thoi, Z. Li, Solidification inside a clean energy storage unit utilizing phase change material with copper oxide nanoparticles, Journal of Cleaner Production (2019), https://doi.org/10.1016/j.jclepro.2019.118888.

[51] M. Sheikholeslami, O. Mahian, Enhancement of PCM solidification using inorganic nanoparticles and an external magnetic field with application in energy storage systems, Journal of Cleaner Production 215 (2019) 963–977.

[52] Y. Qin, M. Zhang, J.E. Hiller, Theoretical and experimental studies on the daily accumulative heat gain from cool roofs, Energy 129 (2017) 138–147.

[53] M. Sheikholeslami, B. Rezaeianjouybari, M. Darzi, A. Shafee, Z. Li, T.K. Nguyen, Application of nano-refrigerant for boiling heat transfer enhancement employing an experimental study, International Journal of Heat and Mass Transfer 141 (2019) 974–980.

[54] M. Sheikholeslami, M. Jafaryar, M. Hedayat, A. Shafee, Z. Li, T.K. Nguyen, M. Bakouri, Heat transfer and turbulent simulation of nanomaterial due to compound turbulator including irreversibility analysis, International Journal of Heat and Mass Transfer 137 (2019) 1290–1300.

[55] M. Sheikholeslami, Application of Darcy law for nanofluid flow in a porous cavity under the impact of Lorentz forces, Journal of Molecular Liquids 266 (2018) 495–503.

[56] M. Sheikholeslami, Finite element method for PCM solidification in existence of CuO nanoparticles, Journal of Molecular Liquids 265 (2018) 347–355.

[57] M. Sheikholeslami, Numerical investigation of nanofluid free convection under the influence of electric field in a porous enclosure, Journal of Molecular Liquids 249 (2018) 1212–1221.

[58] M. Sheikholeslami, CuO-water nanofluid flow due to magnetic field inside a porous media considering Brownian motion, Journal of Molecular Liquids 249 (2018) 921–929.

[59] M. Sheikholeslami, Numerical investigation for CuO-H_2O nanofluid flow in a porous channel with magnetic field using mesoscopic method, Journal of Molecular Liquids 249 (2018) 739–746.

[60] Y. Qin, J.E. Hiller, Understanding pavement-surface energy balance and its implications on cool pavement development, Energy and Buildings 85 (2014) 389–399.

[61] M. Sheikholeslami, H.B. Rokni, Numerical modeling of nanofluid natural convection in a semi annulus in existence of Lorentz force, Computer Methods in Applied Mechanics and Engineering 317 (2017) 419–430.

[62] M. Sheikholeslami, A. Zeeshan, Analysis of flow and heat transfer in water based nanofluid due to magnetic field in a porous enclosure with constant heat flux using CVFEM, Computer Methods in Applied Mechanics and Engineering 320 (2017) 68–81.

[63] M. Sheikholeslami, S.A. Shehzad, Z. Li, Water based nanofluid free convection heat transfer in a three dimensional porous cavity with hot sphere obstacle in existence of Lorenz forces, International Journal of Heat and Mass Transfer 125 (2018) 375–386.

[64] M. Sheikholeslami, M. Jafaryar, Z. Li, Nanofluid turbulent convective flow in a circular duct with helical turbulators considering CuO nanoparticles, International Journal of Heat and Mass Transfer 124 (2018) 980–989.

[65] M. Sheikholeslami, A. Ghasemi, Solidification heat transfer of nanofluid in existence of thermal radiation by means of FEM, International Journal of Heat and Mass Transfer 123 (2018) 418–431.

[66] M. Sheikholeslami, S.A. Shehzad, CVFEM simulation for nanofluid migration in a porous medium using Darcy model, International Journal of Heat and Mass Transfer 122 (2018) 1264–1271.

[67] Y. Qin, Pavement surface maximum temperature increases linearly with solar absorption and reciprocal thermal inertial, International Journal of Heat and Mass Transfer 97 (2016) 391–399.

[68] M. Sheikholeslami, M. Darzi, M.K. Sadoughi, Heat transfer improvement and pressure drop during condensation of refrigerant-based nanofluid; an experimental procedure, International Journal of Heat and Mass Transfer 122 (2018) 643–650.

[69] M. Sheikholeslami, S.A. Shehzad, Simulation of water based nanofluid convective flow inside a porous enclosure via Non-equilibrium model, International Journal of Heat and Mass Transfer 120 (2018) 1200–1212.

[70] M. Sheikholeslami, M. Seyednezhad, Simulation of nanofluid flow and natural convection in a porous media under the influence of electric field using CVFEM, International Journal of Heat and Mass Transfer 120 (2018) 772–781.

[71] Y. Qin, J. Liang, H. Yang, Z. Deng, Gas permeability of pervious concrete and its implications on the application of pervious pavements, Measurement 78 (2016) 104–110.

[72] M. Sheikholeslami, H.B. Rokni, Numerical simulation for impact of Coulomb force on nanofluid heat transfer in a porous enclosure in presence of thermal radiation, International Journal of Heat and Mass Transfer 118 (2018) 823–831.

[73] M. Sheikholeslami, S.A. Shehzad, Numerical analysis of Fe_3O_4 –H_2O nanofluid flow in permeable media under the effect of external magnetic source, International Journal of Heat and Mass Transfer 118 (2018) 182–192.

[74] Y. Qin, J. Liang, K. Tan, F. Li, A side by side comparison of the cooling effect of building blocks with retro-reflective and diffuse-reflective walls, Solar Energy 133 (2016) 172–179.

[75] M. Sheikholeslami, M.K. Sadoughi, Simulation of CuO- water nanofluid heat transfer enhancement in presence of melting surface, International Journal of Heat and Mass Transfer 116 (2018) 909–919.

[76] M. Sheikholeslami, M. Jafaryar, A. Shafee, Z. Li, R.-ul Haq, Heat transfer of nanoparticles employing innovative turbulator considering entropy generation, International Journal of Heat and Mass Transfer 136 (2019) 1233–1240.

[77] M. Sheikholeslami, R.-ul Haq, A. Shafee, Z. Li, Y.G. Elaraki, I. Tlili, Heat transfer simulation of heat storage unit with nanoparticles and fins through a heat exchanger, International Journal of Heat and Mass Transfer 135 (2019) 470–478.

[78] M. Sheikholeslami, R.-ul Haq, A. Shafee, Z. Li, Heat transfer behavior of nanoparticle enhanced PCM solidification through an enclosure with V shaped fins, International Journal of Heat and Mass Transfer 130 (2019) 1322–1342.

[79] M. Sheikholeslami, S.A. Shehzad, Z. Li, A. Shafee, Numerical modeling for alumina nanofluid magnetohydrodynamic convective heat transfer in a permeable medium using Darcy law, International Journal of Heat and Mass Transfer 127 (2018) 614−622.

[80] M. Sheikholeslami, Z. Li, A. Shafee, Lorentz forces effect on NEPCM heat transfer during solidification in a porous energy storage system, International Journal of Heat and Mass Transfer 127 (2018) 665−674.

[81] Y. Qin, H. He, A new simplified method for measuring the albedo of limited extent targets, Solar Energy 157 (Suppl. C) (2017) 1047−1055.

[82] Y. Qin, Y. He, B. Wu, S. Ma, X. Zhang, Regulating top albedo and bottom emissivity of concrete roof tiles for reducing building heat gains, Energy and Buildings 156 (Suppl. C) (2017) 218−224.

[83] M. Sheikholeslami, M. Jafaryar, S. Saleem, Z. Li, A. Shafee, Y. Jiang, Nanofluid heat transfer augmentation and exergy loss inside a pipe equipped with innovative turbulators, International Journal of Heat and Mass Transfer 126 (2018) 156−163.

[84] M. Sheikholeslami, A. Ghasemi, Z. Li, A. Shafee, S. Saleem, Influence of CuO nanoparticles on heat transfer behavior of PCM in solidification process considering radiative source term, International Journal of Heat and Mass Transfer 126 (2018) 1252−1264.

[85] M. Sheikholeslami, M. Darzi, Z. Li, Experimental investigation for entropy generation and exergy loss of nano-refrigerant condensation process, International Journal of Heat and Mass Transfer 125 (2018) 1087−1095.

[86] M. Sheikholeslami, H.B. Rokni, Melting heat transfer influence on nanofluid flow inside a cavity in existence of magnetic field, International Journal of Heat and Mass Transfer 114 (2017) 517−526.

[87] M. Sheikholeslami, M. Sadoughi, Mesoscopic method for MHD nanofluid flow inside a porous cavity considering various shapes of nanoparticles, International Journal of Heat and Mass Transfer 113 (2017) 106−114.

[88] M. Sheikholeslami, M.M. Bhatti, Forced convection of nanofluid in presence of constant magnetic field considering shape effects of nanoparticles, International Journal of Heat and Mass Transfer 111 (2017) 1039−1049.

[89] M. Sheikholeslami, M.M. Bhatti, Active method for nanofluid heat transfer enhancement by means of EHD, International Journal of Heat and Mass Transfer 109 (2017) 115−122.

[90] M. Sheikholeslami, S.A. Shehzad, Magnetohydrodynamic nanofluid convection in a porous enclosure considering heat flux boundary condition, International Journal of Heat and Mass Transfer 106 (2017) 1261−1269.

[91] M. Sheikholeslami, M.A. Sheremet, A. Shafee, I. Tlili, Simulation of nanoliquid thermogravitational convection within a porous chamber imposing magnetic and radiation impacts, Physica A: Statistical Mechanics and its Applications (2020), https://doi.org/10.1016/j.physa.2019.124058.

[92] M. Sheikholeslami, R. Ellahi, Three dimensional mesoscopic simulation of magnetic field effect on natural convection of nanofluid, International Journal of Heat and Mass Transfer 89 (2015) 799−808.

[93] M. Sheikholeslami, A.N. Keshteli, H. Babazadeh, Nanoparticles favorable effects on performance of thermal storage units, Journal of Molecular Liquids 300 (2020) 112329.

[94] M. Sheikholeslami, S.A. Shehzad, Thermal radiation of ferrofluid in existence of Lorentz forces considering variable viscosity, International Journal of Heat and Mass Transfer 109 (2017) 82–92.

[95] M. Sheikholeslami, H.B. Rokni, Nanofluid two phase model analysis in existence of induced magnetic field, International Journal of Heat and Mass Transfer 107 (2017) 288–299.

[96] M. Sheikholeslami, H.B. Rokni, Simulation of nanofluid heat transfer in presence of magnetic field: a review, International Journal of Heat and Mass Transfer 115 (2017) 1203–1233.

[97] M. Sheikholeslami, T. Hayat, A. Alsaedi, On simulation of nanofluid radiation and natural convection in an enclosure with elliptical cylinders, International Journal of Heat and Mass Transfer 115 (2017) 981–991.

[98] A. Shafee, M. Sheikholeslami, M. Jafaryar, H. Babazadeh, Utilizing copper oxide nanoparticles for expedition of solidification within a storage system, Journal of Molecular Liquids 302 (2020) 112371, https://doi.org/10.1016/j.molliq.2019.112371.

[99] F. Li, M. Sheikholeslami, R.N. Dara, M. Jafaryar, A. Shafee, T. Nguyen-Thoi, Z. Li, Numerical study for nanofluid behavior inside a storage finned enclosure involving melting process, Journal of Molecular Liquids 297 (2020) 111939.

[100] M. Sheikholeslami, S.A. Shehzad, CVFEM for influence of external magnetic source on $Fe_3O_4 - H_2O$ nanofluid behavior in a permeable cavity considering shape effect, International Journal of Heat and Mass Transfer 115 (2017) 180–191.

[101] M. Sheikholeslami, M. Seyednezhad, Nanofluid heat transfer in a permeable enclosure in presence of variable magnetic field by means of CVFEM, International Journal of Heat and Mass Transfer 114 (2017) 1169–1180.

[102] M. Sheikholeslami, A. Arabkoohsar, I. Khan, A. Shafee, Z. Li, Impact of Lorentz forces on Fe_3O_4-water ferrofluid entropy and exergy treatment within a permeable semi annulus, Journal of Cleaner Production 221 (2019) 885–898.

[103] M. Sheikholeslami, K. Vajravelu, M.M. Rashidi, Forced convection heat transfer in a semi annulus under the influence of a variable magnetic field, International Journal of Heat and Mass Transfer 92 (2016) 339–348.

[104] M. Sheikholeslami, T. Hayat, A. Alsaedi, Numerical study for external magnetic source influence on water based nanofluid convective heat transfer, International Journal of Heat and Mass Transfer 106 (2017) 745–755.

[105] M. Sheikholeslami, T. Hayat, A. Alsaedi, S. Abelman, Numerical analysis of EHD nanofluid force convective heat transfer considering electric field dependent viscosity, International Journal of Heat and Mass Transfer 108 (2017) 2558–2565.

[106] M. Sheikholeslami, Influence of magnetic field on nanofluid free convection in an open porous cavity by means of Lattice Boltzmann Method, Journal of Molecular Liquids 234 (2017) 364–374.

[107] M. Sheikholeslami, Magnetohydrodynamic nanofluid forced convection in a porous lid driven cubic cavity using Lattice Boltzmann Method, Journal of Molecular Liquids 231 (2017) 555–565.

[108] M. Sheikholeslami, Magnetic field influence on nanofluid thermal radiation in a cavity with tilted elliptic inner cylinder, Journal of Molecular Liquids 229 (2017) 137–147.

[109] M. Sheikholeslami, Influence of Lorentz forces on nanofluid flow in a porous cylinder considering Darcy model, Journal of Molecular Liquids 225 (2017) 903–912.

[110] M. Sheikholeslami, Influence of Coulomb forces on Fe_3O_4-H_2O nanofluid thermal improvement, International Journal of Hydrogen Energy 42 (2017) 821–829.

[111] M. Sheikholeslami, S.A. Shehzad, F.M. Abbasi, Z. Li, Nanofluid flow and forced convection heat transfer due to Lorentz forces in a porous lid driven cubic enclosure with hot obstacle, Computer Methods in Applied Mechanics and Engineering 338 (2018) 491–505.

[112] M. Sheikholeslami, M. Barzegar Gerdroodbary, R. Moradi, A. Shafee, Z. Li, Application of Neural Network for estimation of heat transfer treatment of Al_2O_3- H_2O nanofluid through a channel, Computer Methods in Applied Mechanics and Engineering 344 (2019) 1–12.

CHAPTER

Heat exchanger equipped with various helical turbulators

3

3.1 Convective heat transfer within a tube with twisted tape turbulator

3.1.1 Definition of the problem and governing equation

Fig. 3.1 demonstrates the shape of a twisted tape which is utilized in this simulation. Inner and outer diameters are 5 and 20 mm, respectively. The length of the circular duct is 900 mm [1]. The middle 300 mm is test section in which the modified twisted tube is situated. H_2O-based nanofluid is selected as working fluid. Tables 3.1 and 3.2 illustrate properties of CuO and H_2O. There is no turbolentor in other parts of the pipe to prevent any probable back flow streams.

To simulate this turbulent flow and forced convection heat transfer, following equations should be solved:

$$\frac{\partial}{\partial x_i}(u_i)=0 \tag{3.1}$$

$$\frac{\partial}{\partial x_j}(u_i u_j \rho_{nf})=\frac{\partial}{\partial x_j}\left(-\rho_{nf}\overline{u_j{}'u_i{}'}\right)-\frac{\partial p}{\partial x_i}+\frac{\partial}{\partial x_j}\left(\mu_{nf}\left(\frac{\partial u_i}{\partial x_j}+\frac{\partial u_j}{\partial x_i}\right)\right) \tag{3.2}$$

$$\frac{\partial}{\partial x_i}(\rho_{nf} T u_i)=\frac{\partial}{\partial x_i}\left((\mu_{nf}/Pr_{nf}+\mu_t/Pr_t)\frac{\partial T}{\partial x_i}\right), \tag{3.3}$$

$\rho_{nf}\overline{u_j{}'u_i{}'}$ and μ_t are

$$-\overline{u_j{}'u_i{}'}\,\rho_{nf}=\mu_t\left(\frac{\partial u_i}{\partial x_j}+\frac{\partial u_j}{\partial x_i}\right)-\frac{2}{3}\,\rho_{nf}k\delta_{ij}-\frac{2}{3}\,\mu_t\frac{\partial u_k}{\partial x_k}\,\delta_{ij} \tag{3.4}$$

$$\mu_t=\rho_{nf}C_\mu k^2/\varepsilon \tag{3.5}$$

Nanofluid in Heat Exchangers for Mechanical Systems. https://doi.org/10.1016/B978-0-12-821923-2.00003-8

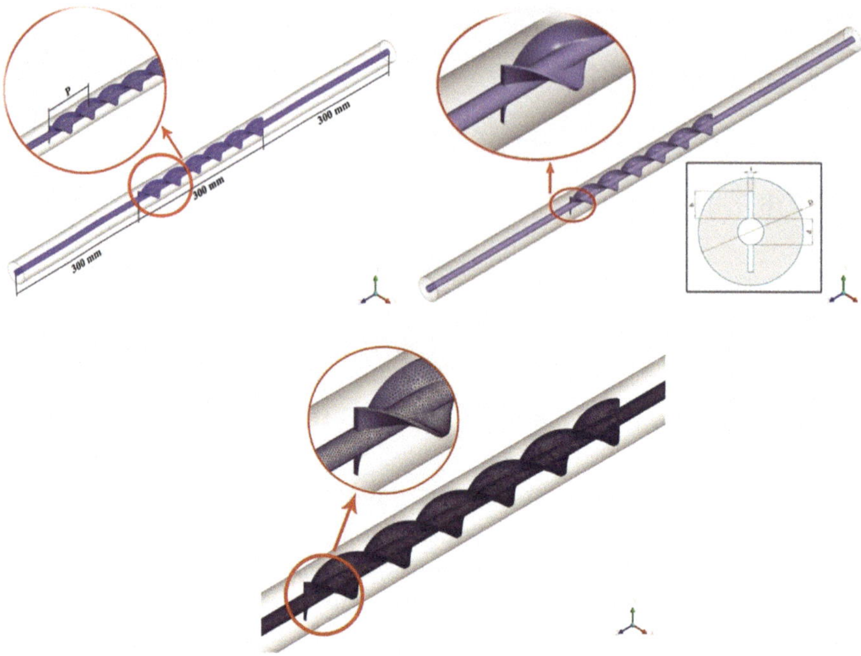

FIGURE 3.1

Geometry of the problem.

Table 3.1 The coefficient values of CuO – Water nanofluid.

Coefficient values	CuO – Water
a_1	−26.5933108
a_2	−0.403818333
a_3	−33.3516805
a_4	−1.915825591
a_5	6.421858E-02
a_6	48.40336955
a_7	−9.787756683
a_8	190.245610009
a_9	10.9285386565
a_{10}	−0.72009983664

Table 3.2 Thermophysical properties of water and nanoparticles.

	$\rho(kg/m^3)$	$C_p(j/kgk)$	k(W/m.k)	μ(Pa.s)
Pure water	997.1	4179	0.613	0.0010003
CuO	6500	540	18	–

k, ε can be calculated as

$$\frac{\partial}{\partial x_j}\left(\left(\frac{\mu_t}{\sigma_k}+\mu_{nf}\right)\frac{\partial k}{\partial x_j}\right)-\rho_{nf}\varepsilon+G_k=\frac{\partial}{\partial x_i}(u_i\rho_{nf}k),$$
$$G_k=-\frac{\partial u_j}{\partial x_i}\rho_{nf}\overline{u_j{}'u_i{}'} \tag{3.6}$$

$$\frac{\partial}{\partial x_i}(u_i\rho_{nf}\varepsilon)=\frac{\partial}{\partial x_j}\left(\left(\frac{\mu_t}{\sigma_\varepsilon}+\mu_{nf}\right)\frac{\partial \varepsilon}{\partial x_j}\right)+\frac{\varepsilon}{k}G_kC_{1\varepsilon}-\rho_{nf}\frac{\varepsilon^2}{k}C_{2\varepsilon} \tag{3.7}$$

$$C_{1\varepsilon}=1.42,\ C_\mu=0.0845,\ C_{2\varepsilon}=1.68, Pr_t=0.85, \sigma_k=1, \sigma_\varepsilon=1.3 \tag{3.8}$$

ANSYS Fluent 14 is used to model this problem. Turbulent flow is modeled by k-ε (RNG) model. Table 3.3 shows the details of simulation setup.
$(\rho C_p)_{nf}, \rho_{nf}, k_{nf}$ and μ_{nf} are [47]

$$(\rho C_p)_{nf}=(\rho C_p)_f(1-\phi)+(\rho C_p)_s\phi \tag{3.9}$$

$$\rho_{nf}=\rho_f(1-\phi)+\rho_s\phi \tag{3.10}$$

$$\frac{k_{nf}}{k_f}=1+3\frac{\phi(-1+k_p/k_f)}{\phi(1-k_p/k_f)+(2+k_p/k_f)}+5\times10^4 g'(\phi,T,d_p)\phi\rho_f c_{p,f}\sqrt{\frac{\kappa_b T}{\rho_p d_p}}$$
$$g'(\phi,T,d_p)=\left(a_1+a_2Ln(d_p)+a_5Ln(d_p)^2+a_3Ln(\phi)+a_4Ln(\phi)\ln(d_p)\right)Ln(T)$$
$$+\left(a_6+a_7Ln(d_p)+a_{10}Ln(d_p)^2+a_8Ln(\phi)+a_9\ln(d_p)Ln(\phi)\right),$$
$$\frac{4\times10^{-8}}{d_p}=-1/k_p+1/k_{p,eff} \tag{3.11}$$

$$\frac{\mu_{nf}}{\mu_f}=\frac{k_{\text{Brownian}}}{Pr\ k_f}+\frac{1}{(1-\phi)^{2.5}} \tag{3.12}$$

Table 3.3 Solver setting of ANSYS FLUENT.

Solver	Pressure-based, steady	
Viscous model	K-ε, RNG, Enhanced wall treatment	
Solution methods	Pressure–velocity coupling	SIMPLE
	Gradient	Green-Gauss node based
	Pressure	Standard
Spatial discretization	Momentum	Second order upwind
	Turbulent kinetic energy	Second order upwind
	Turbulent dissipation rate	Second order upwind

Inlet and outlet boundary conditions are

$$u_i = 0, v_i = 0, w_i = cte, T_i = cte, I = 0.16(\mathrm{Re})^{\frac{-1}{8}} \tag{3.13}$$

$$\frac{\partial v}{\partial z} = \frac{\partial u}{\partial z} = \frac{\partial w}{\partial z} = 0, \frac{\partial T}{\partial z} = 0 \tag{3.14}$$

Thermal and frictional entropy generations are

$$S_{g,h} = \frac{k}{T^2}\left[\left(\frac{\partial T}{\partial x}\right)^2 + \left(\frac{\partial T}{\partial y}\right)^2 + \left(\frac{\partial T}{\partial z}\right)^2\right], \tag{3.15}$$

$$S_{g,f} = \frac{\mu}{T}\left\{\begin{array}{l} 2\left[\left(\frac{\partial v_x}{\partial x}\right)^2 + \left(\frac{\partial v_y}{\partial y}\right)^2 + \left(\frac{\partial v_z}{\partial z}\right)^2\right] + \left(\frac{\partial v_x}{\partial y} + \frac{\partial v_y}{\partial x}\right)^2 \\ + \left(\frac{\partial v_x}{\partial z} + \frac{\partial v_z}{\partial x}\right)^2 + \left(\frac{\partial v_y}{\partial z} + \frac{\partial v_z}{\partial y}\right)^2 \end{array}\right\} \tag{3.16}$$

Nu and f are defined as

$$Nu = \frac{hD_h}{k}, \quad f = \frac{\Delta p}{\frac{\rho v_m^2}{2}\frac{L}{D_h}} \tag{3.17}$$

Different grids are examined to select the good grid size. There is an example in Table 3.4. The grid system of 1,378,654 cells is chosen in numerical simulations. Furthermore, we checked the ranges of Y^+ in current paper (see Table 3.5) to satisfy the limitation of our selected turbulent model ($Y^+ < 5$). To verify the current code, we reproduced the results of previous published papers and show the good accuracy (see Fig. 3.2) [2].

Table 3.4 Grid independence test when $PR = 5$, $BR = 0.5$, $\mathrm{Re} = 20{,}000$.

Mesh refinement	Total element number	*Nu*	*f*
Very coarse mesh	201,134	180.105	0.092
Coarse mesh	310,221	180.024	0.094
Normal mesh	522,892	179.876	0.095
Fine mesh	1,001,312	179.581	0.096
Very fine mesh	1,378,654	179.478	0.096

Table 3.5 Ranges of Y^+ for various cases when Re = 20,000.

Case no.	Surface	Y^+
$PR = 15$, $BR = 0.3$	Outer wall	3.653840085
	Twisted tape wall	3.329417279
$PR = 15$, $BR = 0.5$	Outer wall	3.162444045
	Twisted tape wall	2.958740417
$PR = 5$, $BR = 0.3$	Outer wall	4.252006463
	Twisted tape wall	3.592632247
$PR = 5$, $BR = 0.5$	Outer wall	4.419943403
	Twisted tape wall	3.678451027

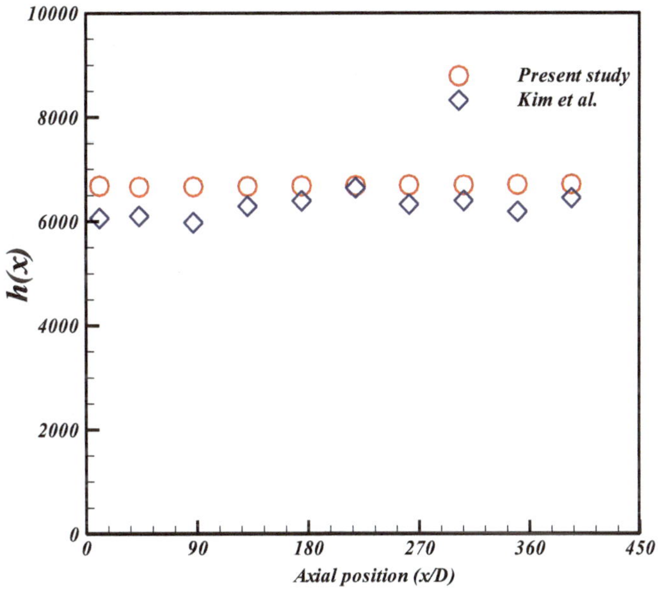

FIGURE 3.2

Comparison of the results obtained for water in a horizontal pipe [48].

3.1.2 Effects of active parameters

In this chapter, influence of inserting twisted tape swirling device on pressure drop, heat transfer augmentation, and entropy generation has been simulated. The working fluid is H_2O–CuO nanofluid, and its properties are estimated according to KKL model. Roles of pitch ratio ($PR = 5$ to 15), height ratio ($BR = 0.3$ to 0.5), and Reynolds number (Re = 5000 to 20000) are illustrated.

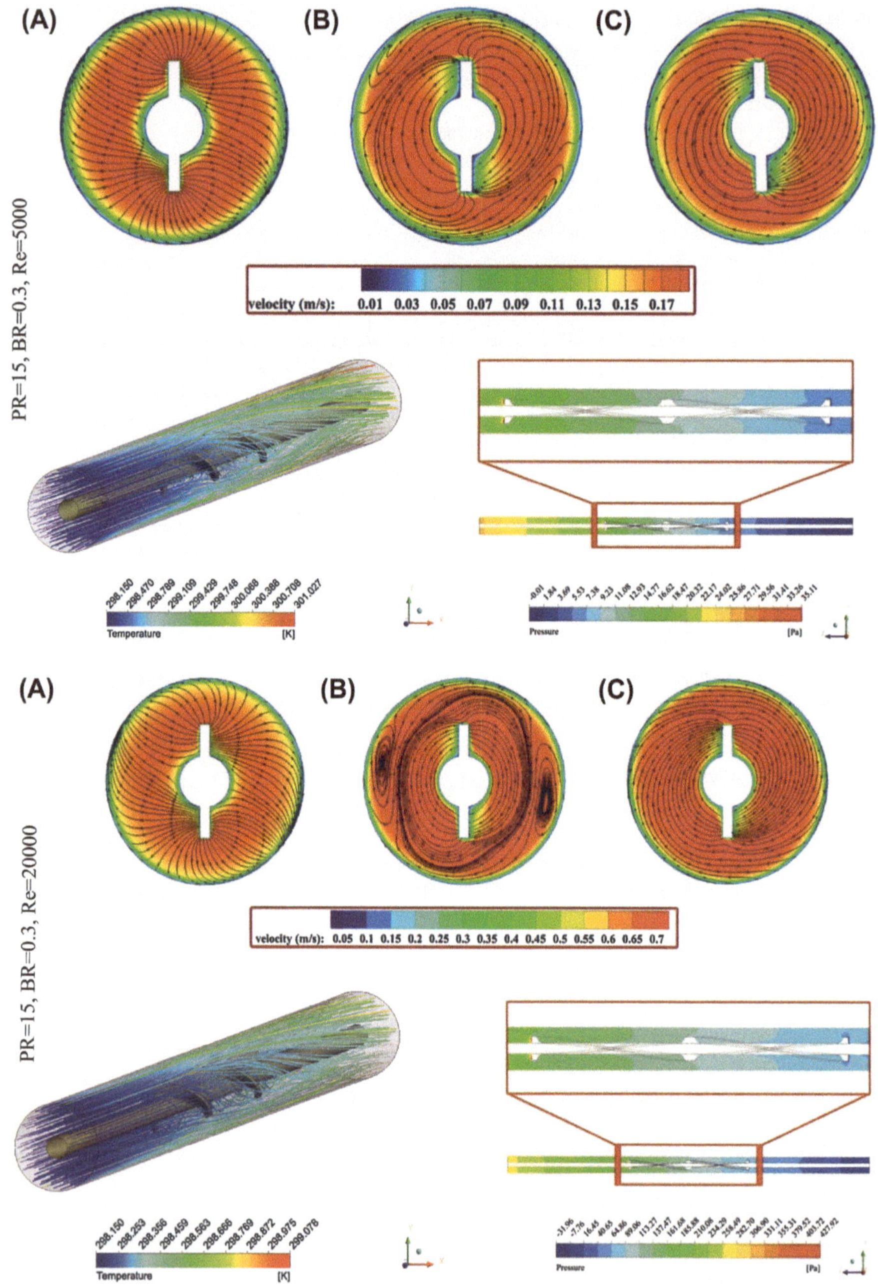

FIGURE 3.3

Effects of *PR*, *BR*, and Re on temperature, velocity, streamline, and pressure contours when $\phi = 0.05$. (*a*) section at inlet of test section; (*b*) section at middle of test section; (*c*) section at outlet of test section.

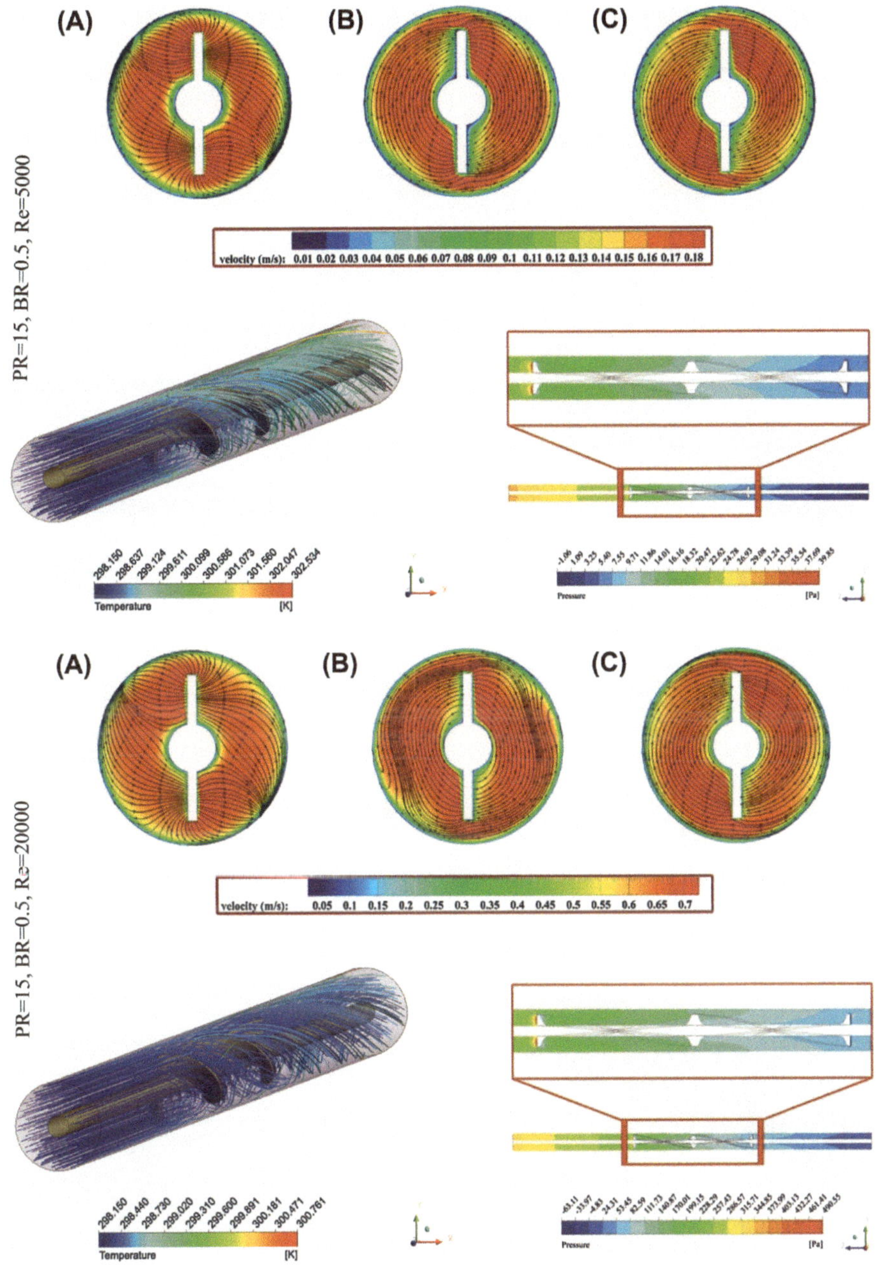

FIGURE 3.3 cont'd.

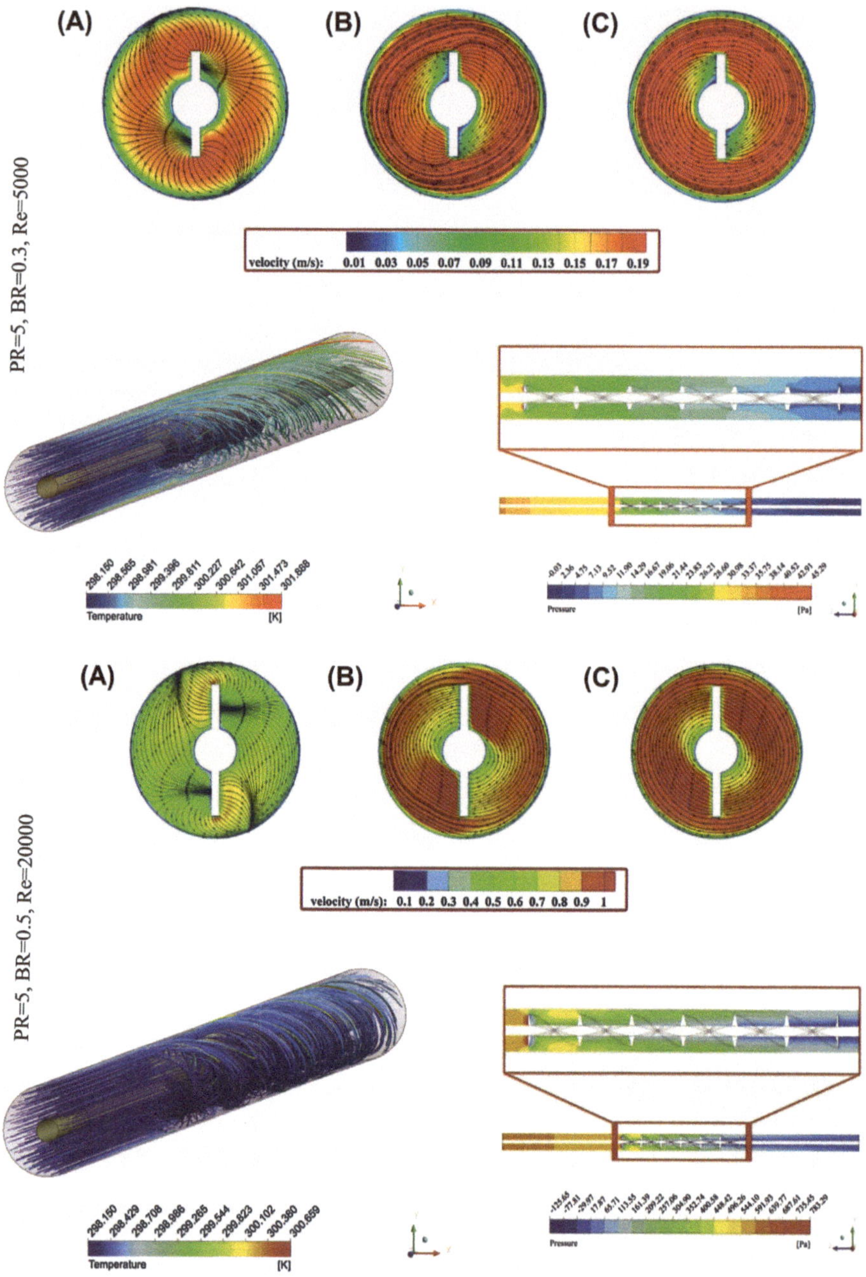

FIGURE 3.3 cont'd.

Fig. 3.3 demonstrates the influence of pitch ratio, Reynolds number, and height ratio on pressure drop and heat transfer behavior. Temperature boundary layer thickness decreases with augment of Reynolds number due to increasing nanofluid mixing with rise of inlet velocity. Augmenting swirl intensity can enhance the convective heat transfer but pressure drop augments. Lower pitch ratio leads to higher revolution of twisted tape. Deceasing pitch ratio makes temperature gradient to enhance but with cost of higher pressure drop. Secondary flow becomes stronger with decrease of pitch ratio. So turbulent intensity has reverse relationship with pitch ratio. Better nanofluid mixing can be occurred for higher values of height ratio. So, disruption of thermal boundary increases with augment of height ratio.

Figs. 3.4 and 3.5 show the impact of PR, BR, and Re on Darcy factor and Nusselt number. Also, according to simulation data, following formulas can be offered:

$$\begin{aligned} Nu^* = &\, 0.78 + 0.46PR^* + 0.51BR + 8.6\mathrm{Re}^* \\ & - 0.61(PR^*)(BR) - 0.76(PR^*)(\mathrm{Re}^*) + 4.01(BR)(\mathrm{Re}^*) \\ & - 0.17(PR^*)^2 - 1.01(BR)^2 - 0.93(\mathrm{Re}^*)^2 \end{aligned} \tag{3.18}$$

where $Nu^* = 0.1Nu$, $PR^* = 0.1PR$, $\mathrm{Re}^* = 10^{-4}\mathrm{Re}$.

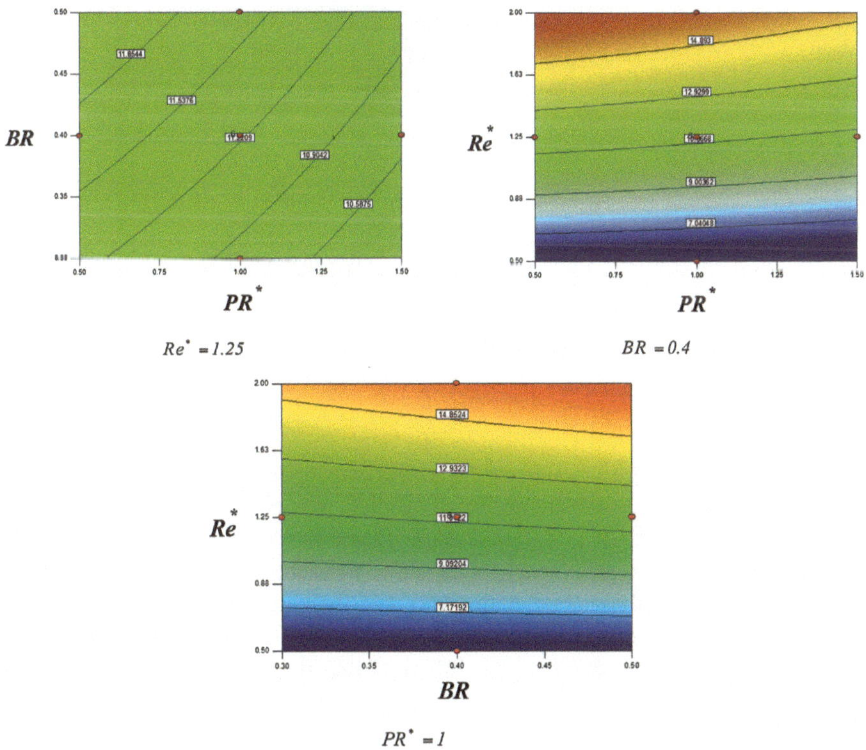

FIGURE 3.4

Effects of Reynolds number, pitch ratio, and height ratio on Nusselt number (Nu^*).

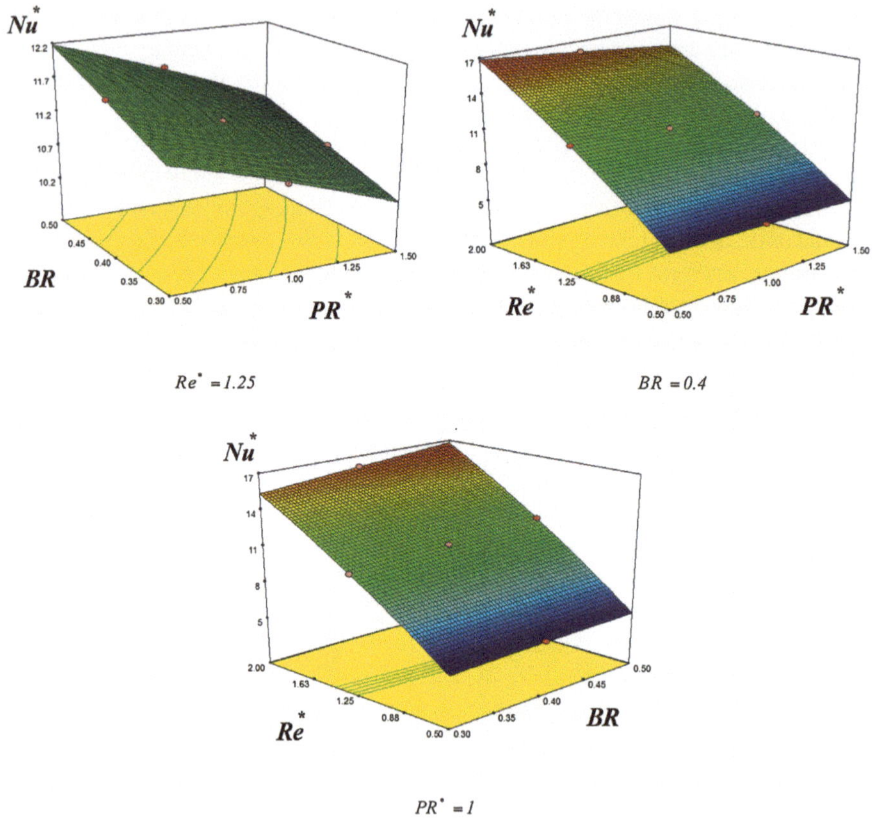

FIGURE 3.4 cont'd.

$$
\begin{aligned}
f^* = {} & 5.24 + 0.2PR^* + 23.2BR - 4.8\mathrm{Re}^* \\
& - 18.36(PR^*)\,(BR) - 0.79(PR^*)(\mathrm{Re}^*) - 2.94(BR)(\mathrm{Re}^*) \\
& + 1.25(PR^*)^2 + 15.43\,(BR)^2 + 1.22\,(\mathrm{Re}^*)^2
\end{aligned}
\tag{3.19}
$$

where $f^* = 100f, PR^* = 0.1PR$, $\mathrm{Re}^* = 10^{-4}\mathrm{Re}$.

Deceasing temperature boundary layer thickness due to increasing Reynolds number and height ratio makes temperature gradient to enhance. So, Nusselt number augments with rise of *BR* and Re. Friction factor has reverse relationship with Reynolds number due to augment in bulk flow. Pressure drop augments with rise of *BR*. So, Darcy factor augments with rise of *BR*. Increasing pitch ratio has reverse treatment in comparison to height ratio. Nusselt number decreases with augment of *PR* and Darcy factor reduces. Fig. 3.6 illustrates the frictional and thermal entropy generation contours for different values of active parameters. Frictional entropy generation enhances with rise of pressure drop. So, $S_{g,f}$ augments with rise of *BR* and Re, while it decreases with rise of *PR*. Thermal entropy generation has direct relationship with pitch ratio. Also it can be found that $S_{g,h}$ reduces with rise of *BR* and Re.

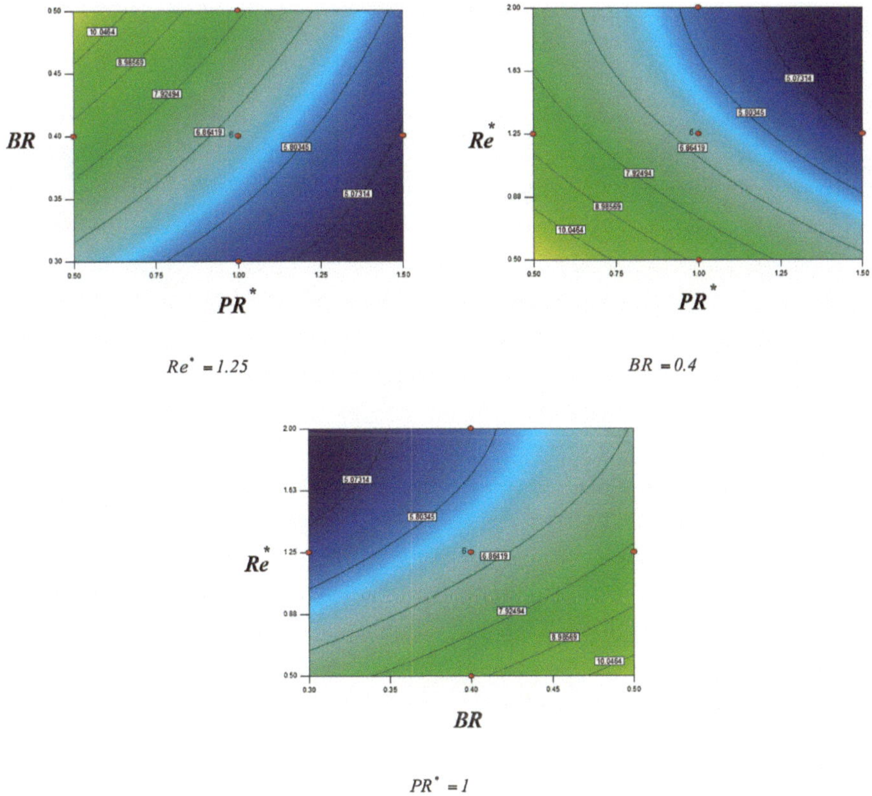

FIGURE 3.5

Effects of Reynolds number, pitch ratio, and height ratio on friction factor (f^*).

3.2 Turbulent nanofluid in appearance of twisted tape with alternate axis

3.2.1 Definition of the problem

In this research, the twisted tape is utilized to enhance heat transfer (Fig. 3.7) [3]. We have three sections: inlet, test, and outlet sections. All the outlet and inlet sections are smooth straight tubes, and the test section is a twisted tube. The flow direction is along z-direction of Cartesian coordinate. The working fluid is CuO–water nanofluid ($\phi = 0.05$). The main geometrical parameters of the twisted tubes are pitch (P), length (b), and thickness (t). In this study, length of the tube is 900 mm and diameter is 20 mm. The middle 300 mm is considered as the test section, and a twisted tape is inserted in this section. In order to ensure that the fully developed flow is formed and avoid any probable back flow streams, the inlet and outlet sections are considered equals to 300 mm. Geometrical parameters for various cases are illustrated in Table 3.6.

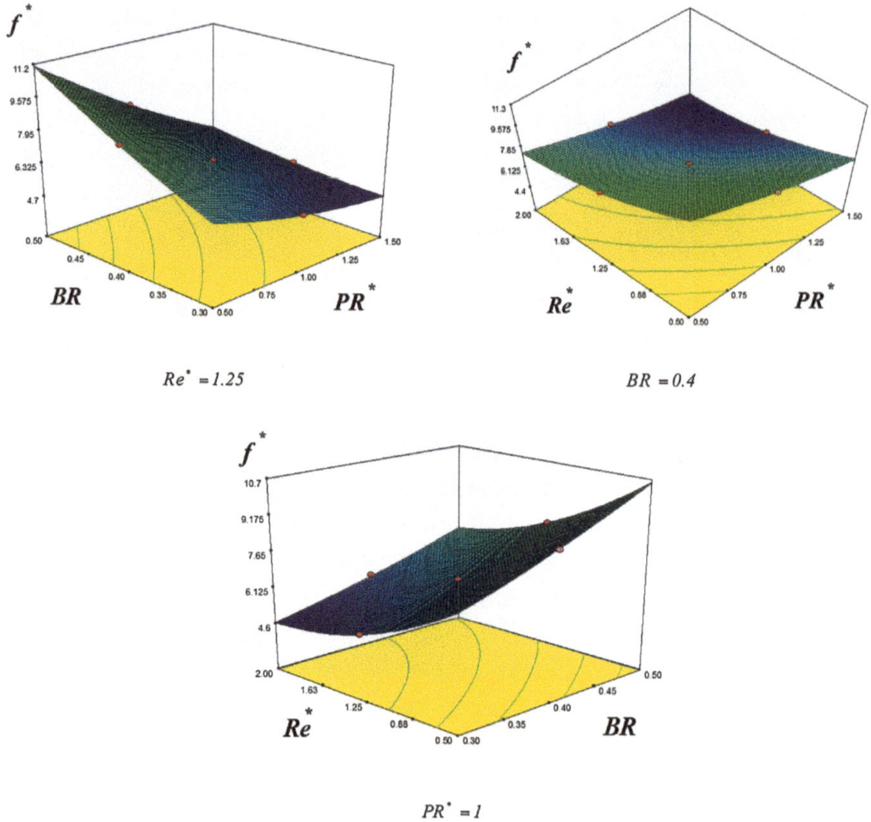

FIGURE 3.5 cont'd.

Fig. 3.8A shows a sample mesh for this heat exchanger. To find grid-independent outputs, different grids have been tested. For instance, according to Table 3.7, the grid system of 1,231,077 cells is chosen in numerical simulations. More details exist in Table 3.8. Moreover, the hydrothermal boundary layers should be captured by the grid to have a reliable grid. Fig. 3.8(B) shows the convergence of various cases. The value of nondimensional y^+ parameter should be less than 5 for our selected model. Fig. 3.8C and D shows that $Y^+ < 5$. So, this model limitation is satisfied.

3.2.2 Effects of active parameters

In this research, the twisted tape with alternate axis is utilized to intensify the thermal performance of heat exchanger filled with CuO−water nanofluid. Brownian motion impact is considered in order to predict nanofluid properties.

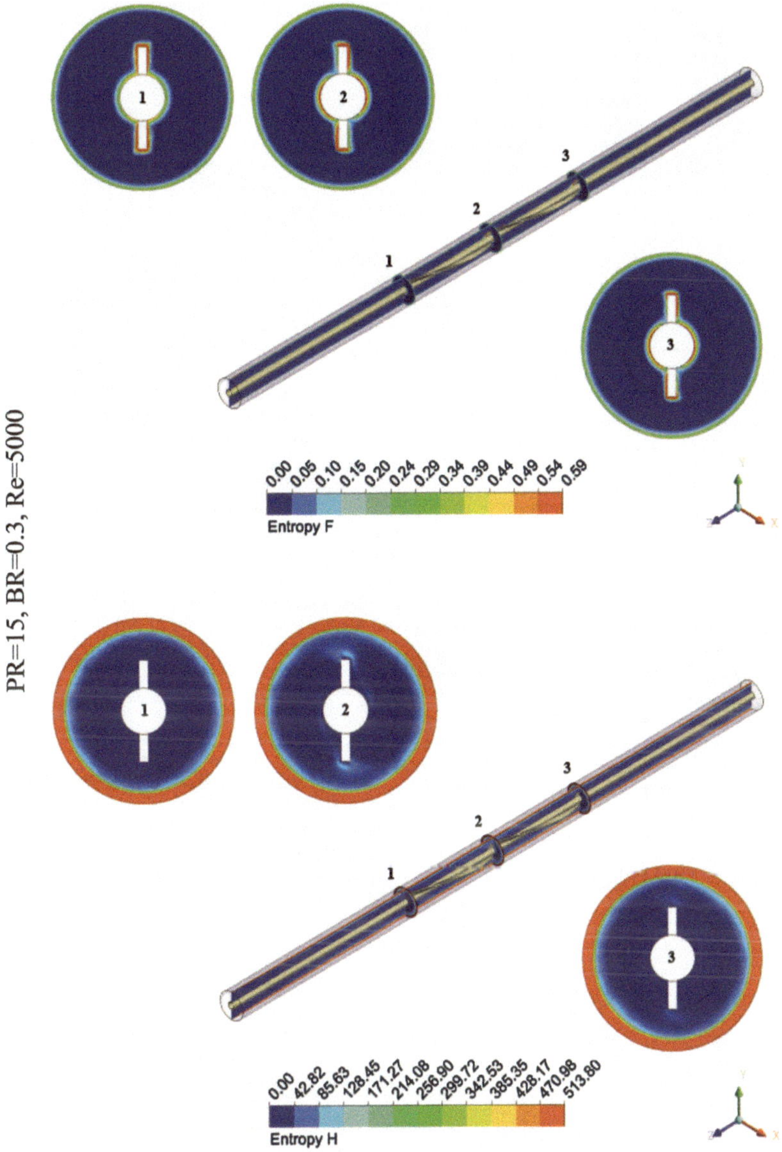

FIGURE 3.6

Effects of *PR*, *BR*, and Re on frictional and thermal entropy generation contours when $\phi = 0.05$.

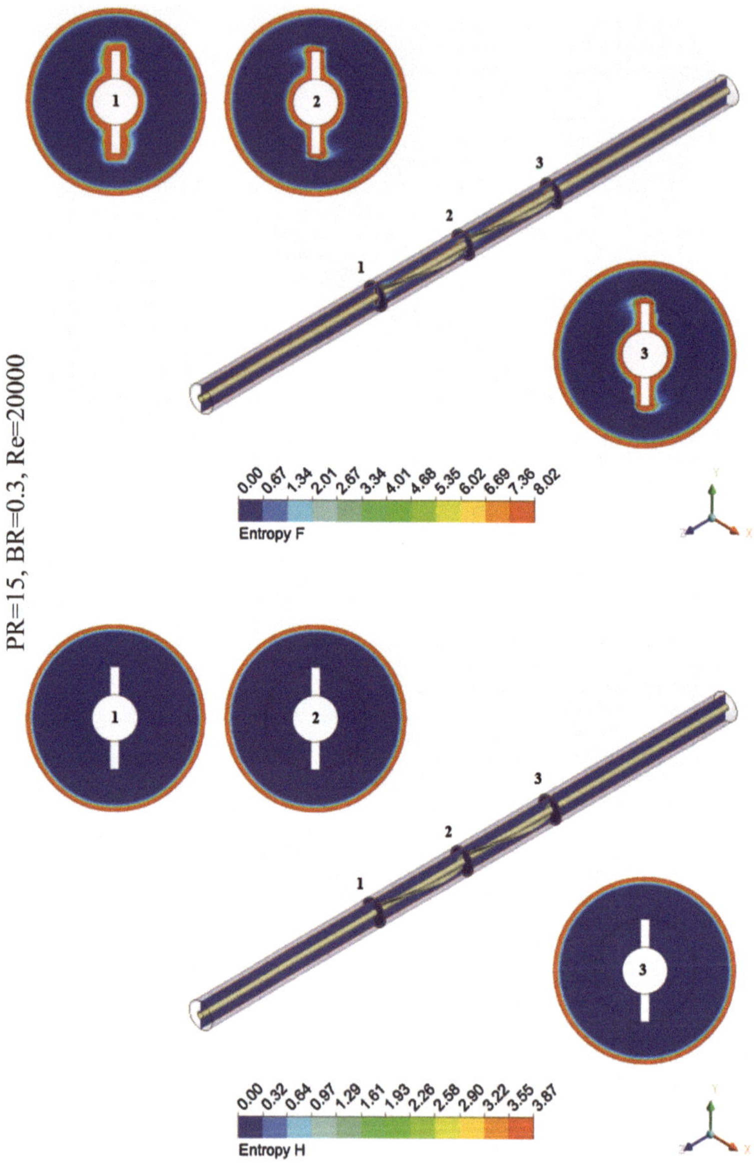

FIGURE 3.6 cont'd.

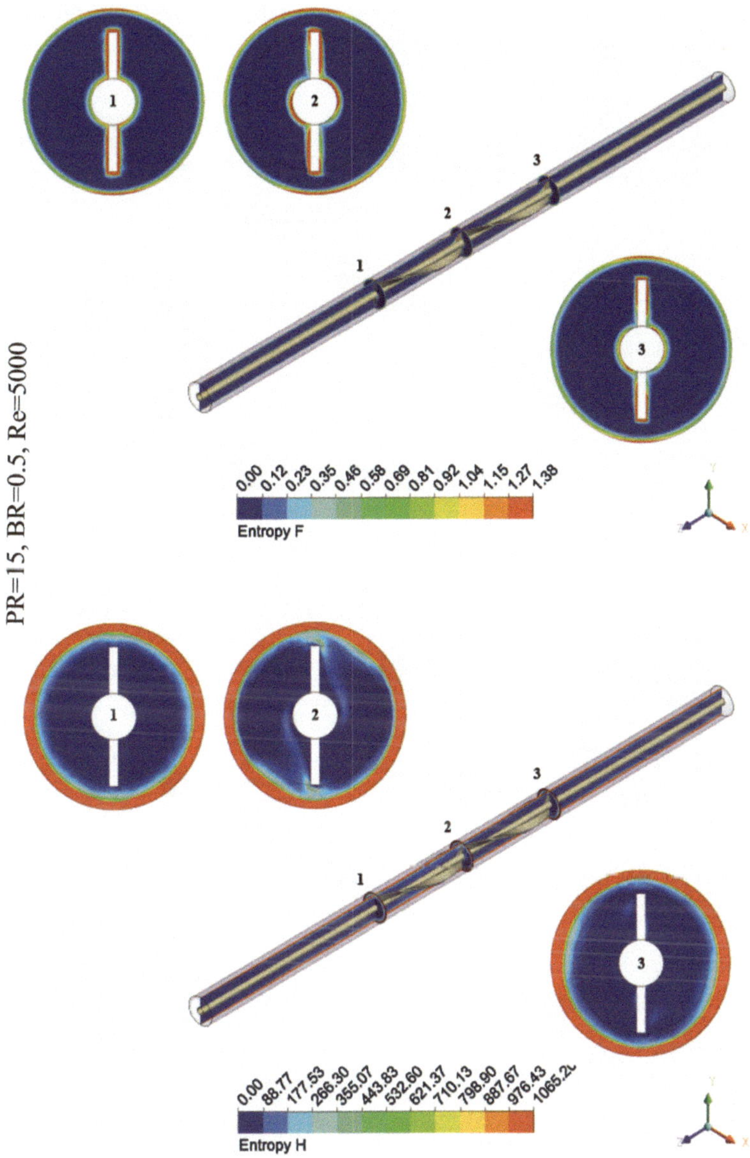

FIGURE 3.6 cont'd.

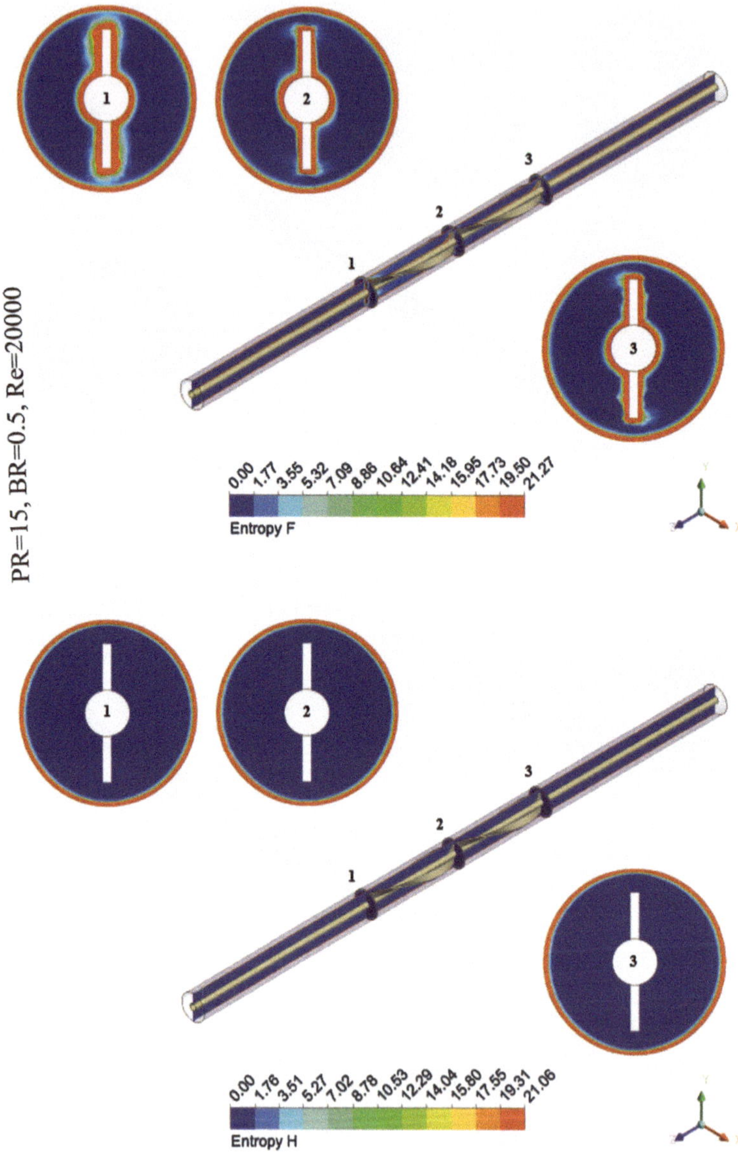

FIGURE 3.6 cont'd.

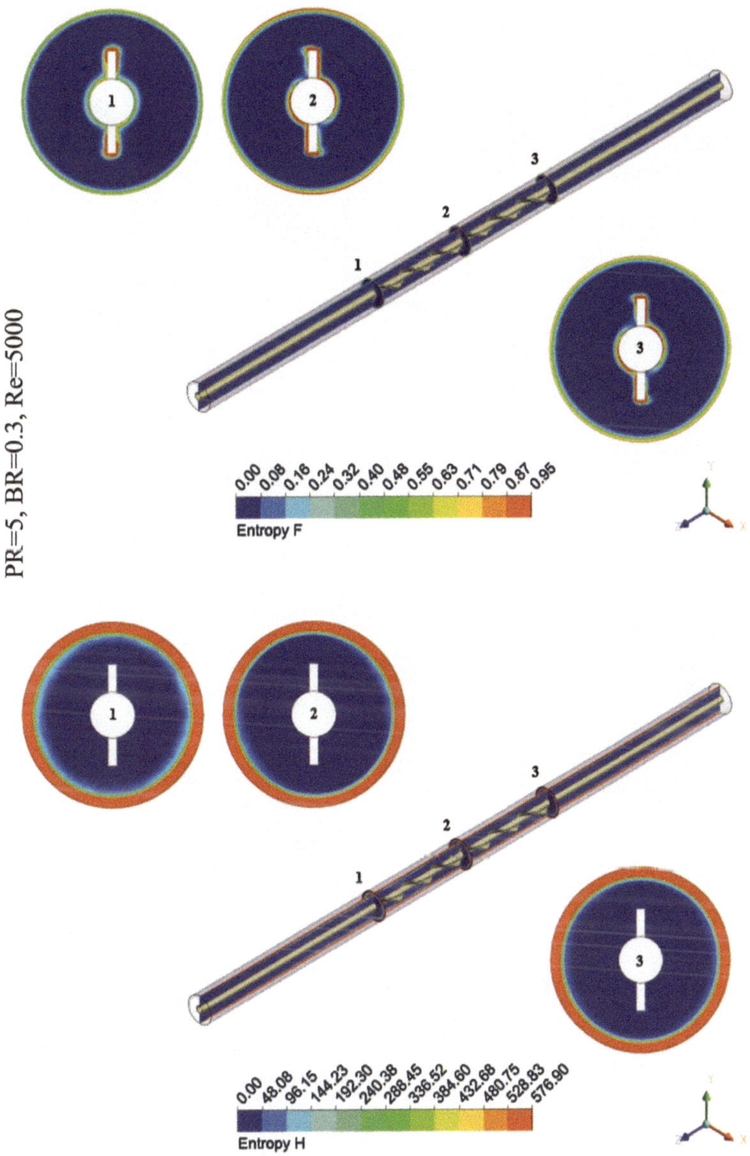

FIGURE 3.6 cont'd.

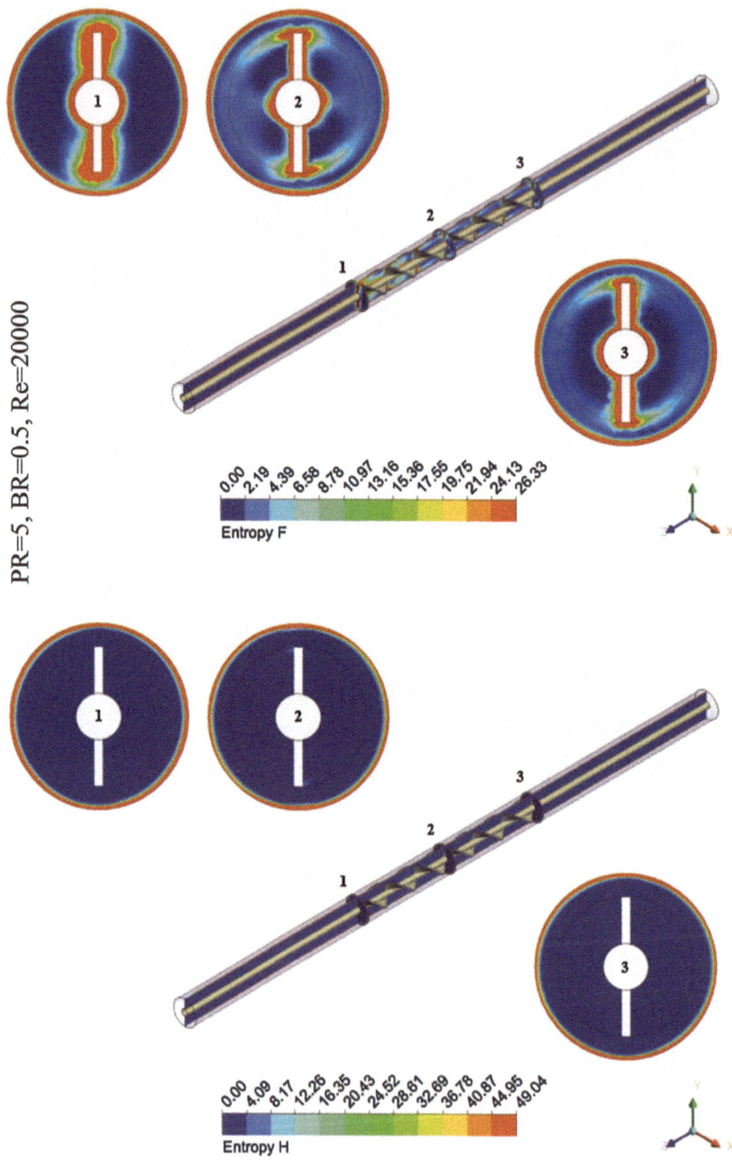

FIGURE 3.6 cont'd.

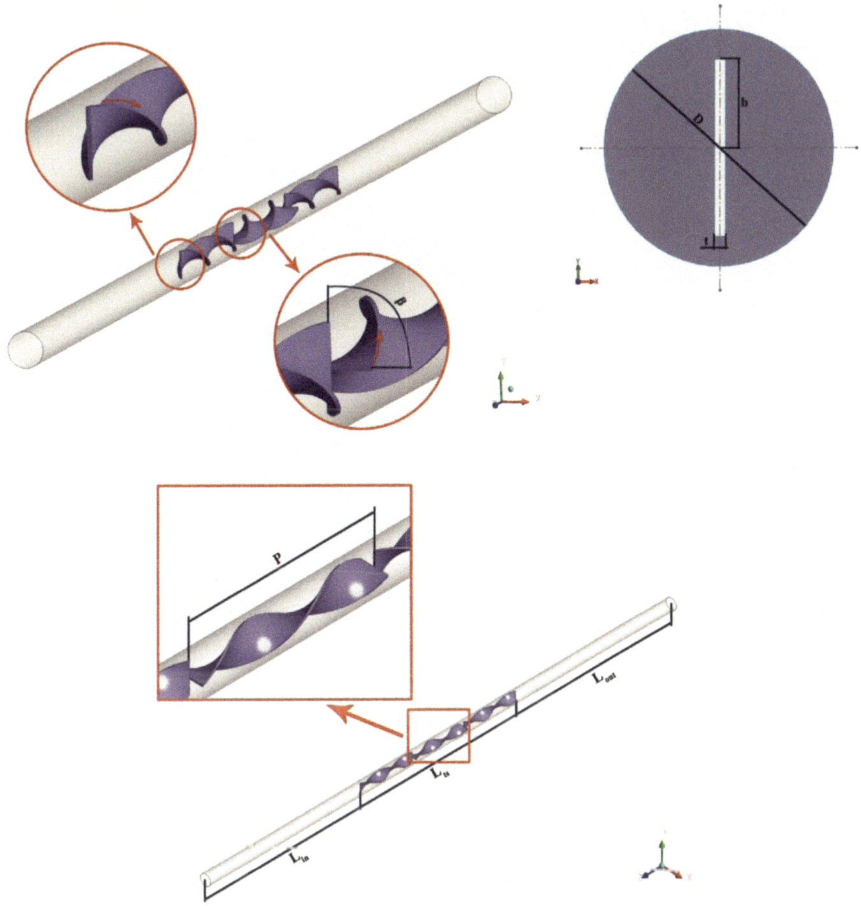

FIGURE 3.7

Geometry of the problem.

Table 3.6 Geometrical parameters for various cases.

Case no.	β	Re
1	0	5000
2	0	12,500
3	0	20,000
4	0.785	5000
5	0.785	12,500
6	0.785	20,000
7	1.57	5000
8	1.57	12,500
9	1.57	20,000

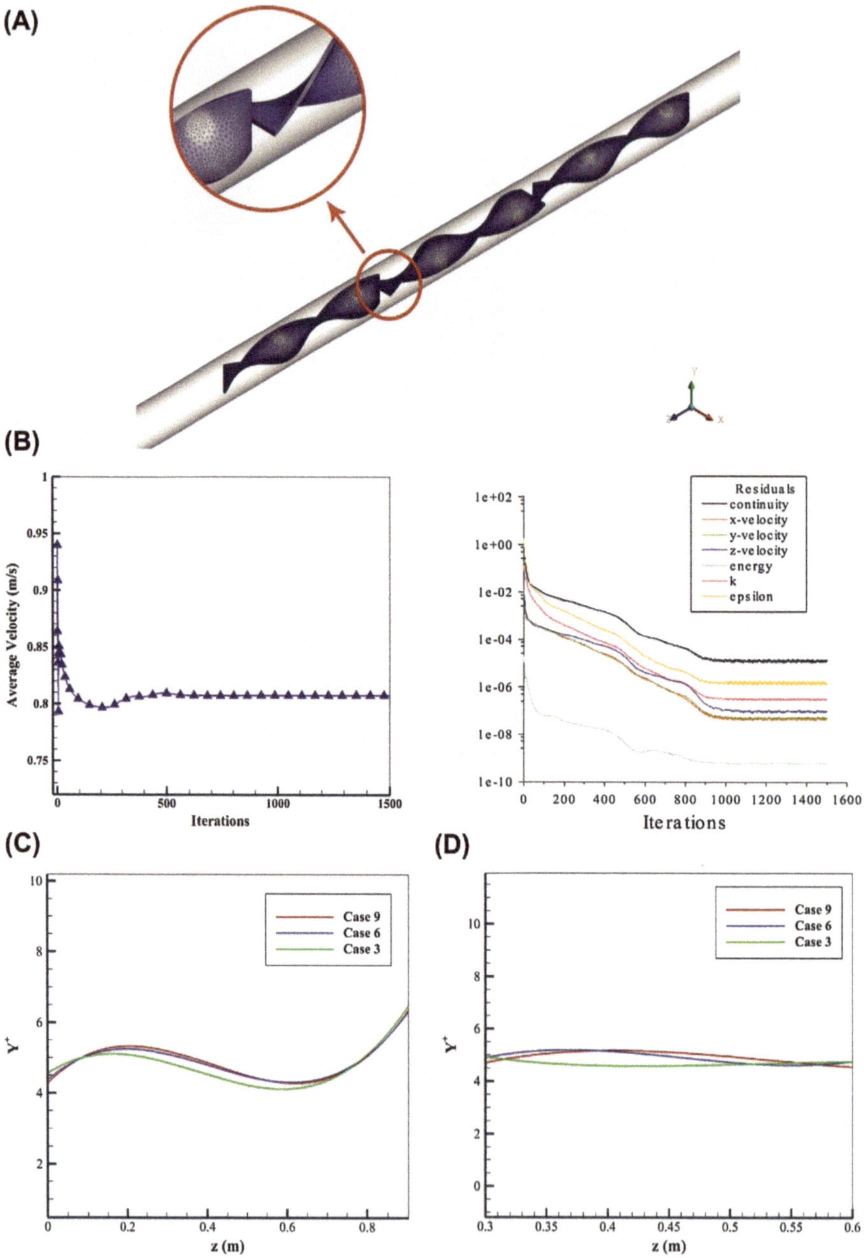

FIGURE 3.8

(A) Detail view of computational mesh, (B) Sample for convergence for case 9, (C) Y^+ on the outer wall, (D) Y^+ on the modified twisted tape.

Table 3.7 Grid independence test for case 9.

Number of element	Number of nodes	*Nu*
842,346	320,782	253.8415
1,053,203	371,067	258.7461
1,231,077	405,753	260.0394
1,482,376	464,125	260.5273

Table 3.8 More details of mesh.

	Number of elements	Number of nodes	Average aspect ratio	Average skewness
Case 1-2-3	1,242,172	408,355	2.8221	0.22992
Case 4-5-6	1,355,657	440,955	2.7689	0.23223
Case 7-8-9	1,231,077	405,753	2.8486	0.23003

Figs. 3.9 and 3.10 demonstrated temperature, velocity, and pressure contours when $\beta = 0$. The Reynolds number is a dimensionless value that measures the ratio of inertial forces to viscous forces. As Reynolds number increases, pressure loss augments and thermal boundary layer thickness decreases. Temperature gradient augments with rise of Reynolds number. This is due to augment in fluid mixing and swirl intensity by increase of inlet velocity. At a given Reynolds number, Nusselt number in the tube with turbolentor is significantly higher than those in the plain tube.

Figs. 3.11 and 3.12 illustrated temperature, velocity, and pressure contours when $\beta = 0.25\pi$. Increasing revolution angle leads to higher temperature gradient but with cost of higher pressure loss. Secondary flow enhances with rotating axis of twisted tape and in turn convective flow enhances. Rate of heat transfer augments with rise of revolution angle. This is attributed to a stronger turbulent intensity and thus a better fluid mixing.

Figs. 3.13 and 3.14 depict temperature, velocity, and pressure contours when $\beta = 0.5\pi$. The effect of revolution angle is more significant at lower Reynolds number. This can be explained that at lower Reynolds number, a thermal boundary becomes thicker; therefore the swirl flows induced by turbolentor possess more significant effect on disruption of thermal boundary. Moreover, using higher revolution angle possessed higher rate of heat transfer.

Fig. 3.15 shows sample for velocity vector for three cases with higher Reynolds number at $z = 0.45$ m. As you can see in this figure, increasing revolution angle leads to increase in pressure drop due to augment in turbulence intensity. The

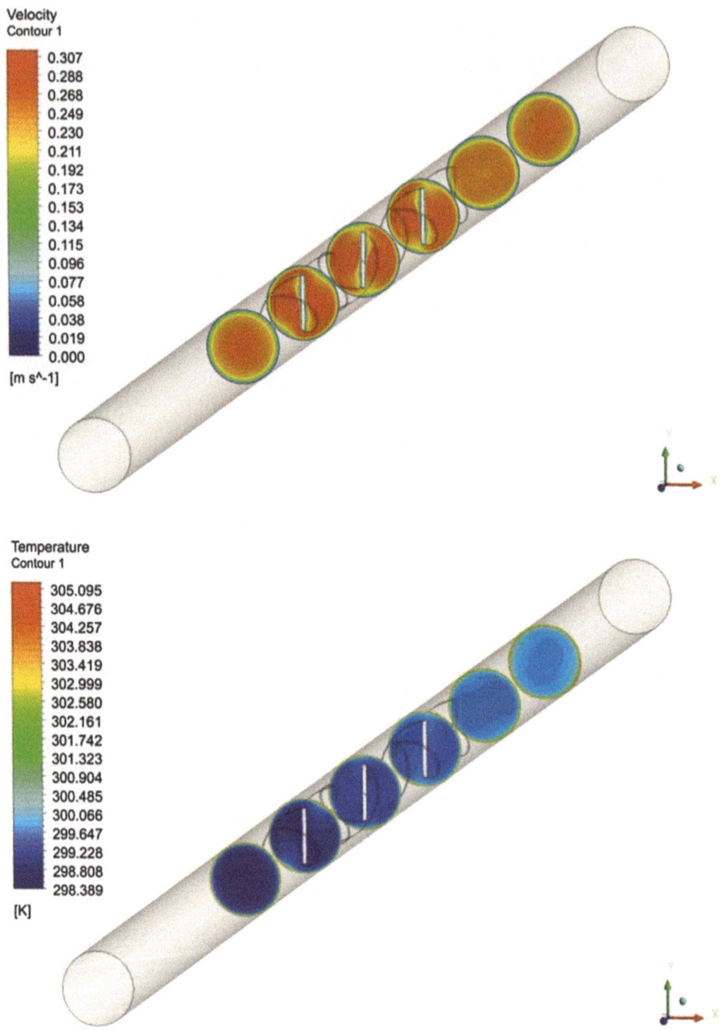

FIGURE 3.9

Temperature, velocity, and pressure contours in various sections for case 1.

strength of secondary vortexes enhances with rise of revolution angle and in turn thermal boundary layer thickness decreases.

Fig. 3.16 and Table 3.9 demonstrate the influence of active parameters on Nu^* and f^*. Also, according to the obtained data, following correlations can be presented:

$$Nu^* = 0.21 + 0.03\beta + 1.3\mathrm{Re}^* + 0.023(\beta)(\mathrm{Re}^*) - 0.025(\beta)^2 - 0.073(\mathrm{Re}^*)^2 \tag{3.20}$$

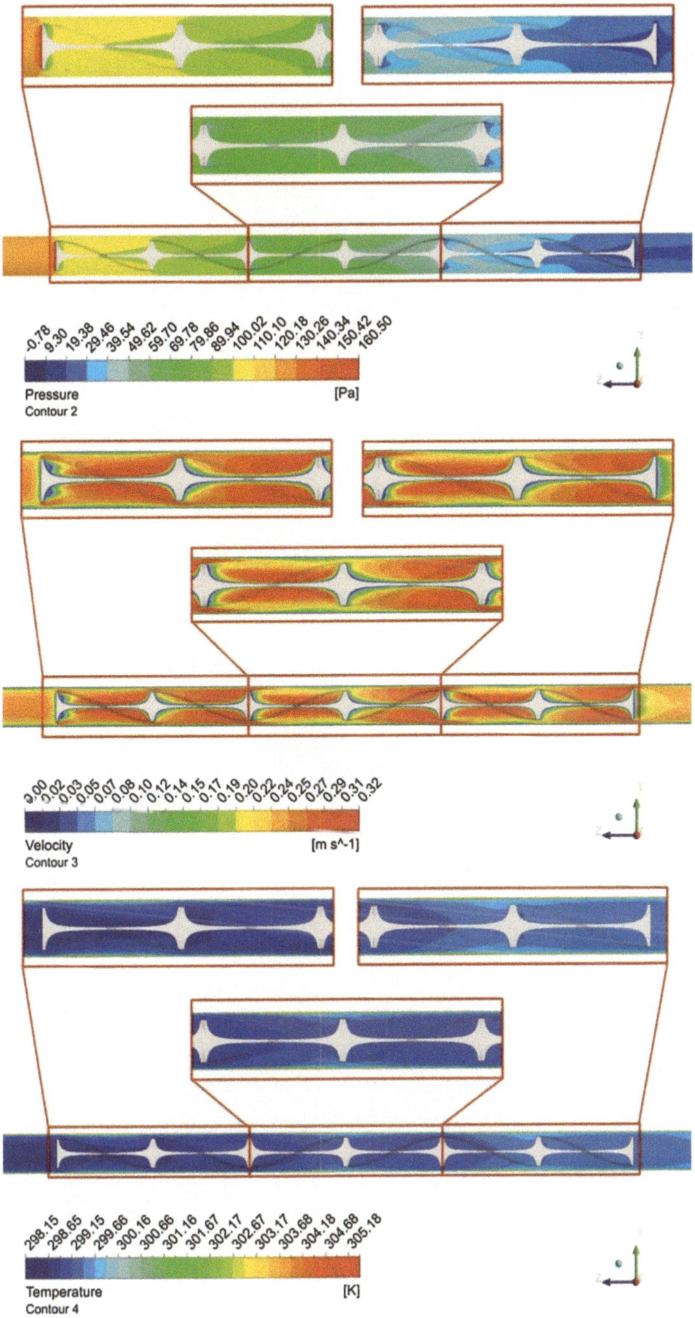

FIGURE 3.9 cont'd.

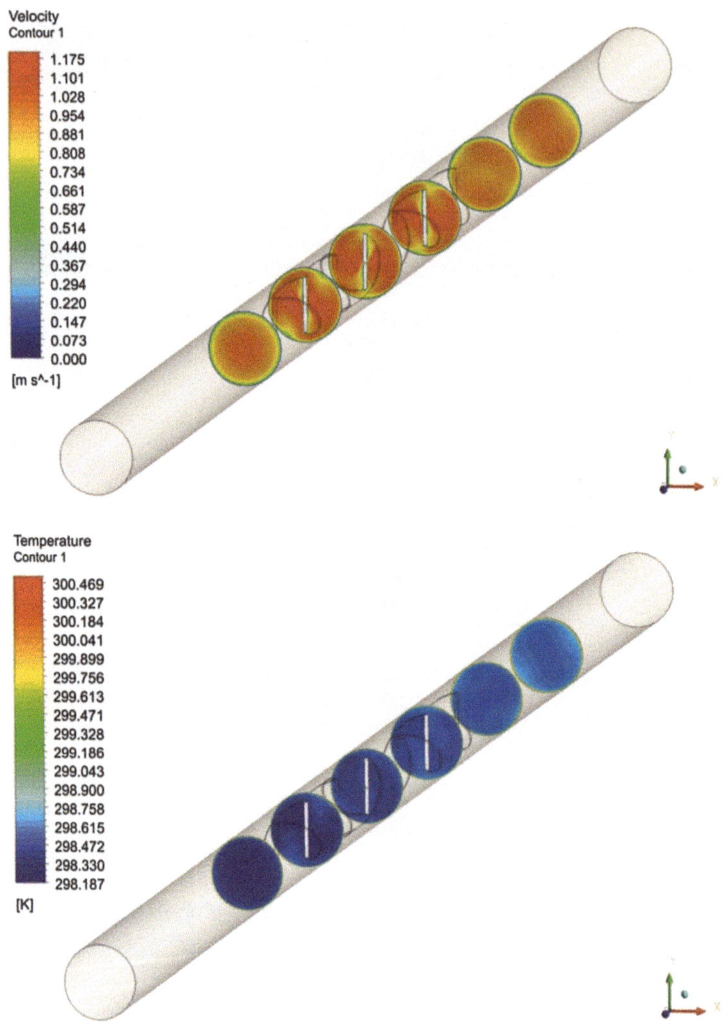

FIGURE 3.10

Temperature, velocity, and pressure contours in various sections for case 3.

where $Nu^* = 0.01Nu, \mathrm{Re}^* = 10^{-4}\mathrm{Re}$.

$$f^* = 1.3 + 0.175\beta - 0.56\mathrm{Re}^* - 2.2\times 10^{-3}(\beta)(\mathrm{Re}^*) - 0.065(\beta)^2 + 0.133(\mathrm{Re}^*)^2 \tag{3.21}$$

where $f^* = 10f, \mathrm{Re}^* = 10^{-4}\mathrm{Re}$.

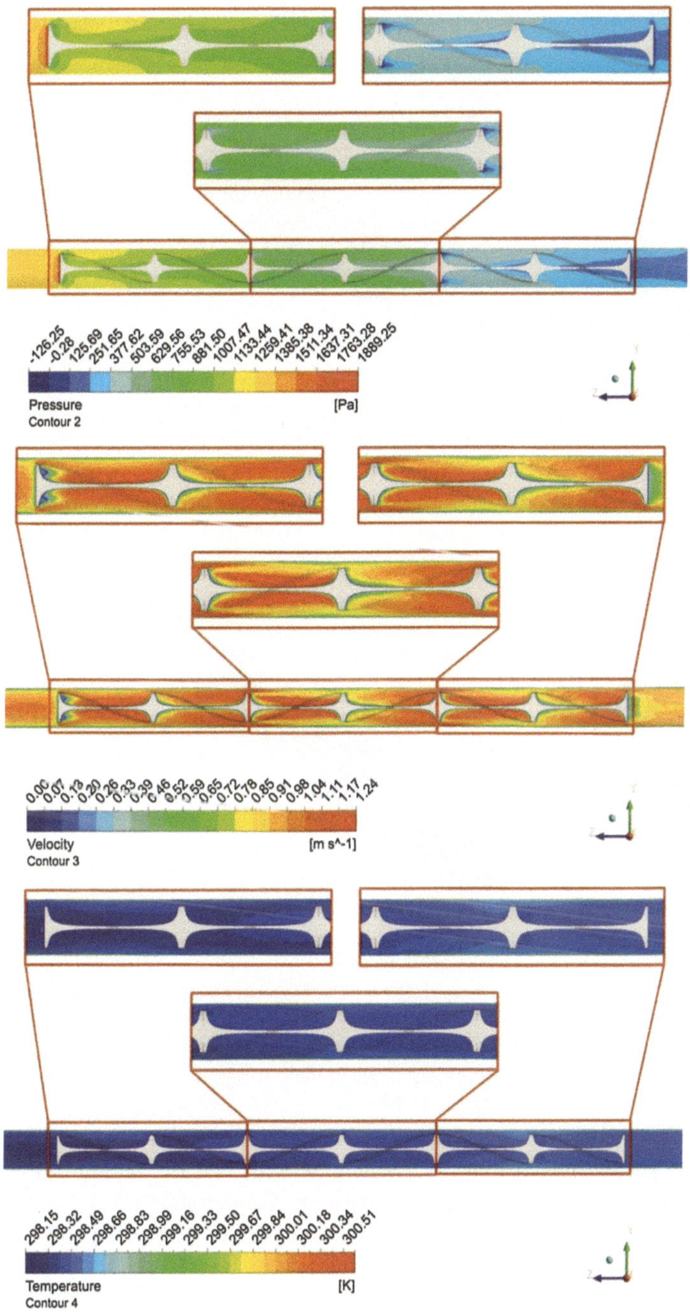

FIGURE 3.10 cont'd.

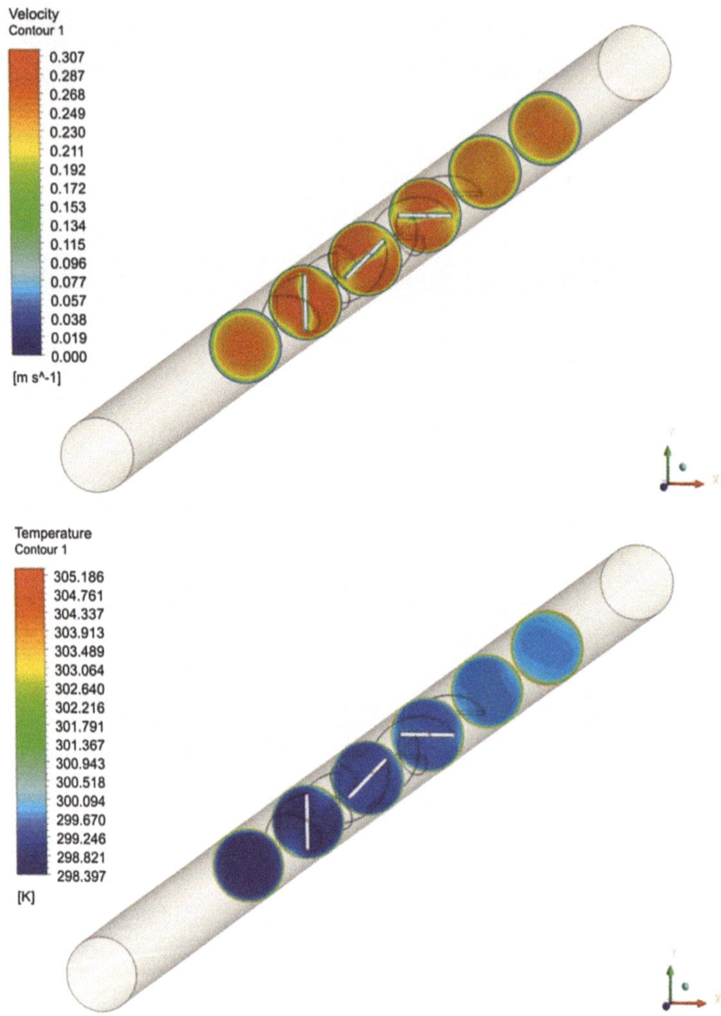

FIGURE 3.11

Temperature, velocity, and pressure contours in various sections for case 4.

These figures showed that Nu^* augments with rise of Reynolds number and revolution angle. As revolution angle increases, the secondary flow becomes stronger and thermal boundary layer thickness decreases. So, temperature gradient enhances with rise of revolution angle and in turn Nusselt number enhances with rise of this parameter. However, friction factor augments with increase of revolution angle. This phenomenon is obtained because of increasing contact surface area,

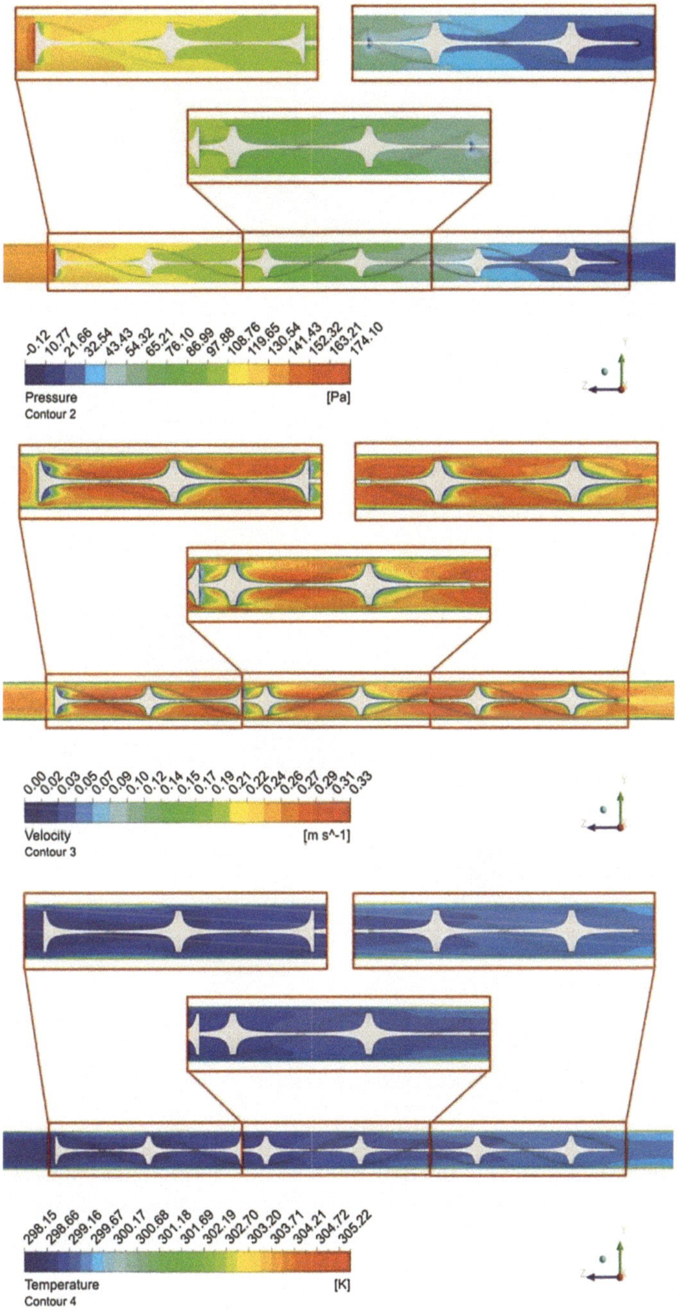

FIGURE 3.11 cont'd.

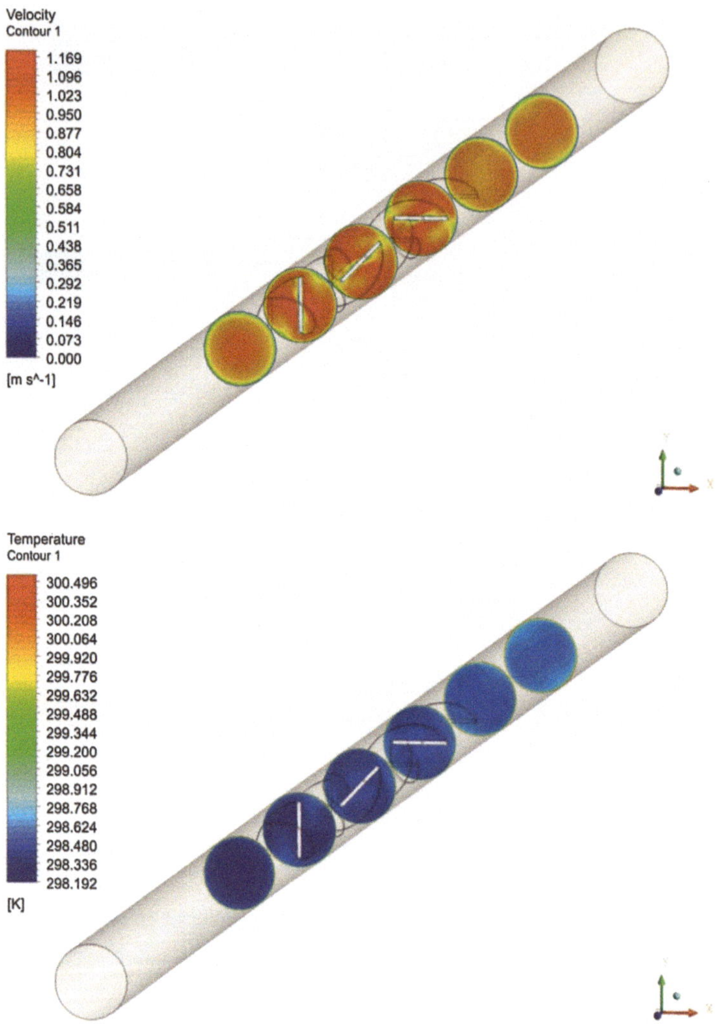

FIGURE 3.12

Temperature, velocity, and pressure contours in various sections for case 6.

residence time, swirl intensity, and fluid mixing with multilongitudinal vortices flow. As Reynolds number increases, the higher turbulence intensities and more chaotic fluid mixing are caused by the interaction of swirl flows by turbolentor. So, Nusselt number increases with rise of Reynolds number. Also pressure drop enhances with rise of this parameter. Friction factor has reverse relationship with Reynolds number due to increase in bulk flow.

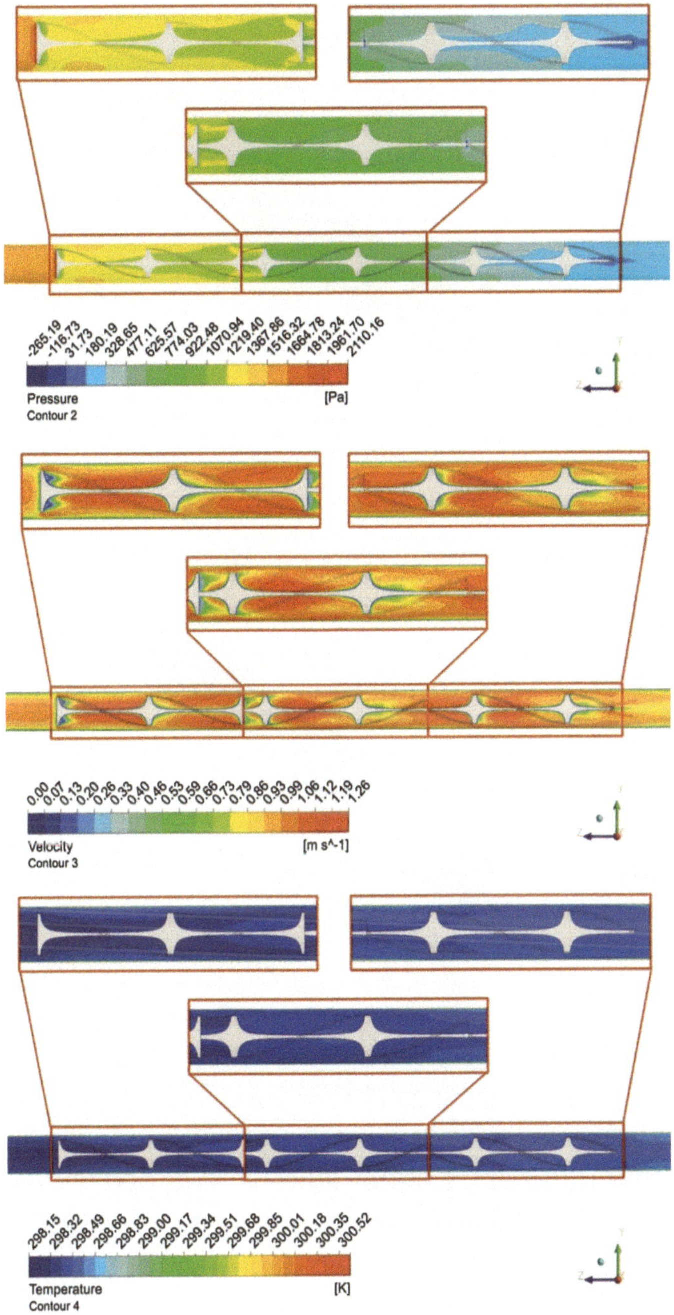

FIGURE 3.12 cont'd.

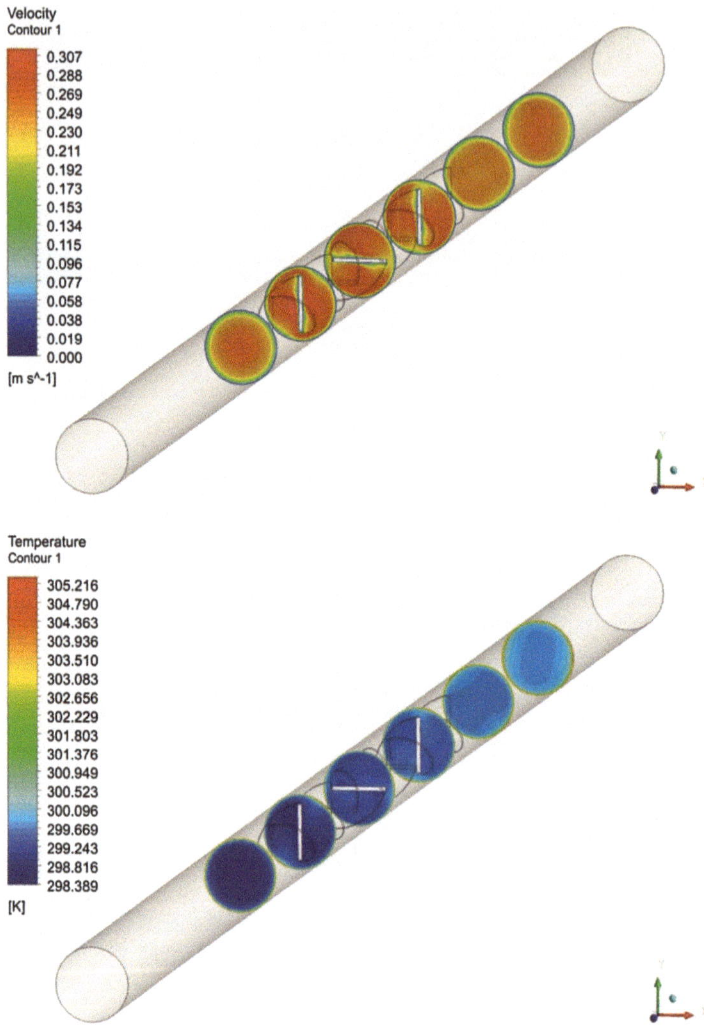

FIGURE 3.13

Temperature, velocity, and pressure contours in various sections for case 7.

3.3 Helical turbulators effect on nanofluid convective

3.3.1 Definition of the problem

Figs. 3.17 and 3.18 depict the geometry of helical twisted tape and sample mesh which is utilized in this simulation [4]. Inner and outer diameters are 5 and 20 mm, respectively. The length of the circular duct is 900 mm. The middle 300 mm is test section in which swirling flow device is situated. H_2O-based

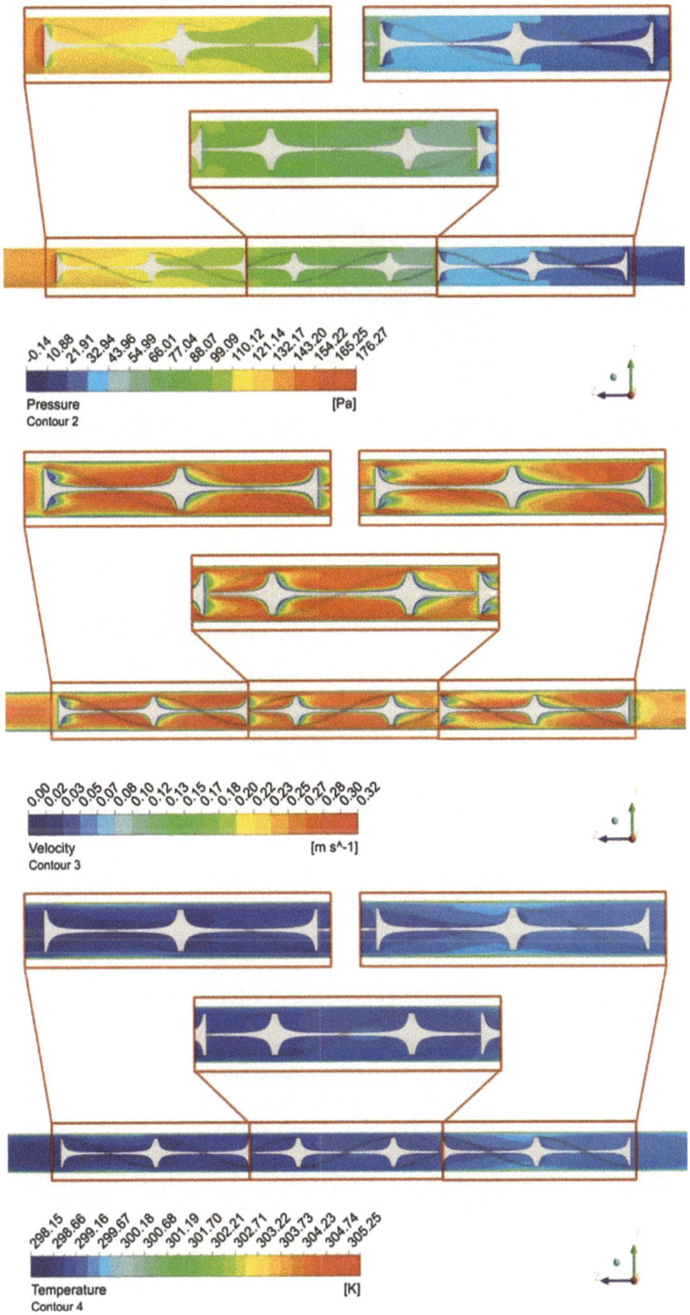

FIGURE 3.13 cont'd.

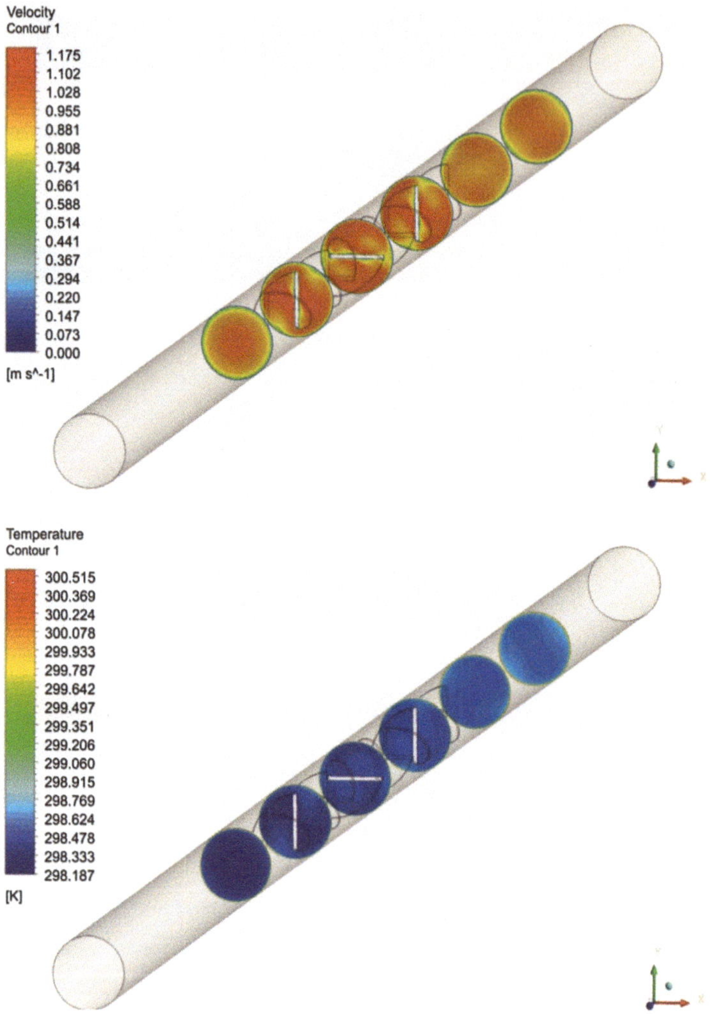

FIGURE 3.14

Temperature, velocity, and pressure contours in various sections for case 9.

nanofluid is selected as working fluid. There is not any turbulators in other parts of the pipe to prevent any probable back flow streams. Table 3.10 shows various cases which are simulated in this chapter.

3.3.2 Effects of active parameters

In this chapter, the helical twisted tape turbulator is inserted in circular tube to intensify the thermal performance. The working fluid is $CuO-H_2O$ nanofluid.

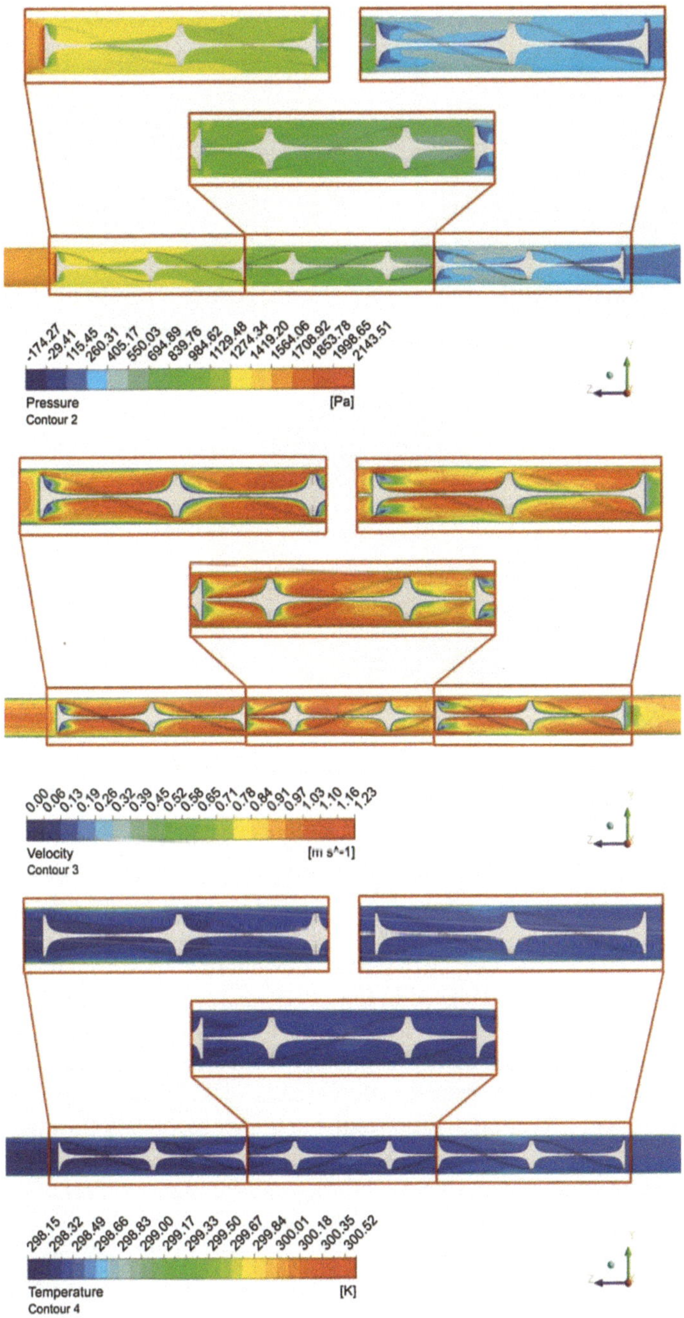

FIGURE 3.14 cont'd.

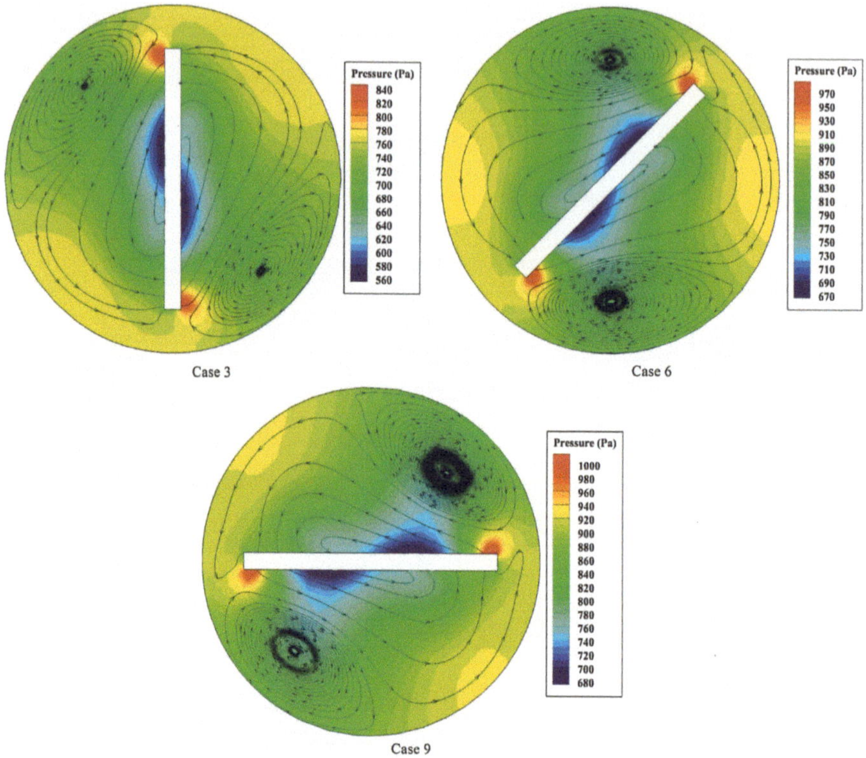

FIGURE 3.15

Velocity Vector and pressure contours at $z = 0.45$ m.

Brownian motion effect is taken into account in order to predict nanofluid properties. Effects of Reynolds number $(\mathrm{Re} = 5000 \text{ to } 15000)$, pitch ratio $(PR = 1.5 \text{ to } 3)$, and width ratio $(BR = 0.25 \text{ to } 0.75)$ on nanofluid treatment are depicted.

Effects of Reynolds number, pitch ratio, and width ratio on stream tracer, temperature, velocity contours are depicted in Figs. 3.19–3.22. As inlet velocity increases, turbulence intensity enhances and temperature boundary layer thickness reduces. Therefore, rate of heat transfer augments with rise of Reynolds number. Convective heat transfer increases with augment of swirl intensity but pressure drop increases, too. Lower pitch ratio results in higher secondary flow but opposite behavior is observed for width ratio. Turbulent intensity is a decreasing function of pitch ratio. More nanofluid mixing can be occurred for higher values of width ratio due to more disruption of thermal boundary layer.

Figs. 3.23 and 3.24 and Table 3.11 depict the effects of PR, BR, and Re on Nu^* and f^*. According to the obtained data, following correlation can be obtained:

$$\begin{aligned} Nu^* = {} & 0.34 + 0.42 \times 10^{-4} PR - 0.26BR + 0.28\mathrm{Re}^* \\ & - 0.017(PR)(BR) - 0.03(PR)(\mathrm{Re}^*) + 0.29(BR)(\mathrm{Re}) \\ & + 3.7 \times 10^{-3}(PR)^2 + 0.38(BR)^2 - 0.18(\mathrm{Re}^*)^2 \end{aligned} \tag{3.22}$$

where $Nu^* = 0.01Nu$, $\mathrm{Re}^* = 10^{-4}\mathrm{Re}$.

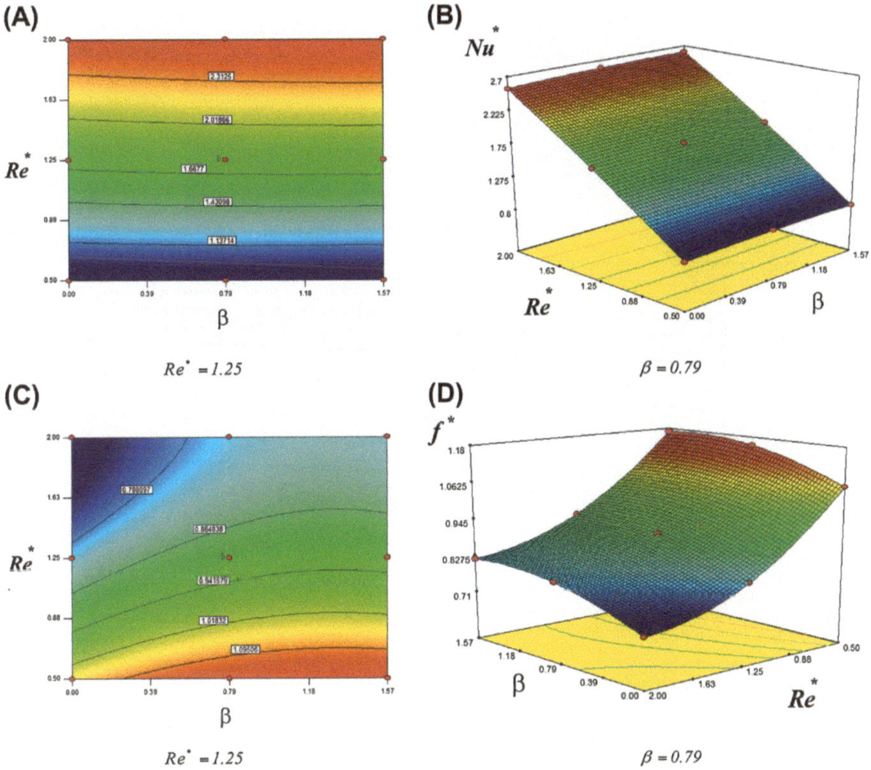

FIGURE 3.16

Effects of Reynolds number, revolution angle on (A, B) Nusselt number; (C, D) Friction factor.

Table 3.9 Values of Nusselt number and friction factor for various cases.

Case no.	*Nu*	*f*
1	84.83086	0.105688
2	173.1856	0.080905
3	253.0332	0.071047
4	85.71245	0.115068
5	176.6266	0.090388
6	259.9045	0.080697
7	86.47866	0.116883
8	177.4868	0.09187
9	260.0394	0.081723

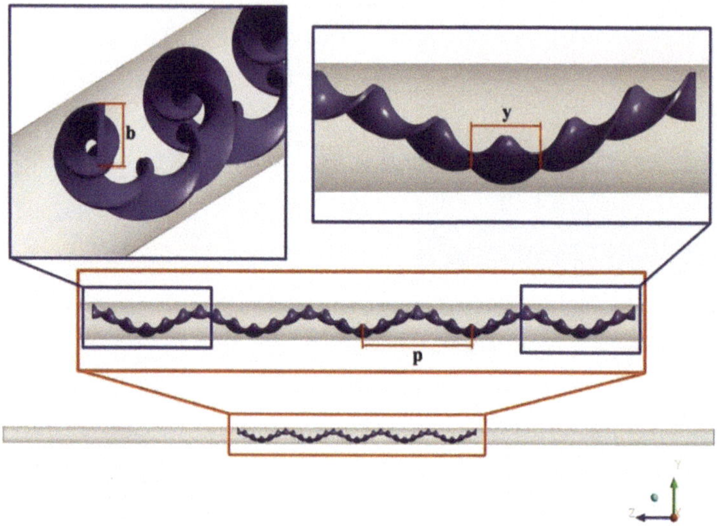

FIGURE 3.17

Geometry of the problem.

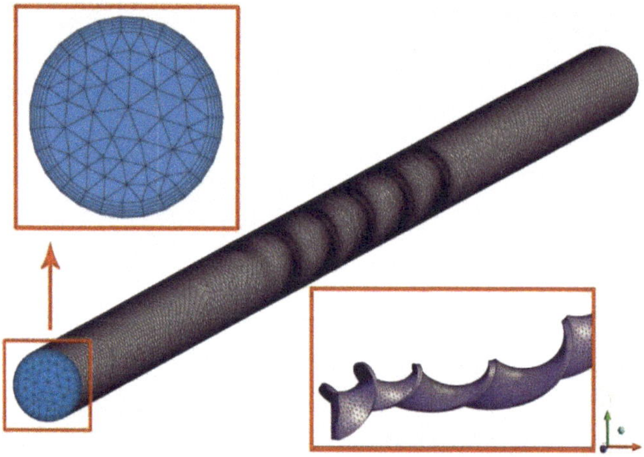

FIGURE 3.18

Detail view of computational mesh.

$$f^* = 1.03 - 0.43PR + 0.78BR - 0.06\text{Re}^* - 0.41(PR)(BR) + 0.01(PR)(\text{Re}^*) - 0.26(BR)(\text{Re}) + 0.12(PR)^2 + 1.36(BR)^2 - 0.04(\text{Re}^*)^2 \tag{3.23}$$

where $f^* = 10f$, $\text{Re}^* = 10^{-4}\text{Re}$.

Table 3.10 Geometrical parameters for various cases.

Case no.	*PR*	*BR*	Re
1	3	0.25	5000
2	3	0.25	15,000
3	3	0.5	5000
4	3	0.5	15,000
5	3	0.75	5000
6	3	0.75	15,000
7	2	0.25	5000
8	2	0.25	15,000
9	2	0.5	5000
10	2	0.5	15,000
11	2	0.75	5000
12	2	0.75	15,000
13	1.5	0.25	5000
14	1.5	0.25	15,000
15	1.5	0.5	5000
16	1.5	0.5	15,000
17	1.5	0.75	5000
18	1.5	0.75	15,000

These figures showed that Nu^* augments with rise of Reynolds number and width ratio, but it reduces with augment of pitch ratio. Secondary flow becomes stronger with rise of Re and *BR*, while it reduces with rise of *PR*. Therefore, temperature gradient augments with augment of Re and *BR*. Pressure drop augments with rise of Re and *BR* due to increasing contact surface area and residence time. Darcy factor has reverse relationship with Reynolds number due to increase in bulk flow.

3.4 Influence of a helical-twisting device on efficiency of nanomaterial

3.4.1 Definition of the problem

Fig. 3.25 presents the geometry of the understudy duct with the designed helical-twisting flow disturber through that. The figure also marks various dimensions of the duct and the flow disturber, which are t: 1 mm, l: 15 mm, y: 23.93 mm, P: 166.67 mm, L_{in}: 500 mm, L_t: 500 mm, L_{out}: 500 mm, and b: ranging from 5 to 15 mm. As such, the diameter of the duct is 50.8 mm. The flow through the pipe is a forced convective water–CuO nanofluid [5].

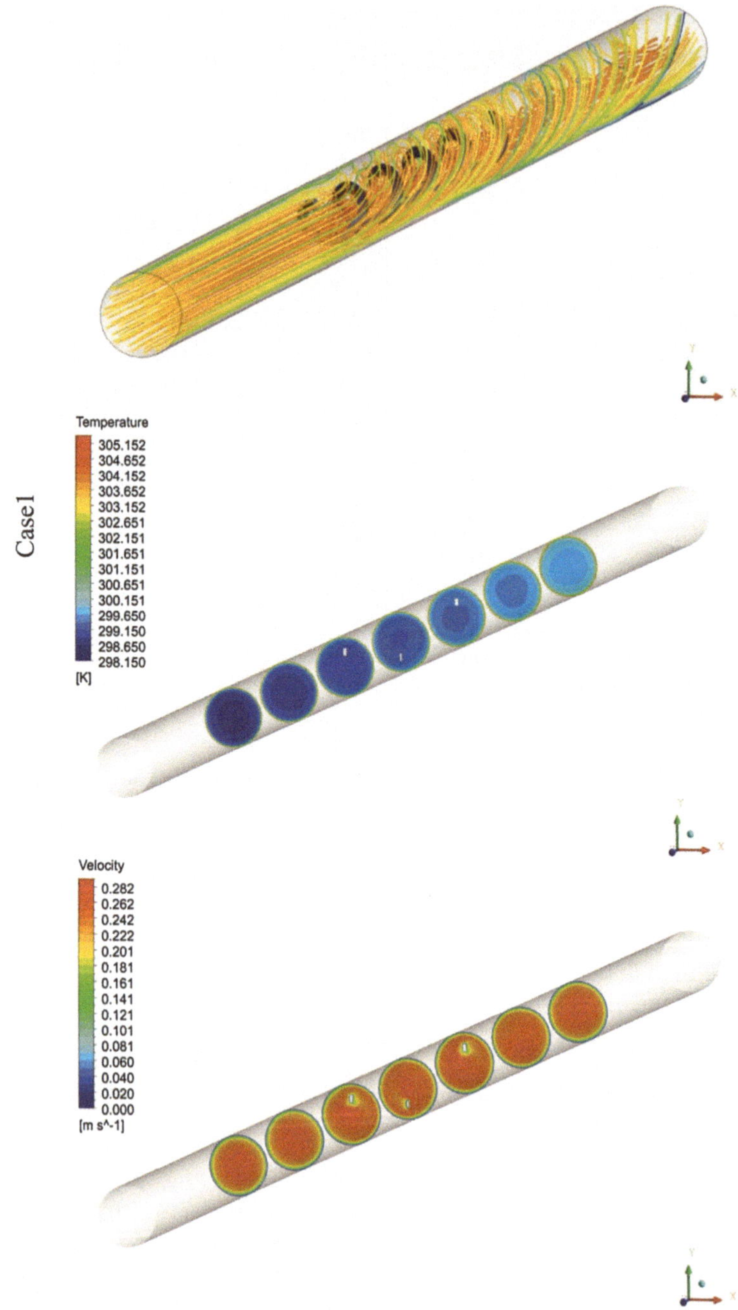

FIGURE 3.19

Stream tracer, temperature, velocity contours for cases 1 and 2 when $\phi = 0.05$.

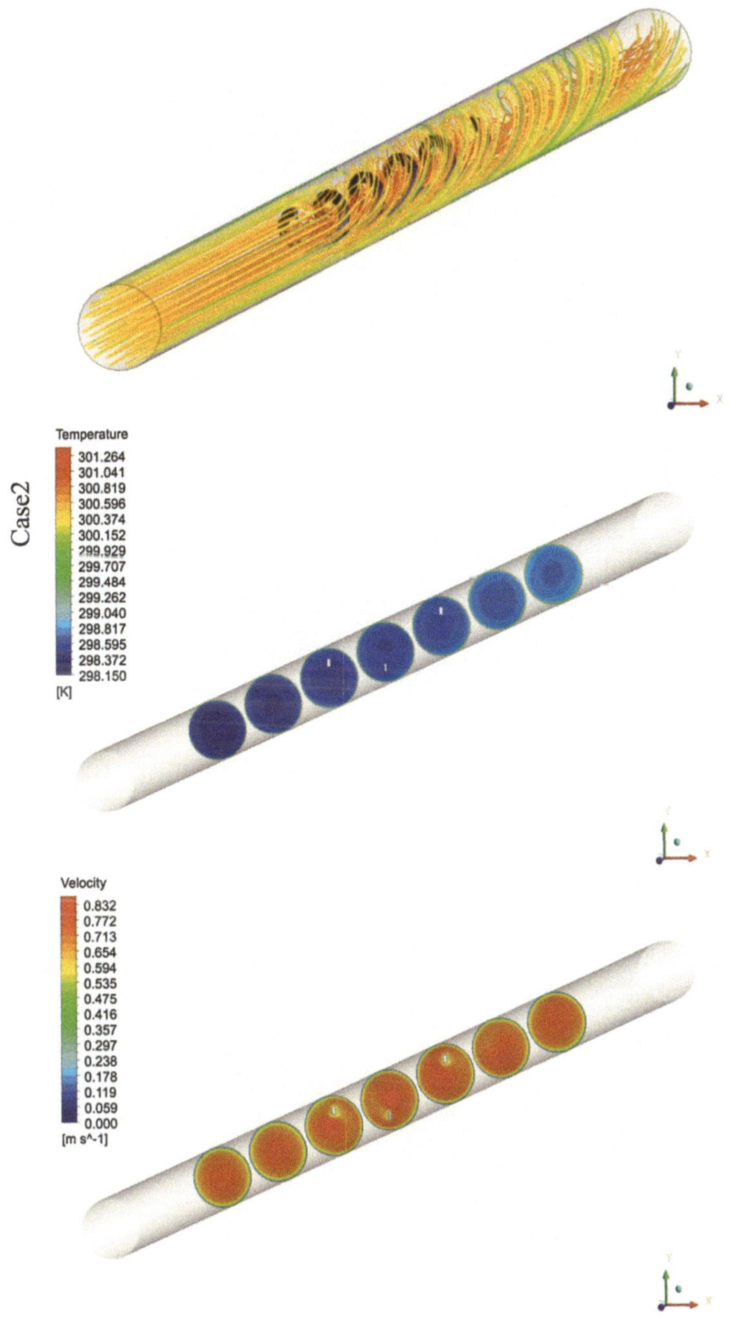

FIGURE 3.19 cont'd.

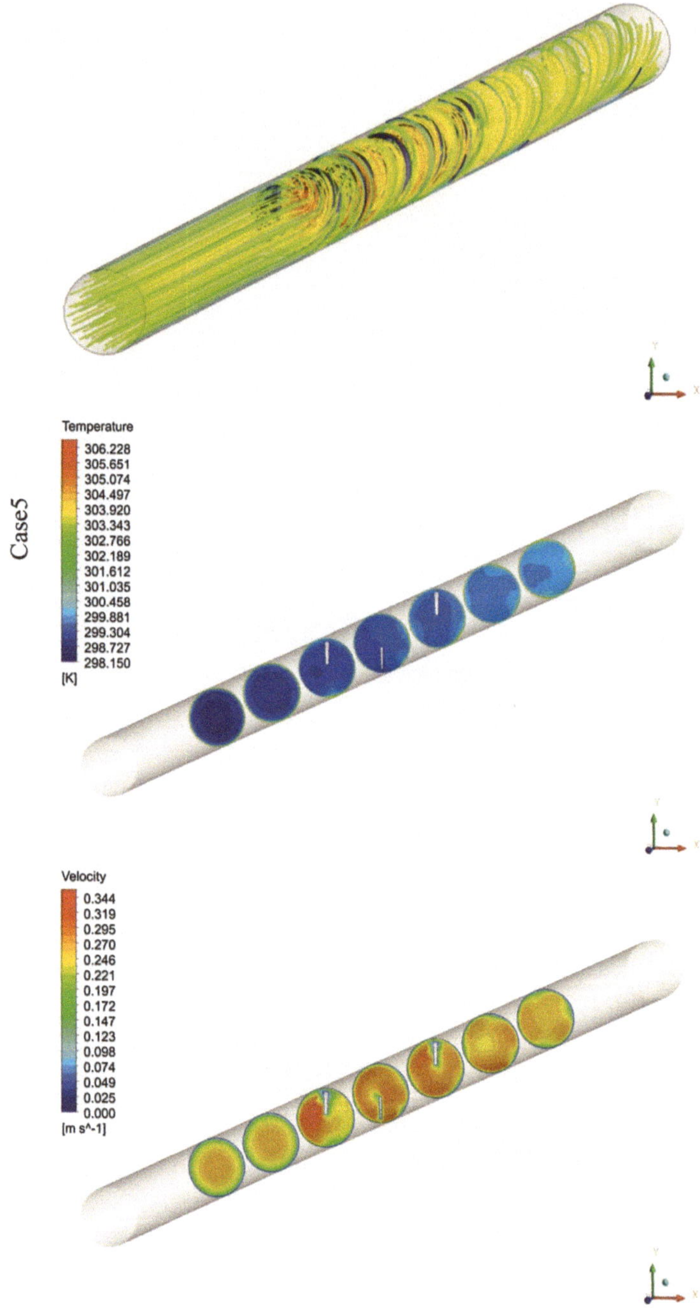

FIGURE 3.20

Stream tracer, temperature, velocity contours for cases 5 and 6 when $\phi = 0.05$.

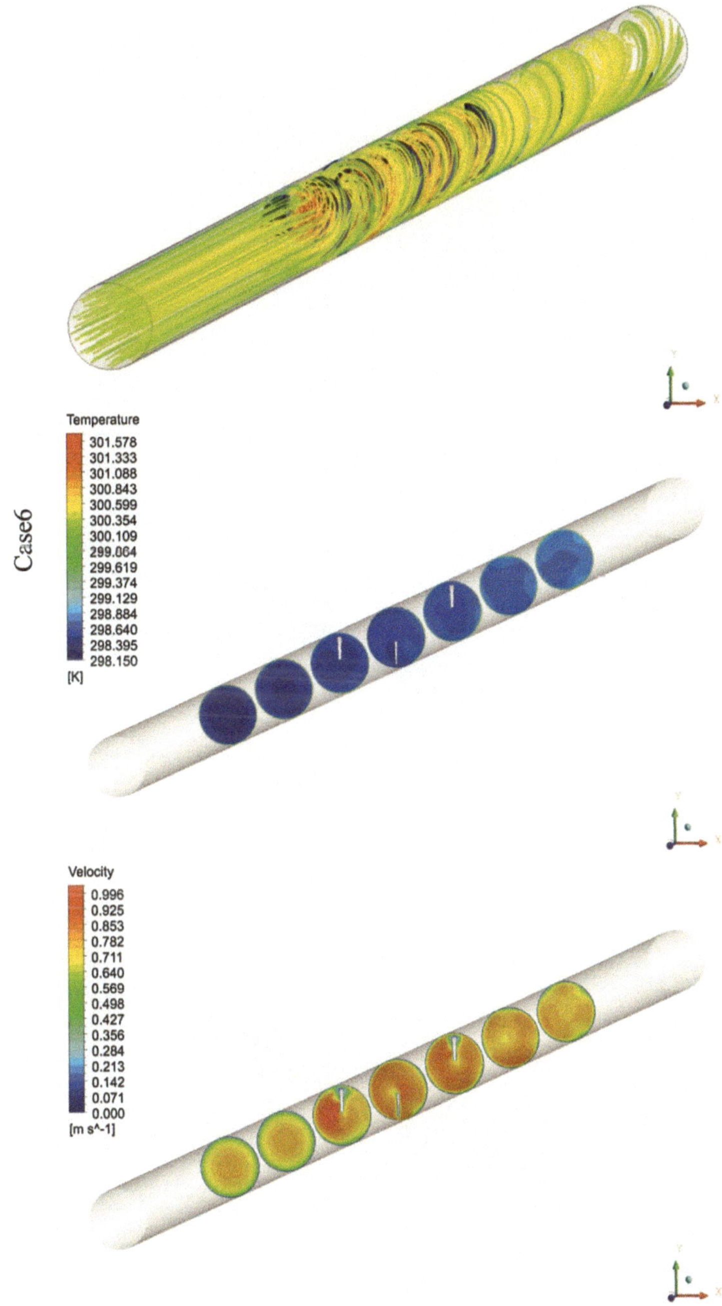

FIGURE 3.20 cont'd.

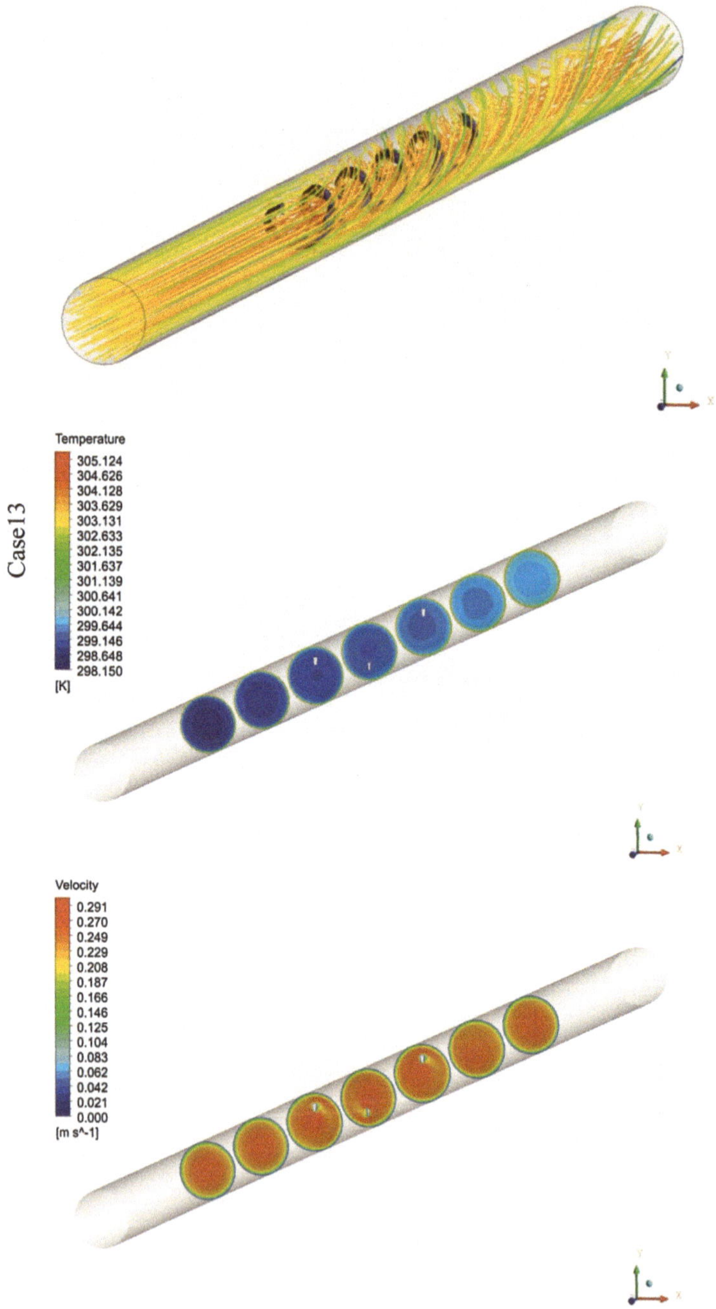

FIGURE 3.21

Stream tracer, temperature, velocity contours for cases 13 and 14 when $\phi = 0.05$.

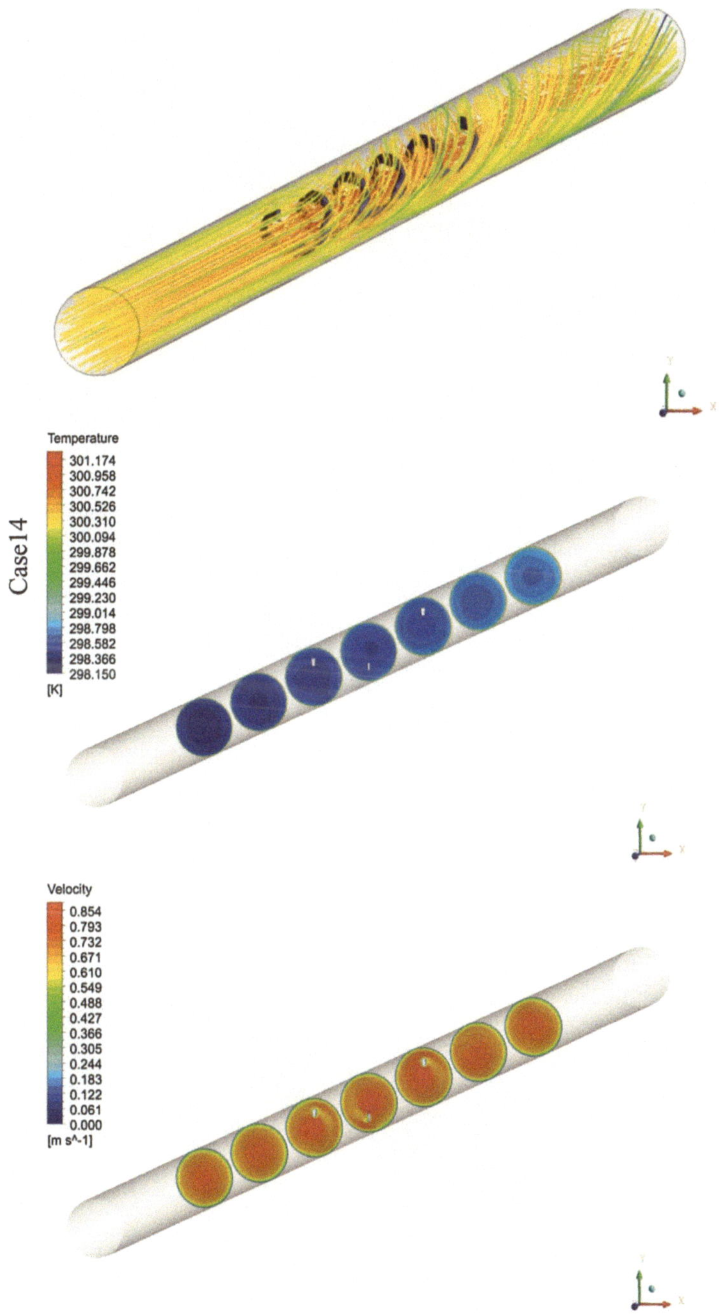

FIGURE 3.21 cont'd.

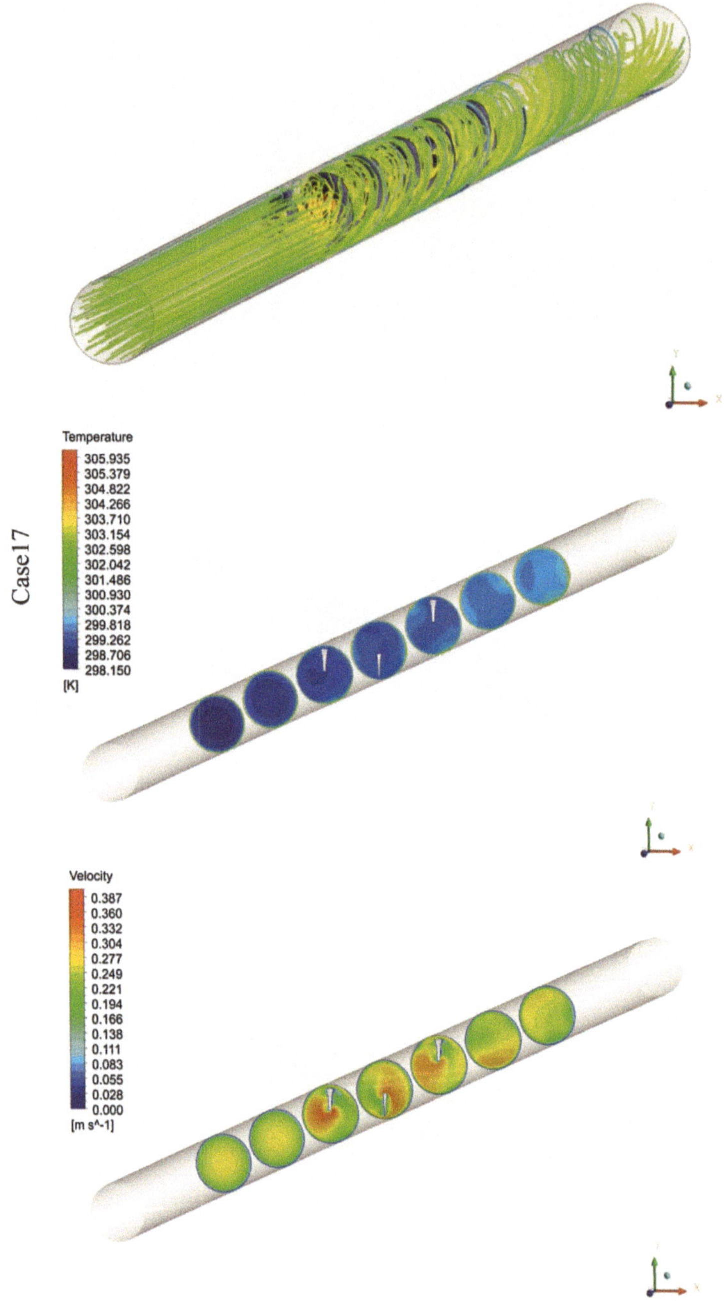

FIGURE 3.22

Stream tracer, temperature, velocity contours for cases 17 and 18 when $\phi = 0.05$.

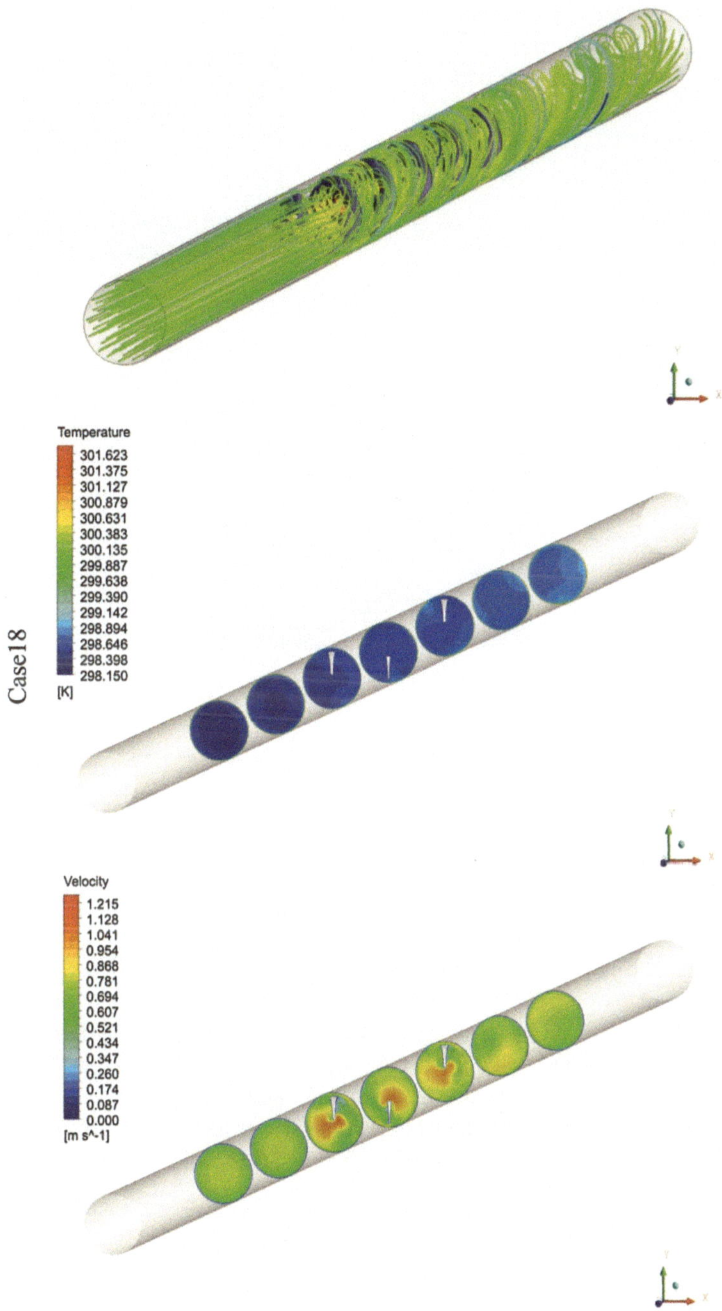

FIGURE 3.22 cont'd.

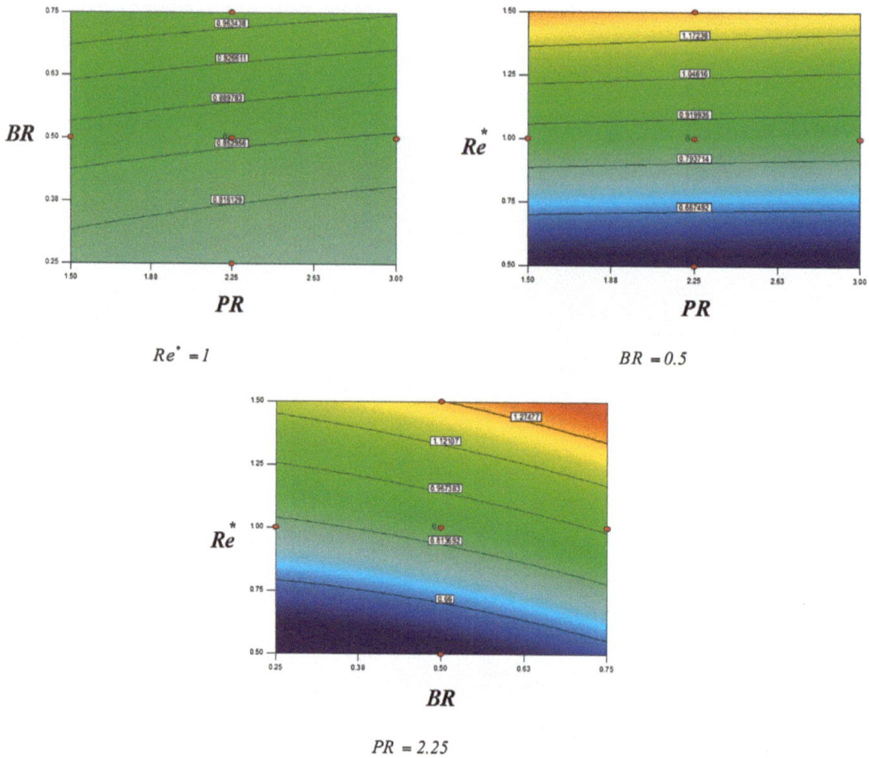

FIGURE 3.23

Effects of Reynolds number, pitch ratio, and width ratio on Nusselt number when $\phi = 0.05$.

As mentioned, the problem is simulated in ANSYS FLUENT software via finite volume method. For any numerical simulation, one needs to create an appropriate mesh grid on the geometry to make the problem understandable/solvable for the software. Fig. 3.26 presents the mesh grid created on the geometry of the duct and the helical-twisting flow disturber. The magnified view of the meshing on a part of the flow disturber is also illustrated in the figure. Several researchers scrutinized various techniques for heat transfer augmentation [6–64].

In the investigations, the studied range of variation for Re* and b (the width of the flow disturber blade) are, respectively, 5000–15,000 and 5–15 mm. R* is 0.001 of Reynolds number.

3.4.2 Effects of active parameters

Nanofluid forced convection in presence of a special type of helical-twisting flow disturber inside a pipe was simulated via finite volume method. Single-phase model was utilized to predict the behavior of water-based nanofluid. The thermophysical

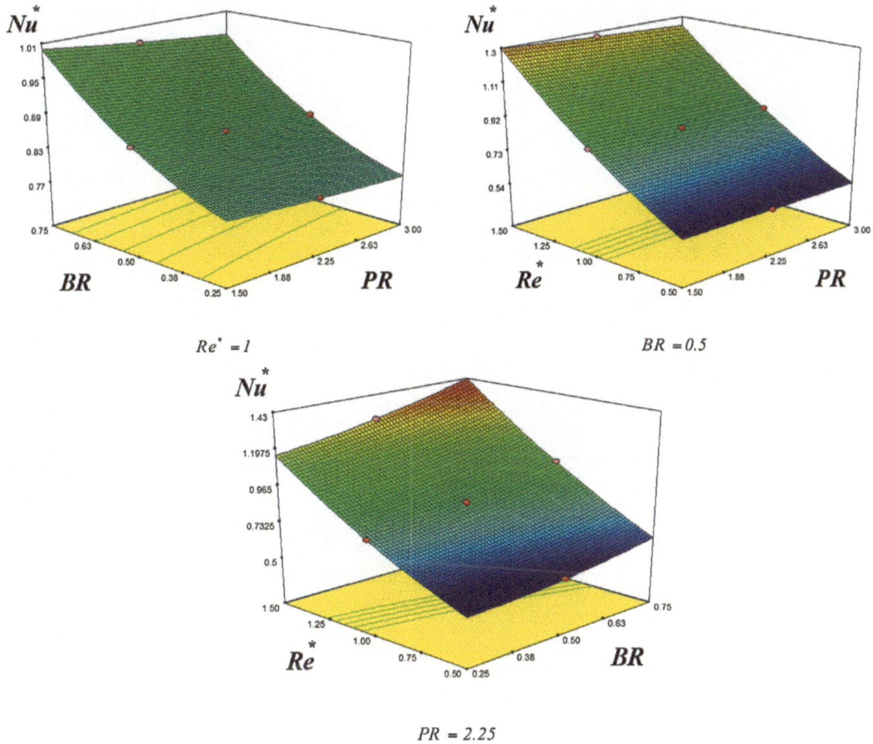

FIGURE 3.23 cont'd.

properties of the nanofluid are estimated considering the Brownian motion effect [40], and the inlet temperature of nanofluid into the duct in all cases is 25 °C. In this section, the results of the simulations are presented. Before presenting the numerical results, one should make sure about the accuracy of the developed numerical solution.

Hereafter, the results corresponding to the contours of velocity, pressure, and temperature through the duct in various conditions of the variable parameters of the problem, i.e., specific cases of Reynolds number and disturber blade width, are presented. The first case of analysis is when Reynolds number is set on 5000 and the blade width changes for the three different sizes of b = 5 mm, 15, and 10 mm. Fig. 3.27 shows the contours of velocity of nanofluid flow through the duct for these three states. As seen, the larger the blade width, the higher velocity values of the nanofluid and a more disturbed pattern of streamlines after the disturber. This will most likely mean a better heat transfer rate. The next figures will directly respond this matter.

Fig. 3.28 exhibits the pressure contours of the fluid through the dust under the same conditions as the previous figure. According to the figure, the larger the value of b, the bigger the pressure drop along the duct. For example, where the internal

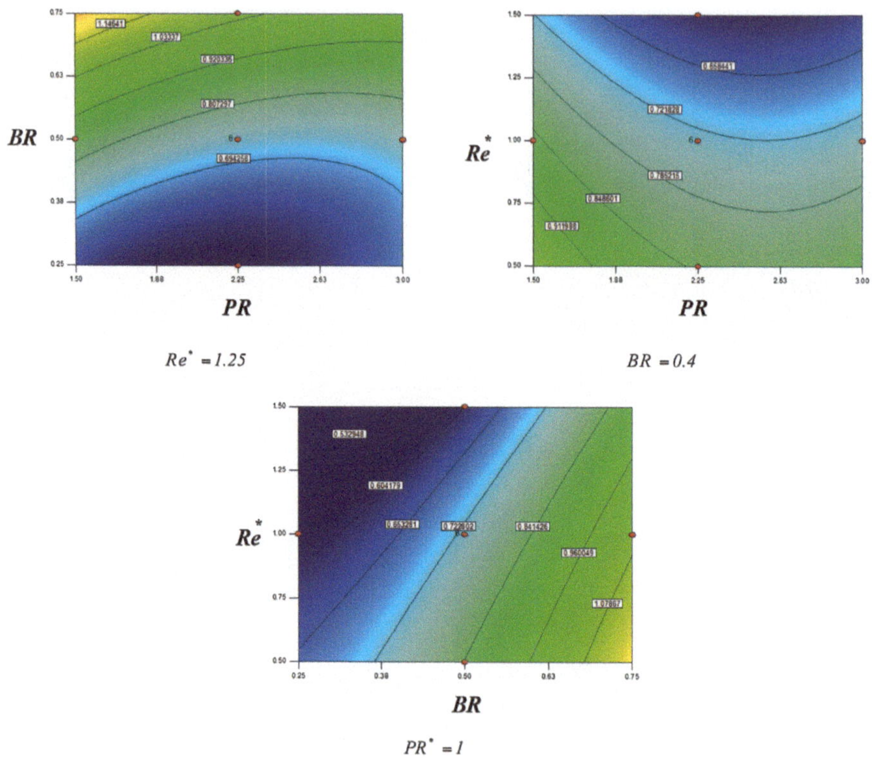

FIGURE 3.24

Effects of Reynolds number, pitch ratio, and width ratio on Darcy factor when $\phi = 0.05$.

pressure of the duct is 11.53 Pa, 17.66 Pa, and 31.55 Pa for the cases of blade width of 5, 10, and 15 mm, respectively, at the end of the duct section, the local gauge pressure is −0.04 Pa for the first case (i.e., b = 5 mm), while this parameter for the same location for the other two cases of b = 10 mm and b = 15 mm are −0.13 Pa and −2.87 Pa, respectively.

By the same token, Fig. 3.29 presents the contours of temperature distribution across the circular duct for the three cases of b = 5, b = 10, and b = 15 mm when the Reynolds number is equal and constant at 5000 for all the three cases. Considering the figures, with an equal and uniform temperature of 298.15 K for all the three cases, a higher temperature gradient and nonuniform temperature profile is observed as the value of b growths.

The next stage is to increase the Reynolds number and then evaluate the effect of the increasing of the blade width in the upper Reynolds level. The Reynolds number of the new condition is the high value of 15,000, and the disturber blade width is again at the three levels of 5, 10 and 15 mm.

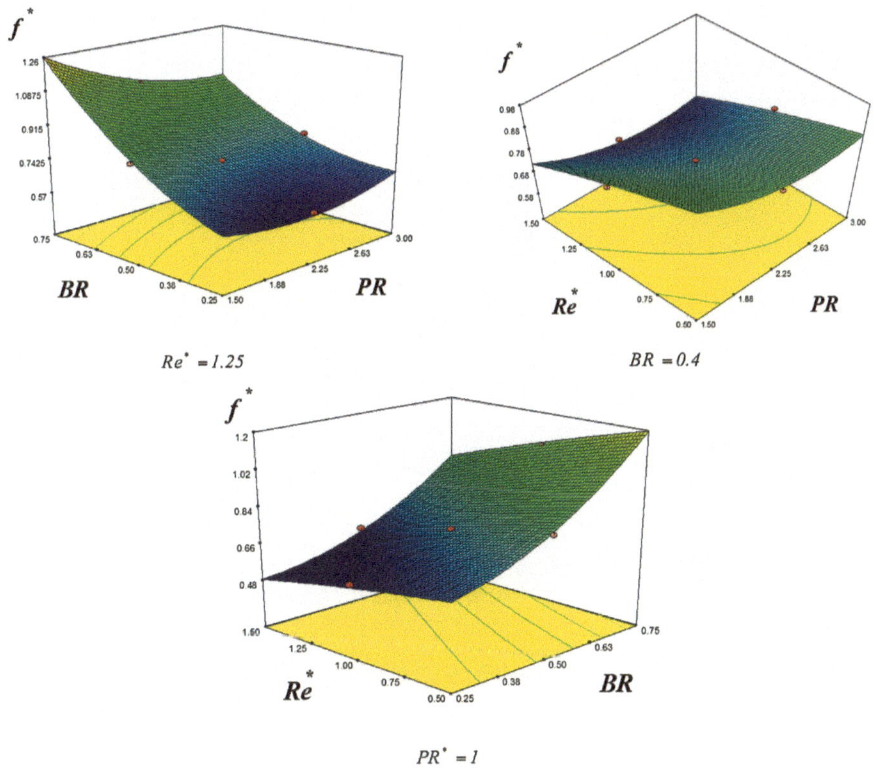

FIGURE 3.24 cont'd.

Table 3.11 Values of Nusselt number and friction factor for various cases.

Case no.	*Nu*	*f*
1	50.45019	0.071159
2	116.3	0.053102
3	54.87257	0.090868
4	126.192	0.068867
5	60.88065	0.112965
6	137.5921	0.08578
7	50.54129	0.068935
8	117.1132	0.05134
9	54.89024	0.08679
10	128.2894	0.06587
11	62.46821	0.117455
12	143.1423	0.089498
13	50.61686	0.068802
14	117.0781	0.050652
15	54.47769	0.093042
16	126.8856	0.069306
17	64.49722	0.146312
18	149.6457	0.109105

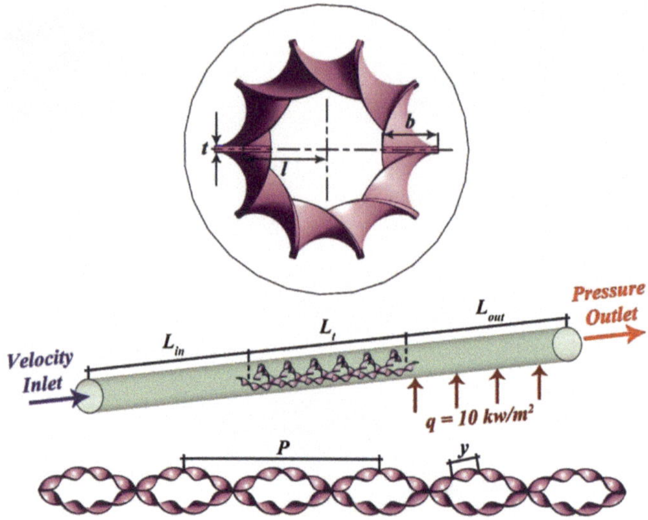

FIGURE 3.25

The geometry of the problem, including the duct and the flow disturber.

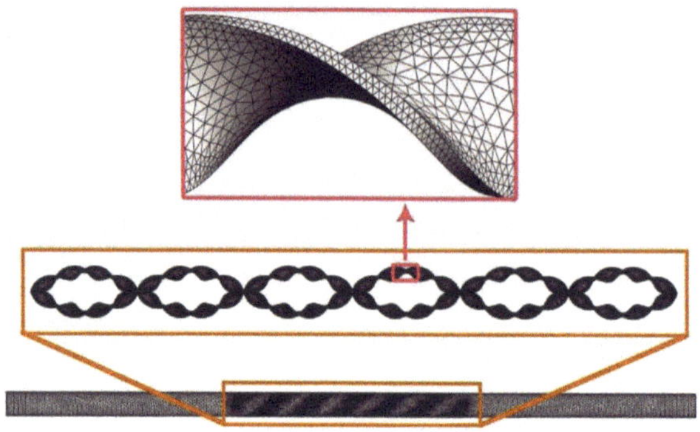

FIGURE 3.26

The structure of the mesh grid created on the geometry of the problem for the numerical solution.

Fig. 3.30 shows the contours of the velocity of the streamlines of the nanofluid for the new Reynolds number and the three considered flow disturber blade widths. Naturally, in this case, the disturbance of the flow is already sharper and the flow disturber makes it more turbulent. Again, the bigger the flow disturber, the more disturb the flow after the blade will be.

Fig. 3.31 illustrates the contours of pressure drop along the duct at Re = 15,000 and the three different blade widths cases. For the case b = 5 00, the inlet pressure

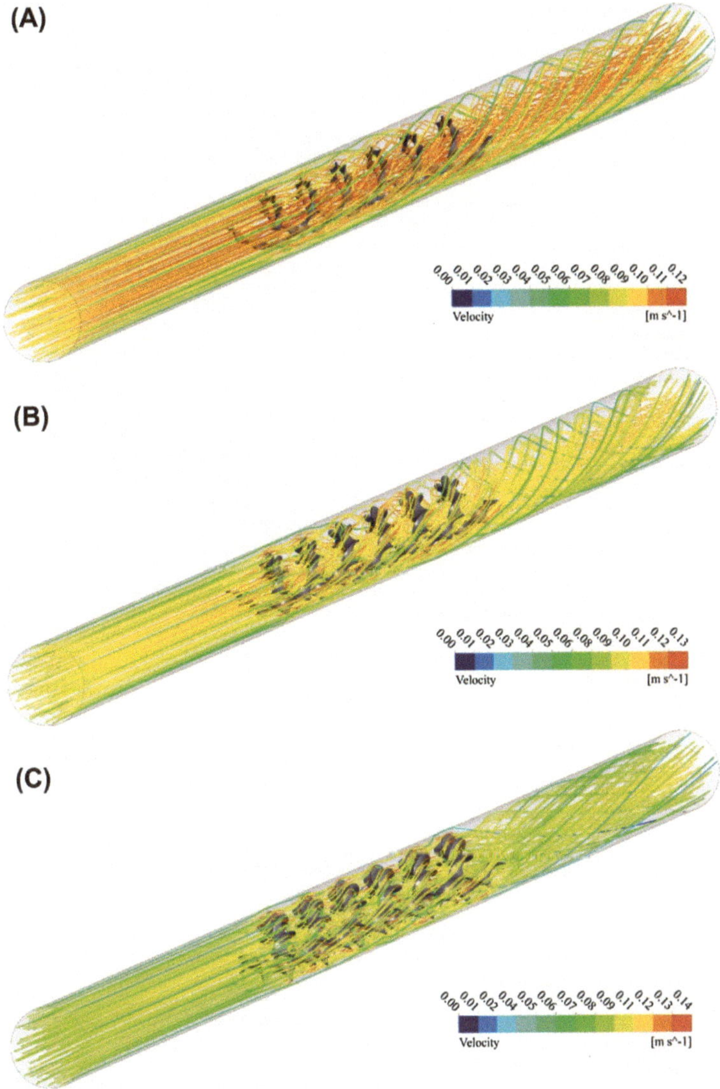

FIGURE 3.27

Velocity contours of the flow through the duct for the cases Re = 5000 and (A) $b = 5$ mm, (B) $b = 10$ mm, and (C) $b = 15$ mm.

of the duct is 75.18 Pa and the outlet pressure is −0.66 Pa. For the other two cases of b = 10 mm and b = 15 mm, the inlet/outlet pressures are, respectively, about 115/-9.9 Pa and 203.6/-20 Pa. Comparing these values with Fig. 3.28, one finds out that the effect of the flow disturber in the same blade width is much stronger at a higher Reynolds number.

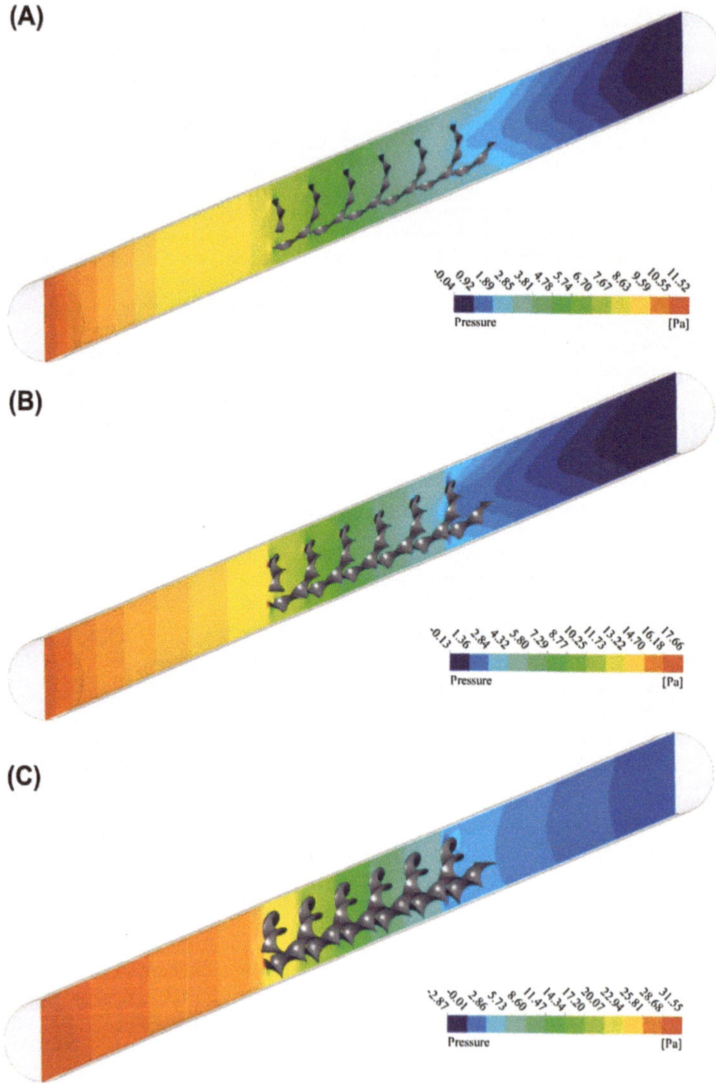

FIGURE 3.28

Pressure contours of the flow through the circular duct for Re = 5000 and (A) $b = 5$ mm, (B) $b = 10$ mm, and (C) $b = 15$ mm.

Fig. 3.32 presents the contours of temperature distributions along the duct. Here also, with the same trend as Fig. 3.29 observed concerning the effect of the blade width (a larger blade, a higher temperature gradient), one realizes the sharper temperature gradient along the duct length at lower Reynolds numbers.

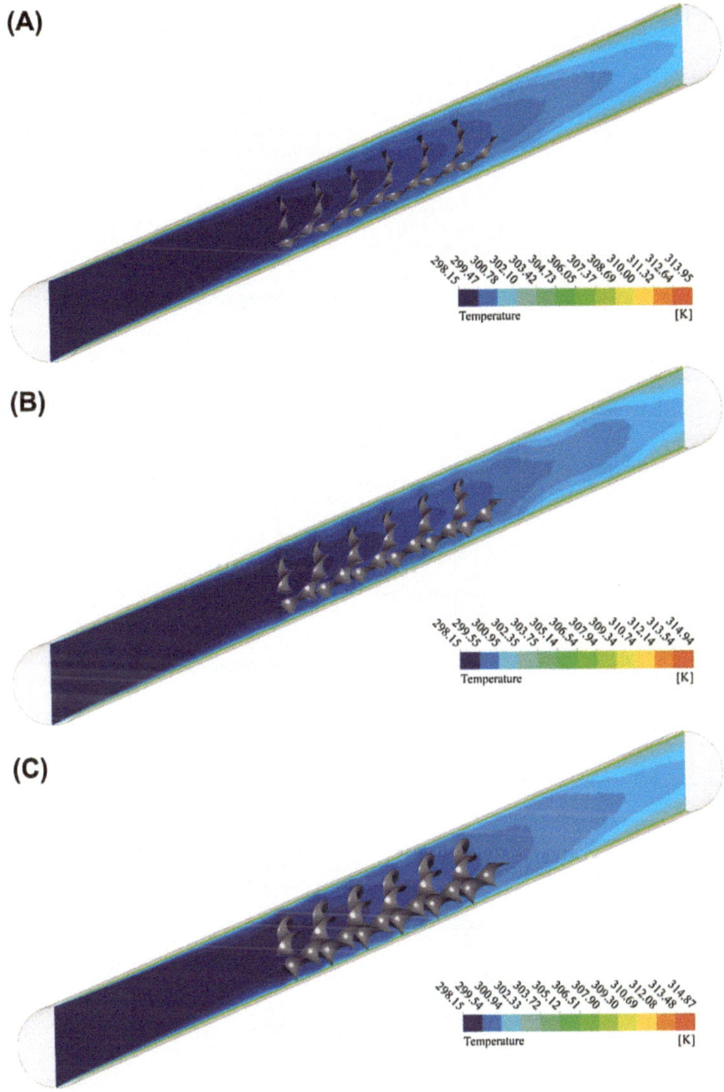

FIGURE 3.29

Contours of temperature distributions of the flow through the circular duct for Re = 5000 and (A) $b = 5$ mm, (B) $b = 10$ mm, and (C) $b = 15$ mm.

Overall, according to the figure sets presented so far, one could conclude that as the flow distracts the nanofluid near the outer surface, the velocity gradient gets larger over the surface because of the high friction. The distortion of vortex augments with the augmentation of the blade width and Reynolds number, accordingly it picks up the velocity gradient resulting in a more friction effect.

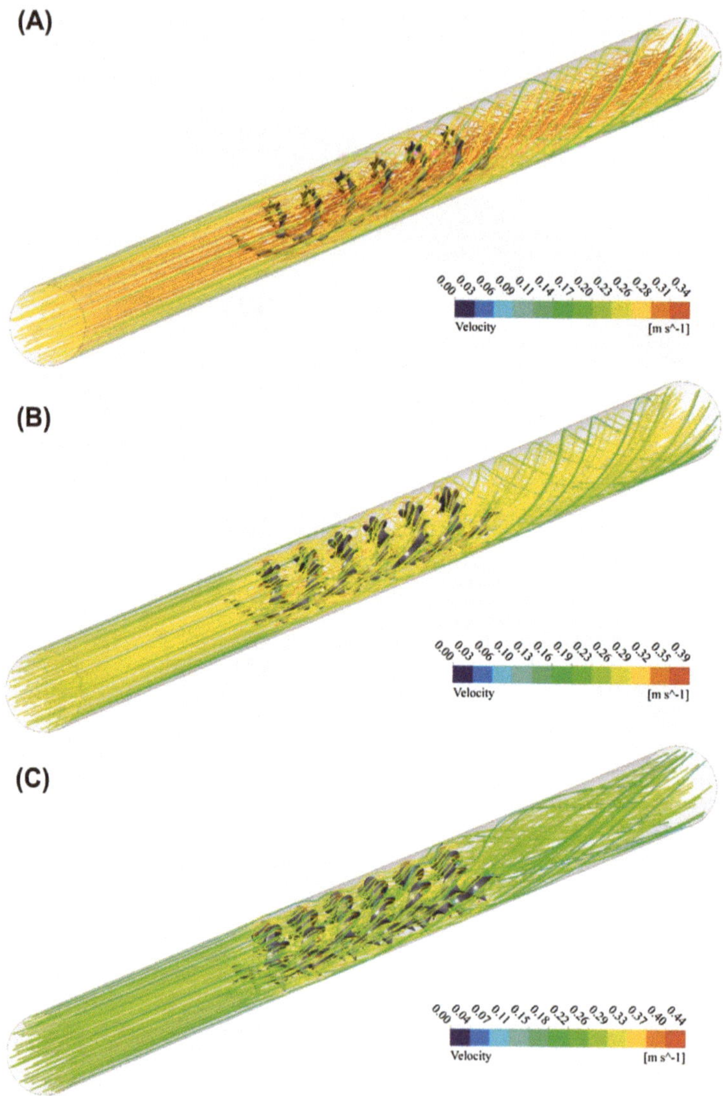

FIGURE 3.30

Contours of velocity of the flow through the circular duct for Re = 15,000 and (A) $b = 5$ mm, (B) $b = 10$ mm, and (C) $b = 15$ mm.

Also, one should note that a thinner boundary layer is obtained as a result of enlarging the values of b and Re due to induction of swirl flows.

In Fig. 3.33, the impacts of changing the values of b and Re on the target parameters of this study, i.e., the *f* and *Nu*, are illustrated. As seen here, the friction factor is reduced with the growth of the Re. Increasing the width of the blade boosts

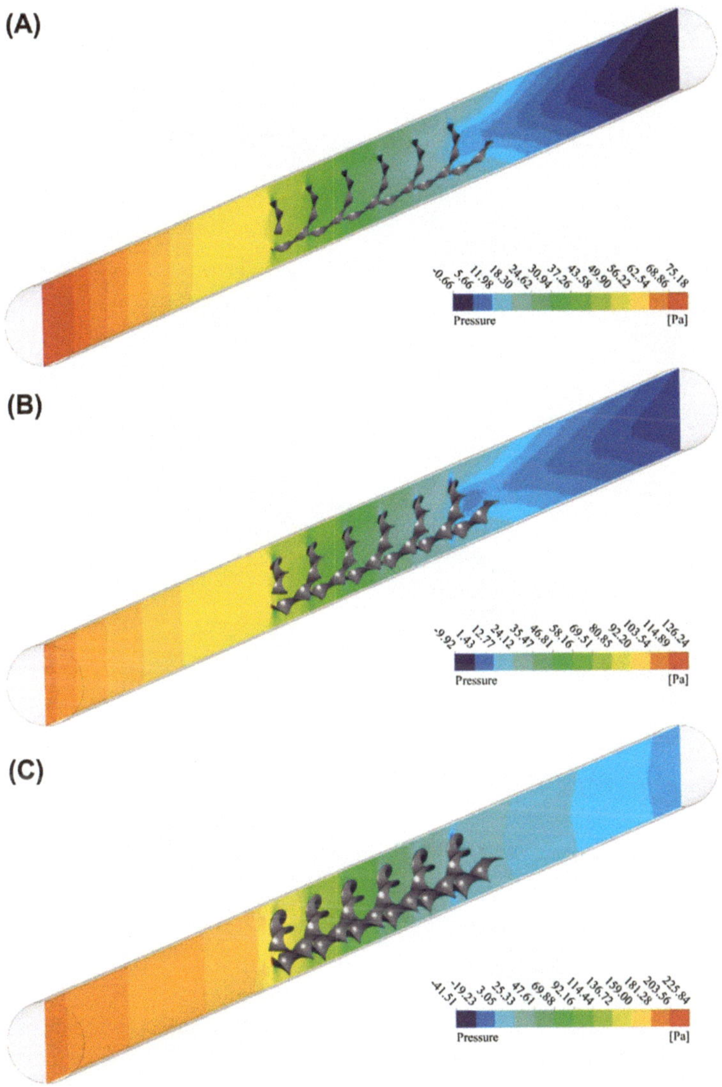

FIGURE 3.31

Pressure contours of the flow through the circular duct for Re = 15,000 and (A) $b = 5$ mm, (B) $b = 10$ mm, and (C) $b = 15$ mm.

the interaction of the inertial forces with pressure forces in the boundary layer. Thus, a greater width of the disturber causes a higher friction factor. On the other hand, a mild change in the Nu has been seen as the blade width varies, whereas the alteration of the Reynolds number largely affects this parameter. The variation of the Nu is in the same way (not reverse) as the change of these two parameters.

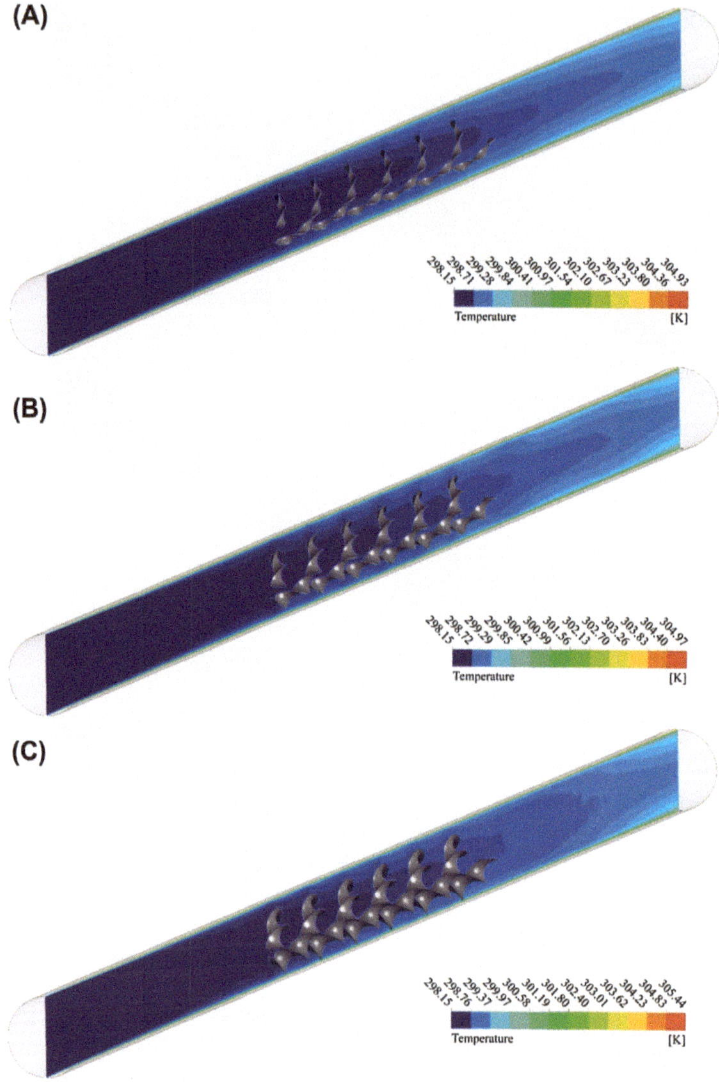

FIGURE 3.32

Contours of temperature distributions of the flow through the circular duct for Re = 15,000 and (A) $b = 5$ mm, (B) $b = 10$ mm, and (C) $b = 15$ mm.

Finally, relying on the results presented and doing curve fitting on the obtained graphs, one could derive the following two equations for the f and Nu as functions of the Reynolds number and the flow disturber blade width as follow;

$$f = 0.22 - 0.039\mathrm{Re}^* + 0.12b - 0.019\, b\, \mathrm{Re}^* + 0.048\, b^2 \qquad (3.24)$$

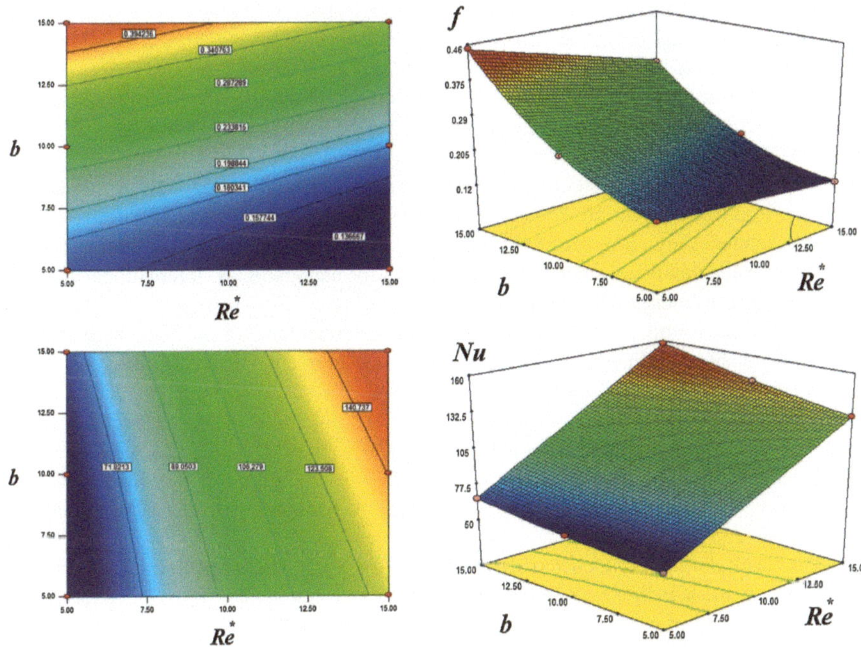

FIGURE 3.33

Variation of friction factor and Nusselt number with Re and *b*

$$Nu = 99 + 41\mathrm{Re}^{*} + 10.6b + 4.38b\,\mathrm{Re}^{*} + 1.98\,b^{2} \tag{3.25}$$

Again, $\mathrm{Re}^{*} = 0.001\mathrm{Re}$. These correlations are valid for $5000 < \mathrm{Re} < 15{,}000$, $5\ \mathrm{mm} < b < 15\ \mathrm{mm}$.

References

[1] M. Jafaryar, M. Sheikholeslami, Z. Li, CuO-water nanofluid flow and heat transfer in a heat exchanger tube with twisted tape turbulator, Powder Technology 336 (2018) 131–143.

[2] D. Kim, Y. Kwon, Y. Cho, C. Li, S. Cheong, Y. Hwang, J. Lee, D. Hong, S. Moona, Convective heat transfer characteristics of nanofluids under laminar and turbulent flow conditions, Current Applied Physics 9 (2009) e119–e123.

[3] M. Jafaryar, M. Sheikholeslami, Z. Li, R. Moradi, Nanofluid turbulent flow in a pipe under the effect of twisted tape with alternate axis, Journal of Thermal Analysis and Calorimetry 135 (1) (2019) 305–323. https://doi.org/10.1007/s10973-018-7093-2.

[4] M. Sheikholeslami, M. Jafaryar, Z. Li, Nanofluid turbulent convective flow in a circular duct with helical turbulators considering CuO nanoparticles, International Journal of Heat and Mass Transfer 124 (2018) 980–989.

[5] M. Sheikholeslami, A. Arabkoohsar, M. Jafaryar, Impact of a helical-twisting device on nanofluid thermal hydraulic performance of a tube, Journal of Thermal Analysis and Calorimetry (2019). https://doi.org/10.1007/s10973-019-08683-x.

[6] M. Sheikholeslami, Effect of uniform suction on nanofluid flow and heat transfer over a cylinder, Journal of the Brazilian Society of Mechanical Sciences and Engineering 37 (2015) 1623–1633.

[7] M. Sheikholeslami, Lattice Boltzmann method simulation of MHD non-Darcy nanofluid free convection, Physica B 516 (2017) 55–71.

[8] M. Sheikholeslami, Magnetic field influence on CuO -H_2O nanofluid convective flow in a permeable cavity considering various shapes for nanoparticles, International Journal of Hydrogen Energy 42 (2017) 19611–19621.

[9] M. Sheikholeslami, Numerical simulation for solidification in a LHTESS by means of nano-enhanced PCM, Journal of the Taiwan Institute of Chemical Engineers 86 (2018) 25–41.

[10] M. Sheikholeslami, Numerical modeling of Nano enhanced PCM solidification in an enclosure with metallic fin, Journal of Molecular Liquids 259 (2018) 424–438.

[11] M. Sheikholeslami, Solidification of NEPCM under the effect of magnetic field in a porous thermal energy storage enclosure using CuO nanoparticles, Journal of Molecular Liquids 263 (2018) 303–315.

[12] M. Sheikholeslami, Influence of magnetic field on Al_2O_3-H_2O nanofluid forced convection heat transfer in a porous lid driven cavity with hot sphere obstacle by means of LBM, Journal of Molecular Liquids 263 (2018) 472–488.

[13] M. Sheikholeslami, New computational approach for exergy and entropy analysis of nanofluid under the impact of Lorentz force through a porous media, Computer Methods in Applied Mechanics and Engineering 344 (2019) 319–333.

[14] M. Sheikholeslami, Numerical approach for MHD Al_2O_3-water nanofluid transportation inside a permeable medium using innovative computer method, Computer Methods in Applied Mechanics and Engineering 344 (2019) 306–318.

[15] X. Ma, M. Sheikholeslami, M. Jafaryar, S. Ahmad, T. Nguyen-Thoi, Z. Li, Solidification inside a clean energy storage unit utilizing phase change material with copper oxide nanoparticles, Journal of Cleaner Production (2019). https://doi.org/10.1016/j.jclepro.2019.118888.

[16] M. Sheikholeslami, Omid Mahian, enhancement of PCM solidification using inorganic nanoparticles and an external magnetic field with application in energy storage systems, Journal of Cleaner Production 215 (2019) 963–977.

[17] M. Sheikholeslami, B. Rezaeianjouybari, M. Darzi, S. Ahmad, Z. Li, T.K. Nguyen, Application of nano-refrigerant for boiling heat transfer enhancement employing an experimental study, International Journal of Heat and Mass Transfer 141 (2019) 974–980.

[18] M. Sheikholeslami, M. Jafaryar, M. Hedayat, S. Ahmad, Z. Li, T.K. Nguyen, M. Bakouri, Heat transfer and turbulent simulation of nanomaterial due to compound turbulator including irreversibility analysis, International Journal of Heat and Mass Transfer 137 (2019) 1290–1300.

[19] M. Sheikholeslami, Application of Darcy law for nanofluid flow in a porous cavity under the impact of Lorentz forces, Journal of Molecular Liquids 266 (2018) 495–503.

[20] M. Sheikholeslami, Finite element method for PCM solidification in existence of CuO nanoparticles, Journal of Molecular Liquids 265 (2018) 347–355.

[21] Y. Qin, J. Luo, Z. Chen, G. Mei, L.-E. Yan, Measuring the albedo of limited-extent targets without the aid of known-albedo masks, Solar Energy 171 (2018) 971–976.

[22] M. Sheikholeslami, M. Jafaryar, S. Ahmad, Z. Li, Rizwan-ul Haq, Heat transfer of nanoparticles employing innovative turbulator considering entropy generation, International Journal of Heat and Mass Transfer 136 (2019) 1233–1240.

[23] Y. Qin, A review on the development of cool pavements to mitigate urban heat island effect, Renewable and Sustainable Energy Reviews 52 (2015) 445–459.

[24] M. Sheikholeslami, H. Rizwan-ul, S. Ahmad, Z. Li, Y.G. Elaraki, I. Tlili, Heat transfer simulation of heat storage unit with nanoparticles and fins through a heat exchanger, International Journal of Heat and Mass Transfer 135 (2019) 470–478.

[25] M. Sheikholeslami, H. Rizwan-ul, S. Ahmad, Z. Li, Heat transfer behavior of Nanoparticle enhanced PCM solidification through an enclosure with V shaped fins, International Journal of Heat and Mass Transfer 130 (2019) 1322–1342.

[26] M. Sheikholeslami, S.A. Shehzad, Z. Li, S. Ahmad, Numerical modeling for Alumina nanofluid magnetohydrodynamic convective heat transfer in a permeable medium using Darcy law, International Journal of Heat and Mass Transfer 127 (2018) 614–622.

[27] Y. Qin, Y. He, J.E. Hiller, G. Mei, A new water-retaining paver block for reducing runoff and cooling pavement, Journal of Cleaner Production 199 (2018) 948–956.

[28] M. Sheikholeslami, Z. Li, S. Ahmad, Lorentz forces effect on NEPCM heat transfer during solidification in a porous energy storage system, International Journal of Heat and Mass Transfer 127 (2018) 665–674.

[29] M. Sheikholeslami, M. Jafaryar, S. Saleem, Z. Li, S. Ahmad, Y. Jiang, Nanofluid heat transfer augmentation and exergy loss inside a pipe equipped with innovative turbulators, International Journal of Heat and Mass Transfer 126 (2018) 156–163.

[30] M. Sheikholeslami, A. Ghasemi, Z. Li, S. Ahmad, S. Saleem, Influence of CuO nanoparticles on heat transfer behavior of PCM in solidification process considering radiative source term, International Journal of Heat and Mass Transfer 126 (2018) 1252–1264.

[31] Y. Qin, Y. Zhao, X. Chen, L. Wang, F. Li, T. Bao, Moist curing increases the solar reflectance of concrete, Construction and Building Materials 215 (2019) 114–118.

[32] M. Sheikholeslami, M. Darzi, M.K. Sadoughi, Heat transfer improvement and pressure drop during condensation of refrigerant-based nanofluid; an experimental procedure, International Journal of Heat and Mass Transfer 122 (2018) 643–650.

[33] M. Sheikholeslami, S.A. Shehzad, Simulation of water based nanofluid convective flow inside a porous enclosure via Non-equilibrium model, International Journal of Heat and Mass Transfer 120 (2018) 1200–1212.

[34] M. Sheikholeslami, M. Seyednezhad, Simulation of nanofluid flow and natural convection in a porous media under the influence of electric field using CVFEM, International Journal of Heat and Mass Transfer 120 (2018) 772–781.

[35] M. Sheikholeslami, H.B. Rokni, Numerical simulation for impact of Coulomb force on nanofluid heat transfer in a porous enclosure in presence of thermal radiation, International Journal of Heat and Mass Transfer 118 (2018) 823–831.

[36] M. Sheikholeslami, S.A. Shehzad, Numerical analysis of $Fe_3O_4-H_2O$ nanofluid flow in permeable media under the effect of external magnetic source, International Journal of Heat and Mass Transfer 118 (2018) 182–192.

[37] M. Sheikholeslami, M.K. Sadoughi, Simulation of CuO- water nanofluid heat transfer enhancement in presence of melting surface, International Journal of Heat and Mass Transfer 116 (2018) 909–919.

[38] M. Sheikholeslami, H.B. Rokni, Simulation of nanofluid heat transfer in presence of magnetic field: a review, International Journal of Heat and Mass Transfer 115 (2017) 1203–1233.

[39] M. Sheikholeslami, T. Hayat, A. Alsaedi, On simulation of nanofluid radiation and natural convection in an enclosure with elliptical cylinders, International Journal of Heat and Mass Transfer 115 (2017) 981–991.

[40] M. Sheikholeslami, S.A. Shehzad, CVFEM for influence of external magnetic source on $Fe_3O_4 - H_2O$ nanofluid behavior in a permeable cavity considering shape effect, International Journal of Heat and Mass Transfer 115 (2017) 180–191.

[41] M. Sheikholeslami, M. Seyednezhad, Nanofluid heat transfer in a permeable enclosure in presence of variable magnetic field by means of CVFEM, International Journal of Heat and Mass Transfer 114 (2017) 1169–1180.

[42] M. Sheikholeslami, B. Houman, Rokni, Melting heat transfer influence on nanofluid flow inside a cavity in existence of magnetic field, International Journal of Heat and Mass Transfer 114 (2017) 517–526.

[43] M. Sheikholeslami, M. Sadoughi, Mesoscopic method for MHD nanofluid flow inside a porous cavity considering various shapes of nanoparticles, International Journal of Heat and Mass Transfer 113 (2017) 106–114.

[44] M. Sheikholeslami, M.M. Bhatti, Forced convection of nanofluid in presence of constant magnetic field considering shape effects of nanoparticles, International Journal of Heat and Mass Transfer 111 (2017) 1039–1049.

[45] M. Sheikholeslami, M.M. Bhatti, Active method for nanofluid heat transfer enhancement by means of EHD, International Journal of Heat and Mass Transfer 109 (2017) 115–122.

[46] M. Sheikholeslami, M.A. Sheremet, A. Shafee, I. Tlili, Simulation of nanoliquid thermogravitational convection within a porous chamber imposing magnetic and radiation impacts, Physica A: Statistical Mechanics and its Applications (2020). https://doi.org/10.1016/j.physa.2019.124058.

[47] Y. Qin, Y. He, B. Wu, S. Ma, X. Zhang, Regulating top albedo and bottom emissivity of concrete roof tiles for reducing building heat gains, Energy and Buildings 156 (Suppl. C) (2017) 218–224.

[48] M. Sheikholeslami, S.A. Shehzad, Thermal radiation of ferrofluid in existence of Lorentz forces considering variable viscosity, International Journal of Heat and Mass Transfer 109 (2017) 82–92.

[49] M. Sheikholeslami, H.B. Rokni, Nanofluid two phase model analysis in existence of induced magnetic field, International Journal of Heat and Mass Transfer 107 (2017) 288–299.

[50] M. Sheikholeslami, A.N. Keshteli, H. Babazadeh, Nanoparticles favorable effects on performance of thermal storage units, Journal of Molecular Liquids 300 (2020) 112329.

[51] M. Sheikholeslami, S.A. Shehzad, Magnetohydrodynamic nanofluid convection in a porous enclosure considering heat flux boundary condition, International Journal of Heat and Mass Transfer 106 (2017) 1261–1269.

[52] M. Sheikholeslami, R. Ellahi, Three dimensional mesoscopic simulation of magnetic field effect on natural convection of nanofluid, International Journal of Heat and Mass Transfer 89 (2015) 799–808.

[53] M. Sheikholeslami, Numerical investigation of nanofluid free convection under the influence of electric field in a porous enclosure, Journal of Molecular Liquids 249 (2018) 1212–1221.

[54] M. Sheikholeslami, CuO-water nanofluid flow due to magnetic field inside a porous media considering Brownian motion, Journal of Molecular Liquids 249 (2018) 921–929.

[55] A. Shafee, M. Sheikholeslami, M. Jafaryar, H. Babazadeh, Utilizing copper oxide nanoparticles for expedition of solidification within a storage system, Journal of Molecular Liquids 302 (2020) 112371. https://doi.org/10.1016/j.molliq.2019.112371.

[56] M. Sheikholeslami, Numerical investigation for CuO-H_2O nanofluid flow in a porous channel with magnetic field using mesoscopic method, Journal of Molecular Liquids 249 (2018) 739–746.

[57] Y. Qin, M. Zhang, J.E. Hiller, Theoretical and experimental studies on the daily accumulative heat gain from cool roofs, Energy 129 (2017) 138–147.

[58] Y. Qin, H. He, A new simplified method for measuring the albedo of limited extent targets, Solar Energy 157 (Suppl. C) (2017) 1047–1055.

[59] M. Sheikholeslami, H.B. Rokni, Numerical modeling of nanofluid natural convection in a semi annulus in existence of Lorentz force, Computer Methods in Applied Mechanics and Engineering 317 (2017) 419–430.

[60] M. Sheikholeslami, A. Zeeshan, Analysis of flow and heat transfer in water based nanofluid due to magnetic field in a porous enclosure with constant heat flux using CVFEM, Computer Methods in Applied Mechanics and Engineering 320 (2017) 68–81.

[61] M. Sheikholeslami, M. Darzi, Z. Li, Experimental investigation for entropy generation and exergy loss of nano-refrigerant condensation process, International Journal of Heat and Mass Transfer 125 (2018) 1087–1095.

[62] M. Sheikholeslami, S.A. Shehzad, Z. Li, Water based nanofluid free convection heat transfer in a three dimensional porous cavity with hot sphere obstacle in existence of Lorenz forces, International Journal of Heat and Mass Transfer 125 (2018) 375–386.

[63] M. Sheikholeslami, A. Ghasemi, Solidification heat transfer of nanofluid in existence of thermal radiation by means of FEM, International Journal of Heat and Mass Transfer 123 (2018) 418–431.

[64] M. Sheikholeslami, S.A. Shehzad, CVFEM simulation for nanofluid migration in a porous medium using Darcy model, International Journal of Heat and Mass Transfer 122 (2018) 1264–1271.

CHAPTER

Entropy generation due to inserting various helical swirl flow devices

4

4.1 Second law analysis for nanofluid flow in appearance of twisted tape turbulators

4.1.1 Definition of the problem

Geometry of turbolentor and sample mesh is shown in Figs. 4.1 and 4.2. Lengths, outer and inner diameters of the pipe are 900, 20, and 5 mm, respectively. We separate the length to three regions: inlet section, test section, and outlet section [1]. Twisted tape is situated in second section, and two other sections are considered to prevent any probable back flow streams. Tables 4.1 and 4.2 illustrate properties of CuO and water.

To simulate this chapter, the governing equations for turbulent flow and forced convection heat transfer of nanofluid are

$$\frac{\partial(u_i)}{\partial x_i}=0 \tag{4.1}$$

$$\frac{\partial}{\partial x_j}(u_i u_j \rho_{nf})=-\frac{\partial p}{\partial x_i}+\frac{\partial}{\partial x_j}\left(-\rho_{nf}\overline{u_j' u_i'}\right)+\frac{\partial}{\partial x_j}\left(\mu_{nf}\left(\frac{\partial u_i}{\partial x_j}+\frac{\partial u_j}{\partial x_i}\right)\right) \tag{4.2}$$

$$\begin{gathered}\frac{\partial}{\partial x_i}(\rho_{nf} T u_i)=\frac{\partial}{\partial x_i}\left((\Gamma+\Gamma_t)\frac{\partial T}{\partial x_i}\right),\\ (\Gamma=\mu_{nf}/\mathrm{Pr}_{nf}),\quad (\Gamma_t=\mu_t/\mathrm{Pr}_t)\end{gathered} \tag{4.3}$$

$\rho_{nf}\overline{u_j' u_i'}$ and μ_t are

$$-\overline{u_j' u_i'}\,\rho_{nf}=\mu_t\left(\frac{\partial u_i}{\partial x_j}+\frac{\partial u_j}{\partial x_i}\right)-\frac{2}{3}\rho_{nf} k\delta_{ij}-\frac{2}{3}\mu_t\frac{\partial u_k}{\partial x_k}\delta_{ij} \tag{4.4}$$

$$\mu_t=\rho_{nf} C_\mu k^2/\varepsilon \tag{4.5}$$

k, ε can be predicted as

Nanofluid in Heat Exchangers for Mechanical Systems. https://doi.org/10.1016/B978-0-12-821923-2.00004-X

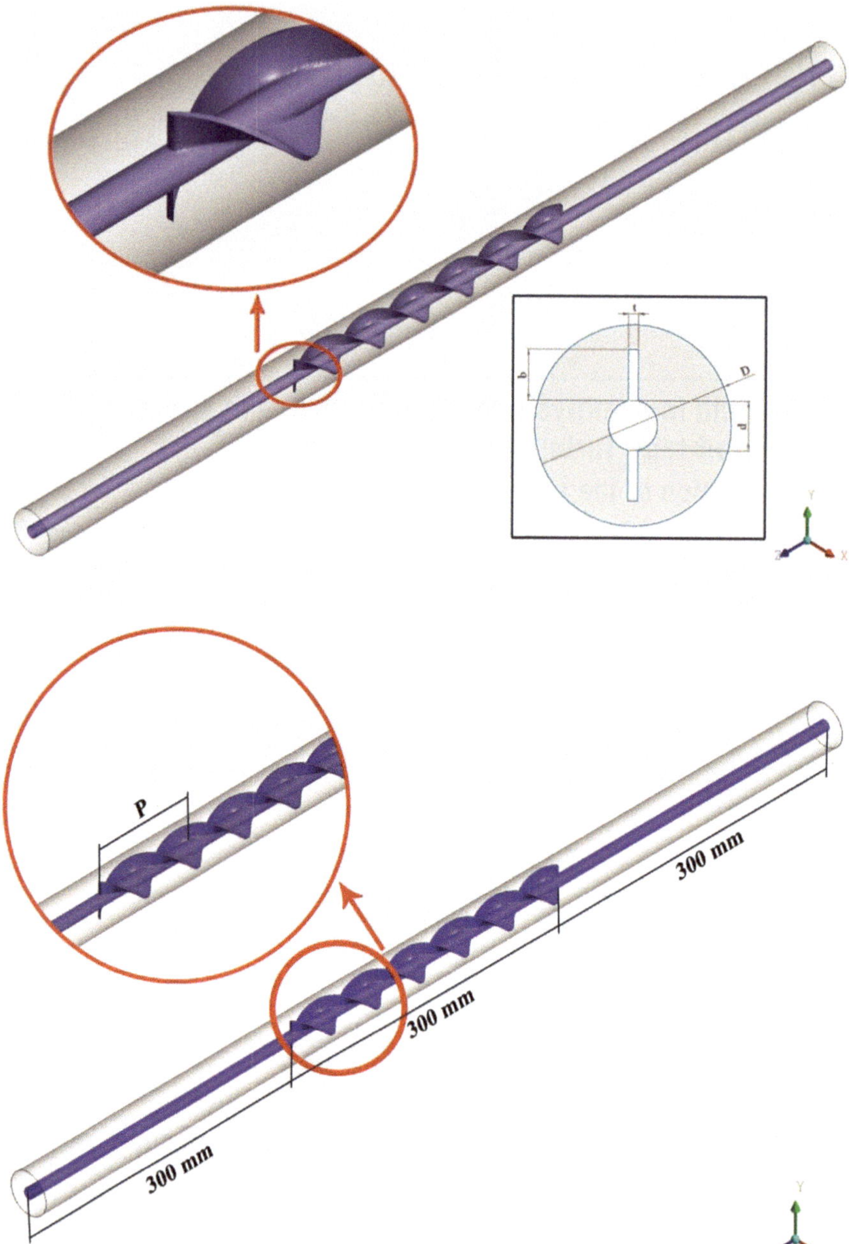

FIGURE 4.1

Geometry of the problem.

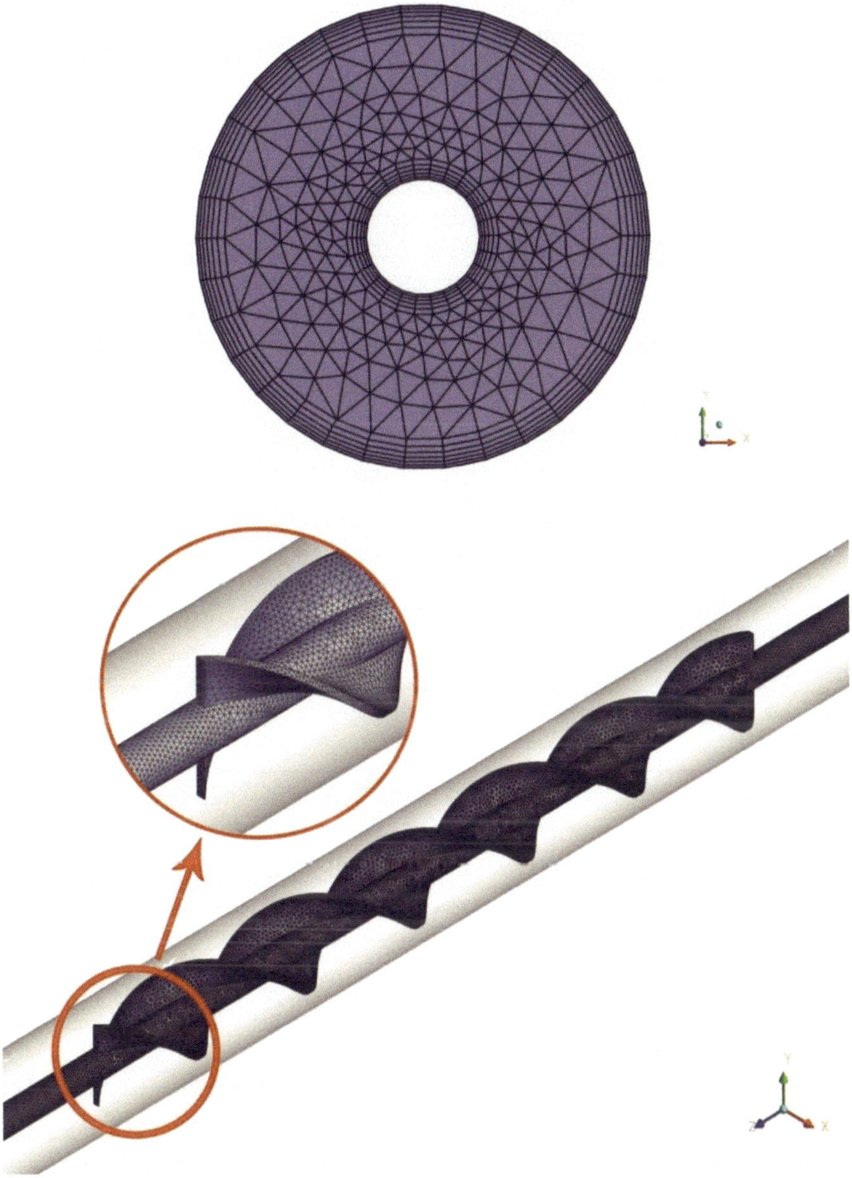

FIGURE 4.2

Detail view of computational mesh.

Table 4.1 The coefficient values of CuO – Water nanofluid.

Coefficient values	CuO-Water
a_1	−26.5933108
a_2	−0.403818333
a_3	−33.3516805
a_4	−1.915825591
a_5	6.421858E-02
a_6	48.40336955
a_7	−9.787756683
a_8	190.245610009
a_9	10.9285386565
a_{10}	−0.72009983664

Table 4.2 Thermophysical properties of water and nanoparticles.

	ρ(kg/m^3)	C_p(j/kgk)	k(W/m.k)	μ(Pa.s)
Pure water	997.1	4179	0.613	0.0010003
CuO	6500	540	18	–

$$\frac{\partial}{\partial x_j}\left(\left(\frac{\mu_t}{\sigma_k}+\mu_{nf}\right)\frac{\partial k}{\partial x_j}\right)-\rho_{nf}\varepsilon+G_k=\frac{\partial}{\partial x_i}(u_i\rho_{nf}k),\quad G_k=-\frac{\partial u_j}{\partial x_i}\rho_{nf}\,\overline{u_j'u_i'} \tag{4.6}$$

$$\frac{\partial}{\partial x_i}(u_i\rho_{nf}\varepsilon)=\frac{\partial}{\partial x_j}\left(\left(\frac{\mu_t}{\sigma_\varepsilon}+\mu_{nf}\right)\frac{\partial \varepsilon}{\partial x_j}\right)+\frac{\varepsilon}{k}G_kC_{1\varepsilon}-\rho_{nf}\frac{\varepsilon^2}{k}C_{2\varepsilon} \tag{4.7}$$

$$C_{1\varepsilon}=1.42,\quad C_\mu=0.0845,\quad C_{2\varepsilon}=1.68,\mathrm{Pr}_t=0.85,\sigma_k=1,\sigma_\varepsilon=1.3 \tag{4.8}$$

ANSYS Fluent software is selected as simulation tool. k-ε (RNG) model was chosen as turbulent model.

ρ_{nf}, $(\rho C_p)_{nf}$, k_{nf} and μ_{nf} are

$$\rho_{nf}=\rho_f(1-\phi)+\rho_s\phi \tag{4.9}$$

$$(\rho C_p)_{nf}=-(\phi-1)(\rho C_p)_f+(\rho C_p)_s\phi \tag{4.10}$$

$$\frac{k_{nf}}{k_f}=1+3\frac{\phi(k_p/k_f-1)}{-\phi(k_p/k_f-1)+(2+k_p/k_f)}+5\times10^4 g'(\phi,T,d_p)\phi\rho_f c_{p,f}\sqrt{\frac{\kappa_b T}{\rho_p d_p}}$$

$$g'(\phi,T,d_p)=\left(a_1+a_2Ln(d_p)+a_5Ln(d_p)^2+a_3Ln(\phi)+a_4Ln(\phi)\ln(d_p)\right)Ln(T)$$
$$+\left(a_6+a_7Ln(d_p)+a_{10}Ln(d_p)^2+a_8Ln(\phi)+a_9\ln(d_p)Ln(\phi)\right),$$
$$\frac{4\times10^{-8}}{d_p}=-1/k_p+1/k_{p,eff} \tag{4.11}$$

$$\frac{\mu_{nf}}{\mu_f}=\frac{k_{Brownian}}{\Pr k_f}+\frac{1}{(1-\phi)^{2.5}} \tag{4.12}$$

Boundary conditions at z = 0 and z = 900 are

$$v_i=0, u_i=0, w_i=cte, T_i=cte, I=0.16(Re)^{\frac{-1}{8}} \tag{4.13}$$

$$\frac{\partial v}{\partial z}=\frac{\partial u}{\partial z}=\frac{\partial w}{\partial z}=0, \frac{\partial T}{\partial z}=0 \tag{4.14}$$

In this chapter, the hydraulic diameter, Re, Nu, and f can be defined as

$$D_h=\frac{4A}{C} \tag{4.15}$$

$$Re=\frac{\rho v_i D_h}{\mu} \tag{4.16}$$

$$Nu=\frac{hD_h}{k} \tag{4.17}$$

$$f=\frac{\Delta p}{\frac{\rho v_m^2}{2}\frac{L}{D_h}} \tag{4.18}$$

The entropy generation and Bejan number can be defined as [2]

$$S_{gen,total}\cong\frac{q''^2 PD_h L}{Nu.k.T_1T_2}+\frac{8\dot{m}^3 fL}{\rho^2 T_{avg}D_h^3P^2}=S_{gen,th}+S_{gen,f},$$
$$T_{avg}=\frac{(T_1-T_2)}{\ln(T_1/T_2)} \tag{4.19}$$

$$Be=\frac{S_{gen,th}}{S_{gen,f}+S_{gen,th}} \tag{4.20}$$

4.1.2 Effects of active parameters

CuO−water turbulent forced convection inside a pipe with turbulator is studied. Entropy generations of nanofluid are investigated, and obtained results are shown as contours and related correlations. Brownian motion impact on thermal conductivity of CuO−water is taken into account. Influence pitch ratio $(PR = 5\ to\ 15)$, Reynolds number (Re $=$ 5000 to 20000), and height ratio $(BR = 0.3\ to\ 0.5)$ on nanofluid entropy generation are illustrated.

Fig. 4.3 depicts the stream tracer contours for various values of active parameters. Using lower pitch ratio leads to better nanofluid mixing due to stronger secondary flows. Besides, similar behavior is observed for Reynolds number and height ratio. As Reynolds number augments, pressure loss augments and temperature

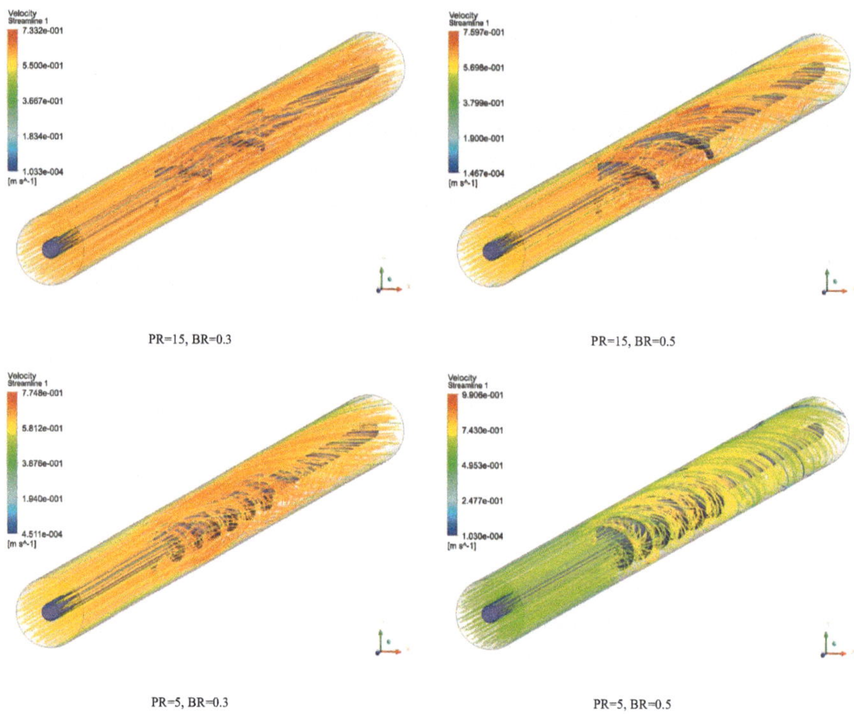

FIGURE 4.3

Stream tracer contours when Re = 20,000.

boundary layer thickness reduces. Increasing height ratio and decreasing pitch ratio lead to higher temperature gradient but with cost of higher pressure loss.

Related formulas for entropy generation and Bejan number are

$$S^*_{gen,f} = 10^5 \times S_{gen,f} = 0.91 - 0.33PR + 0.26BR + 1.55\mathrm{Re} \\ -0.17(PR)(BR) - 0.19(PR)(\mathrm{Re}) + 0.32(\mathrm{Re})(BR) + 0.73\mathrm{Re}^2 \tag{4.21}$$

$$S_{gen,th} = 0.032 + 1.8 \times 10^{-3}PR - 1.3 \times 10^{-3}BR - 0.022\mathrm{Re} \\ +6 \times 10^{-4}(PR)(BR) - 8.54 \times 10^{-4}(PR)(\mathrm{Re}) + 5.4 \times 10^{-4}(\mathrm{Re})(BR) + 0.014\mathrm{Re}^2 \tag{4.22}$$

$$S_{gen,total} = 0.032 + 1.81 \times 10^{-3}PR - 1.3 \times 10^{-3}BR - 0.022\mathrm{Re} \\ +5.9 \times 10^{-4}(PR)(BR) - 8.58 \times 10^{-4}(PR)(\mathrm{Re}) + 5.5 \times 10^{-4}(\mathrm{Re})(BR) + 0.014\mathrm{Re}^2 \tag{4.23}$$

$$\mathrm{Be} = 1 + 1.61 \times 10^{-3}PR - 1.3 \times 10^{-4}BR - 7.04 \times 10^{-4}\mathrm{Re} \\ +9 \times 10^{-5}(PR)(BR) + 2.1 \times 10^{-4}(PR)(\mathrm{Re}) - 1.7 \times 10^{-4}(\mathrm{Re})(BR) - 4.42 \times 10^{-4}\mathrm{Re}^2 \tag{4.24}$$

Effects of Reynolds number, pitch ratio, and height ratio on $S_{gen,f}$, $S_{gen,th}$, $S_{gen,total}$ and Bejan number are shown in Figs. 4.4–4.7. $S_{gen,f}$ is an increasing function of Reynolds number and height ratio. This is due to increase in velocity gradient with increase if Re, BR. Opposite behavior is reported for$S_{gen,th}$. Effect of active parameters on $S_{gen,total}$ is similar to those of $S_{gen,th}$ because $S_{gen,f}$ is obtained to be too lower than $S_{gen,th}$. The ratio of the thermal and total entropy generation rates has been introduced as Bejan number. Bejan number at low Reynolds number is close to 1 because the thermal entropy generation is prominent. Bejan number enhances with augment of pitch ratio. Also, it decreases with rise of Reynolds number and height ratio.

4.2 Nanofluid entropy generation through a pipe considering a new turbulator

4.2.1 Definition of the problem

Two passive techniques have been employed in current study. Using nanofluid and turbulator can improve heat transfer. Single phase was utilized for working fluid with employing experimental correlations. New shape of turbulator has been selected (see

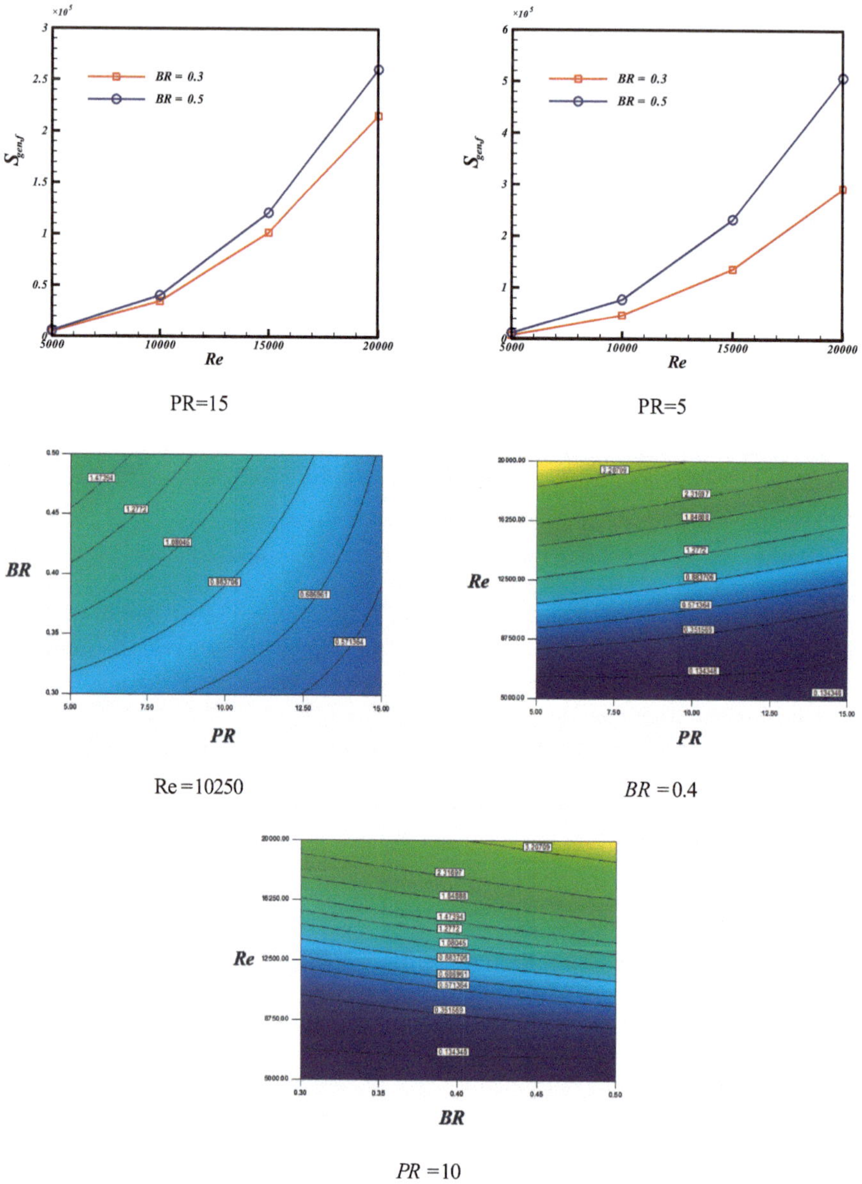

FIGURE 4.4

Effect of Re, *PR*, and *BR* on $S_{gen,f}$.

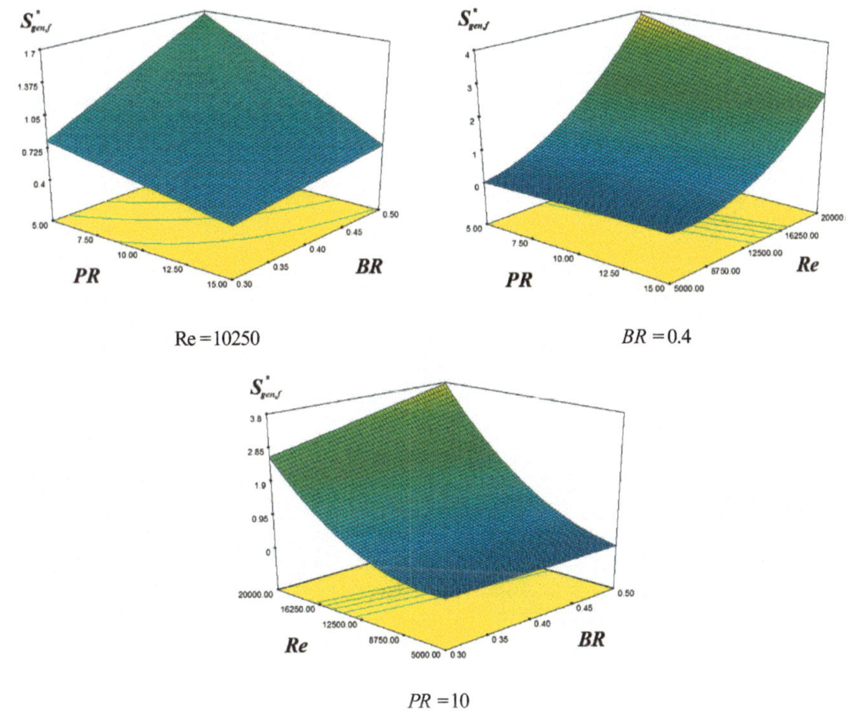

FIGURE 4.4 cont'd.

Fig. 4.8) [3]. Revolution angle was considered as geometric variable. Length of pipe is 900 mm and the test section is central 300 mm. The reason of selecting such test section is because of this fact that test section should not appear any back flow. The inlet boundary was considered as velocity inlet and outlet one was pressure outlet. Heat exchanger is heated by constant heat flux.

$S_{gen,th}$ and $S_{gen,f}$ were calculated as

$$S_{gen,th}=\frac{k_{nf}}{T^2}\left[\left(\frac{\partial T}{\partial z}\right)^2+\left(\frac{\partial T}{\partial y}\right)^2+\left(\frac{\partial T}{\partial x}\right)^2\right], \tag{4.25}$$

$$S_{gen,f}=\frac{\mu_{nf}}{T}\left\{\begin{array}{l}\left(\frac{\partial v}{\partial x}+\frac{\partial u}{\partial y}\right)^2+2\left[\left(\frac{\partial u}{\partial x}\right)^2+\left(\frac{\partial w}{\partial z}\right)^2+\left(\frac{\partial v}{\partial y}\right)^2\right]\\+\left(\frac{\partial v}{\partial z}+\frac{\partial w}{\partial y}\right)^2+\left(\frac{\partial u}{\partial z}+\frac{\partial w}{\partial x}\right)^2\end{array}\right\} \tag{4.26}$$

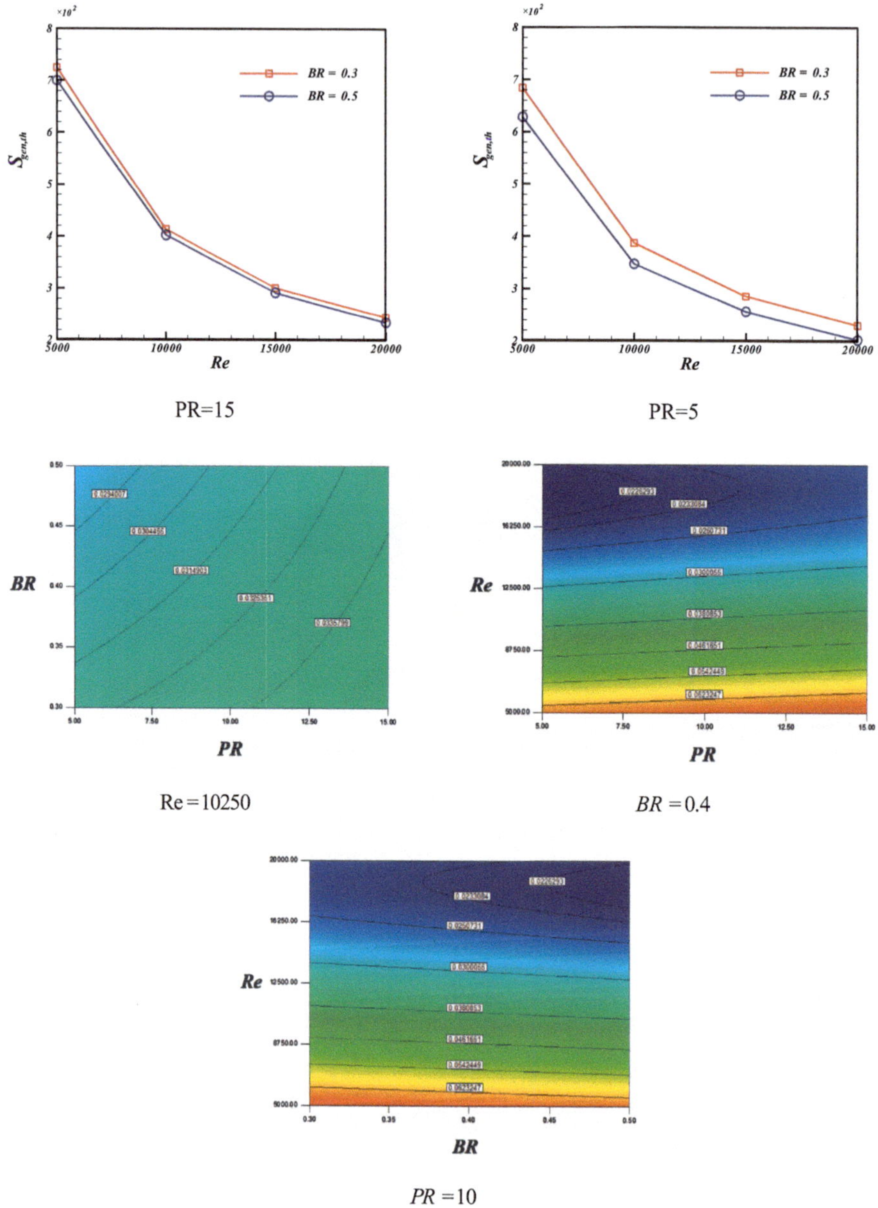

FIGURE 4.5

Effect of Re, *PR*, and *BR* on $S_{gen,th}$.

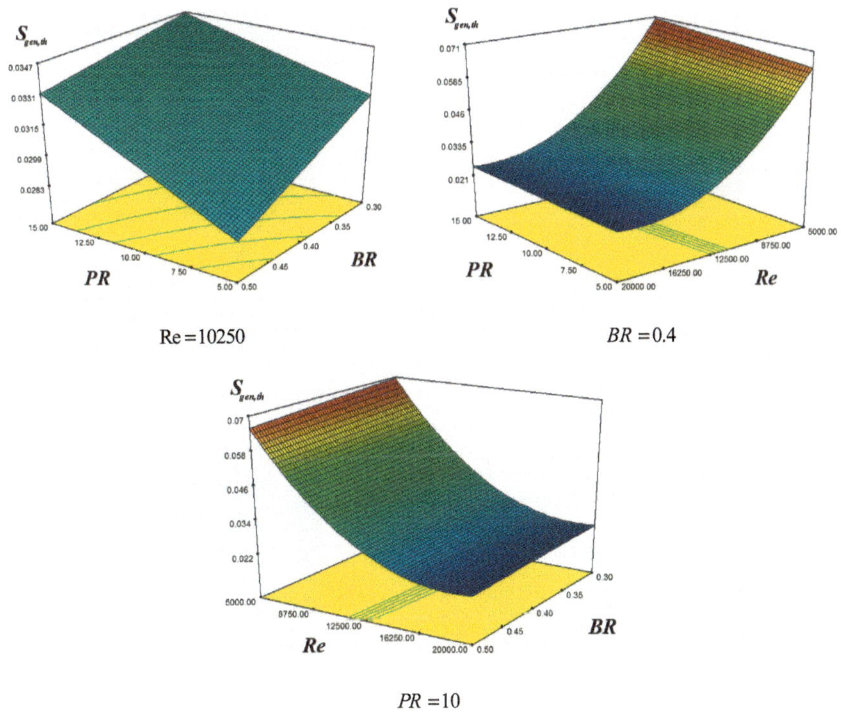

FIGURE 4.5 cont'd.

4.2.2 Effects of active parameters

Entropy generation of turbulent nanofluid flow within a duct with innovative swirling inserts is investigated in current chapter. Homogenous model was considered. We selected two parameters as variables (inlet velocity and revolution angle). Finding the best performance in view of thermal and second law behaviors was our main purpose.

Impacts of inlet velocity and β on velocity contours, thermal and viscous entropy generation contours are depicted in Figs. 4.9–4.14. As inlet velocity enhances, pressure drop and temperature gradient increase. As inlet velocity increases, reduction in the thermal boundary layer thickness can be observed, and it makes convective flow stronger. Pressure loss enhances due to more swirling flow.

Secondary flow rises with augment of β and in turn convective flow rises. Stronger turbulent intensity occurs for higher values of β. So, greater revolution angle makes to reach better nanofluid fluid mixing. The impact of β is more significant at lower inlet velocity because of thicker boundary layer in low Reynolds number. Also, disruption of thermal boundary is more pronounced in lower Re. ΔP augments with rise of β because of intensification in secondary flow.

Viscous entropy generation augments with augment of ΔP. So, $S_{gen,f}$ intensifies with augment of inlet velocity and revolution angle. Thermal entropy generated act

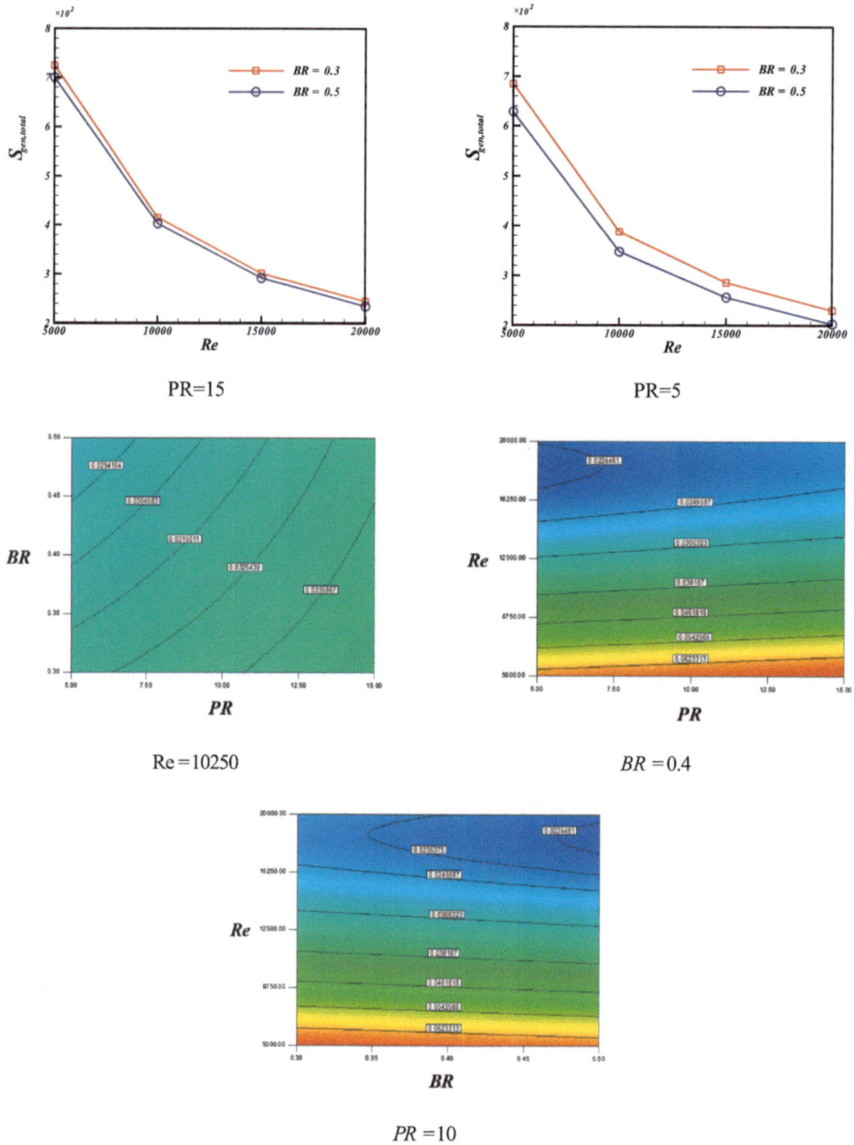

FIGURE 4.6

Effect of Re, *PR*, and *BR* on $S_{gen,total}$.

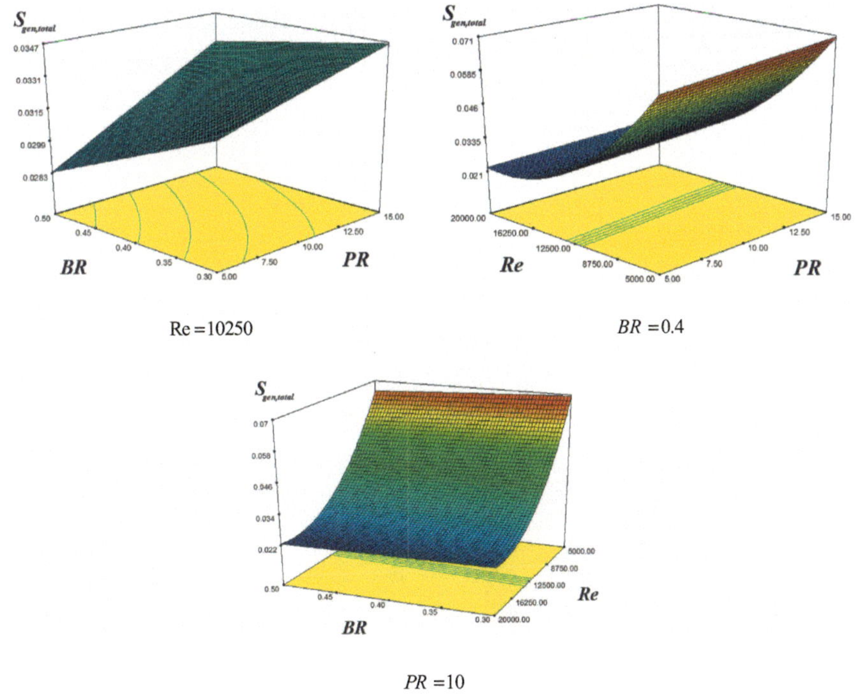

FIGURE 4.6 cont'd.

against pressure loss. $S_{gen,th}$ intensifies with reduction of β and Re. Following formulas are derived for $S_{gen,th}$, Be, and $S_{gen,f}$:

$$S_{gen,th} = 22.26 - 0.73\beta - 21.62\text{Re}^* + 0.64\beta\,\text{Re}^* + 0.57\beta^2 \quad (4.27)$$

$$Be = 0.63 - 0.012\beta - 0.35\text{Re}^* - 0.012\beta\,\text{Re}^* + 0.012\beta^2 \quad (4.28)$$

$$S_{gen,f} = 1.4 + 0.049\beta + 1.16\text{Re}^* + 0.039\beta\,\text{Re}^* - 0.048\beta^2 \quad (4.29)$$

In the above equations, $\text{Re}^* = 0.001\text{Re}$ and mean square error is 0.98. Fig. 4.15 demonstrates the impacts of β and Re on entropy parameters. Secondary vortexes become stronger with rise of both parameters. Therefore, $S_{gen,th}$ is an augmenting function of Re and β. Roles of significant parameters on $S_{gen,f}$ are different from $S_{gen,th}$. So, $S_{gen,th}$ reduces with augment of Re andβ. Because of higher values of $S_{gen,th}$ in comparison to $S_{gen,f}$, it can be concluded that Be detracts with increase of Re and β.

4.3 Helical twisted tapes effect on entropy generation of nanofluid

4.3.1 Definition of the problem

In this research, the helical twisted tape is inserted to enhance heat transfer (Fig. 4.16). We have three sections: inlet section, test section, and outlet. All the outlet and inlet sections are smooth straight tubes [4]. The flow direction is along z-direction of Cartesian coordinate. The working fluid is Al_2O_3-water nanofluid ($\phi = 0.04$). The main geometrical parameters of the twisted tubes are pitch (P), length (b), and thickness (t). In this study, length of the tube is 900 mm and the outer

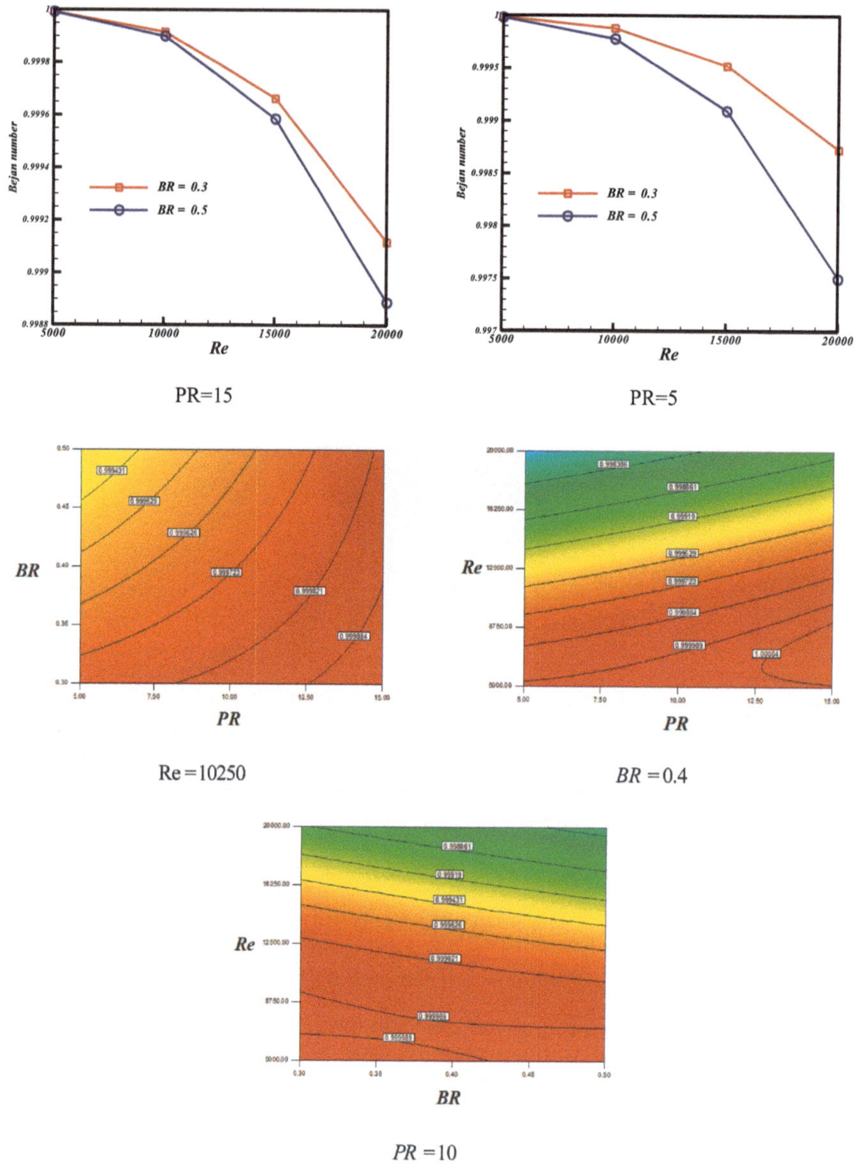

FIGURE 4.7

Effect of Re, *PR*, and *BR* on Bejan number.

and inner diameters are 20 and 5 mm, respectively. The middle 300 mm is considered as test section and helical twisted tape insert in this section. In order to ensure that the fully developed flow is formed and avoid any probable back flow streams, the inlet and outlet sections are considered equals to 300 mm.

4.3.2 Effects of active parameters

In this chapter, turbulent nanofluid flow and entropy generation inside a pipe with helical-shaped twisted tape is investigated. Brownian motion influence is considered in order to estimate Al_2O_3–water nanofluid properties. Stream tracer, temperature,

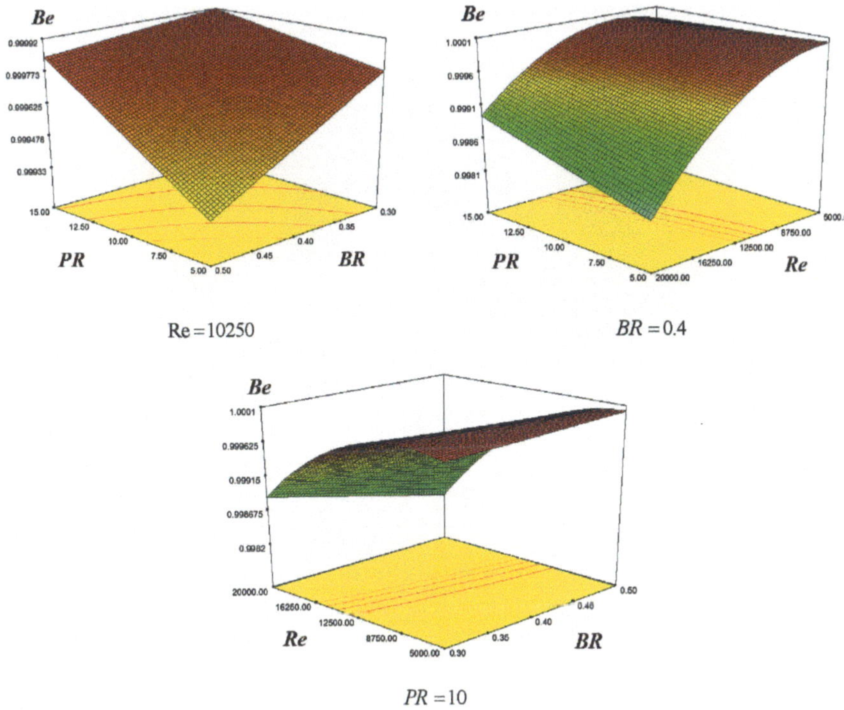

FIGURE 4.7 cont'd.

and entropy generation contours for various cases are shown in Figs. 4.17–4.20. As Reynolds number augments, pressure loss augments and thermal boundary layer thickness decreases. Increasing height ratio and decreasing pitch ratio lead to higher temperature gradient but with cost of higher pressure loss.

Variations of frictional and thermal entropy generation respect to active parameters are depicted in Figs. 4.21 and 4.22. Suitable formulas for entropy generation are

$$S_{gen,f} = 1.19 - 0.32PR + 0.68BR + 1.04\text{Re}^* - 0.021(PR)(BR) - 0.013(PR)(\text{Re}^*) + 0.57(BR)(\text{Re}^*) + 0.09(BR)^2 \tag{4.30}$$

$$S_{gen,th} = 30.17 + 16.2PR + 0.79BR - 27.5\text{Re}^* + 1.14(PR)(BR) - 13.25(PR)(\text{Re}^*) + 1.54(BR)(\text{Re}^*) + 3.61(BR)^2 \tag{4.31}$$

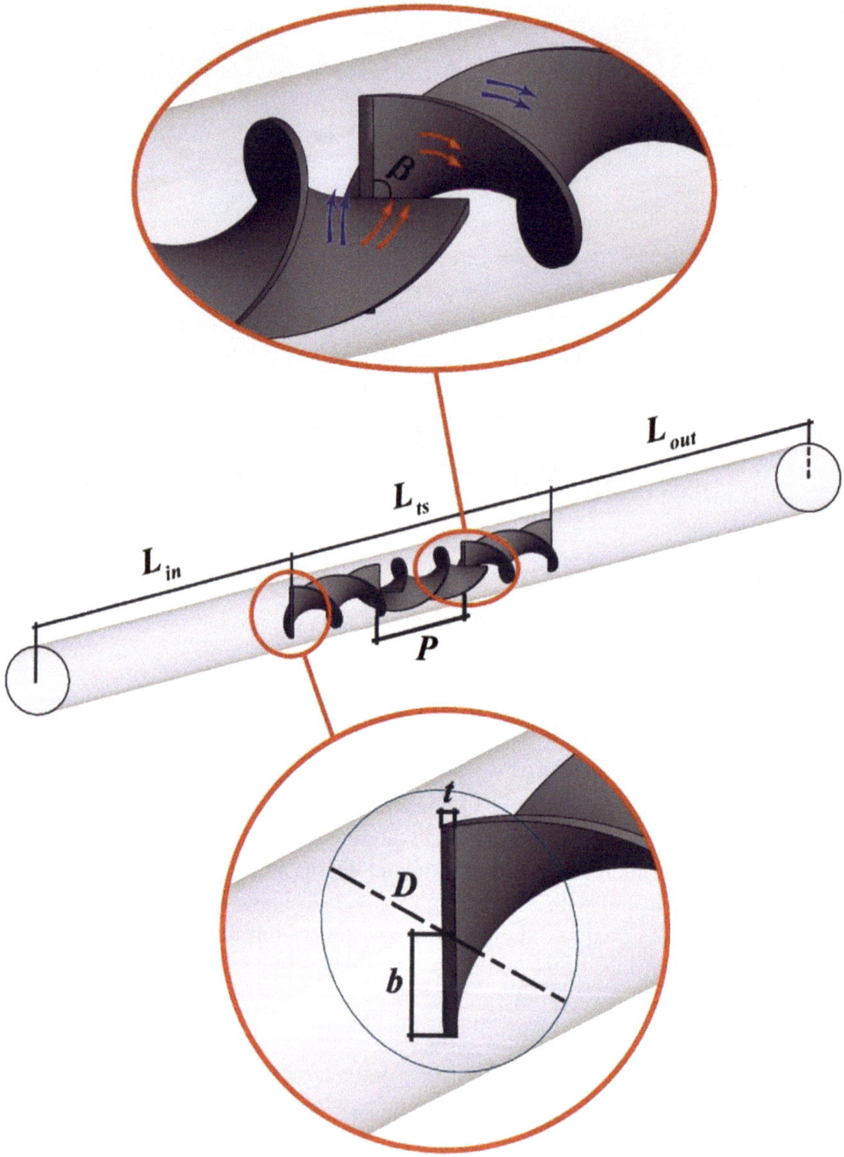

FIGURE 4.8

Present heat exchanger.

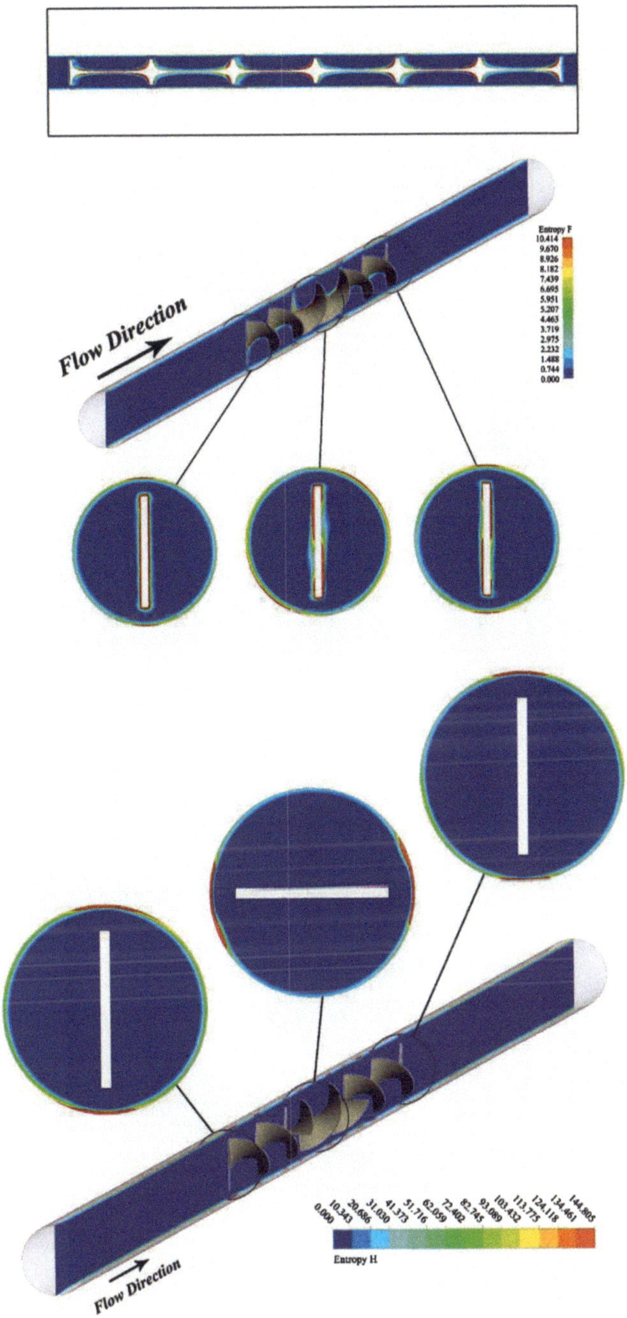

FIGURE 4.9

Frictional and thermal entropy generation contours at $\beta = 0^{\circ}$, Re $= 5000$.

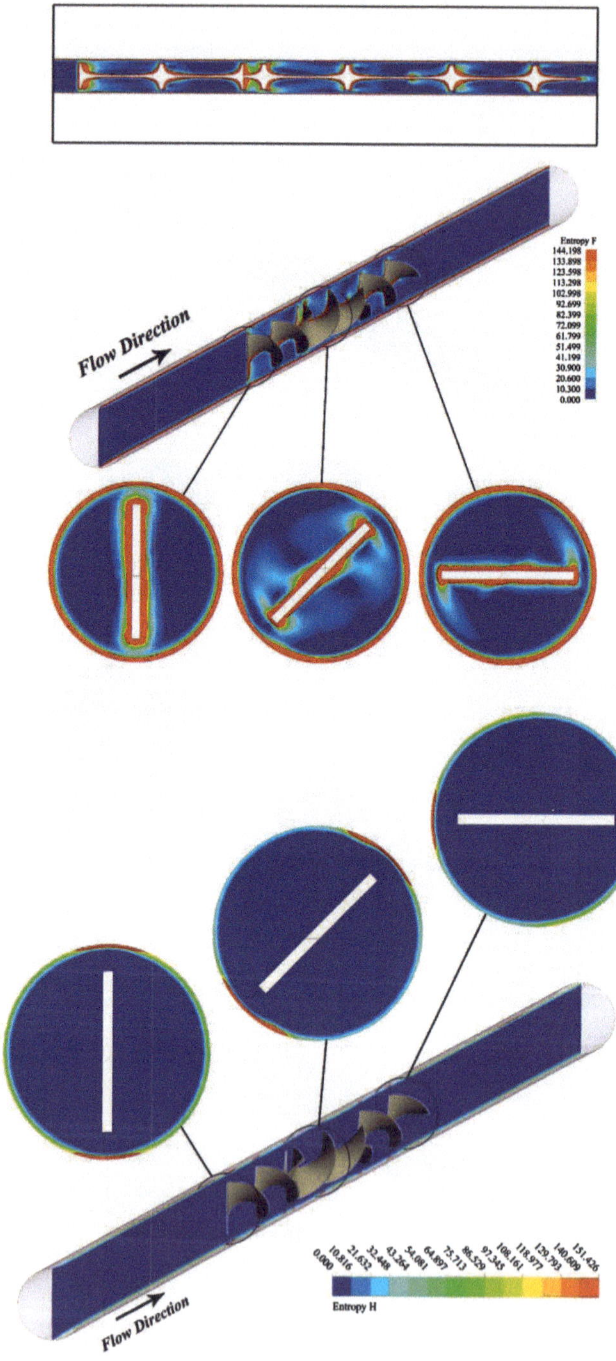

FIGURE 4.10

Frictional, thermal entropy generation and velocity contours at $\beta = 0°$, Re $= 20000$. (A) inlet of test section; (B) middle of test section; (C) outlet of test section.

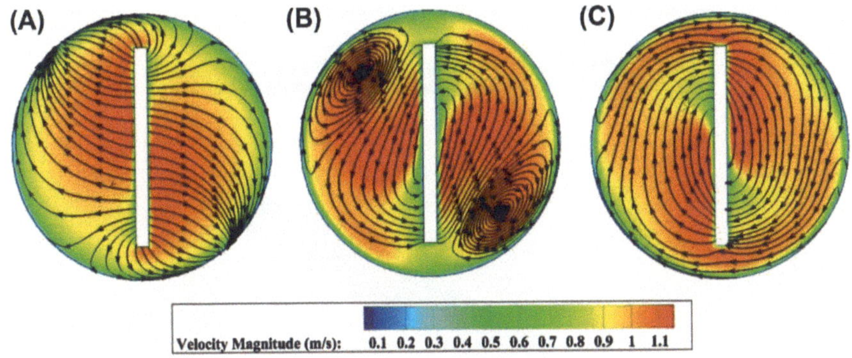

FIGURE 4.10 cont'd.

where $Re^* = Re/10000$. As Reynolds number and height ratio augment, velocity gradient enhances and in turn frictional entropy generation enhances. Opposite treatment is obtained for $S_{gen,th}$. Effect of active parameters on $S_{gen,total}$ is similar to those of $S_{gen,th}$ because $S_{gen,f}$ is found to be too lower than $S_{gen,th}$.

4.4 Heat transfer and entropy generation of nanoparticles employing innovative helical turbulator

4.4.1 Definition of the problem

Layout of the scrutinized heat exchanger has been demonstrated in Fig. 4.23. Complex turbulators have been inserted in test section. To prevent back flow in outlet zone and reach to fully develop flow in inlet zone, L_{out} and L_{in} have been considered. Copper oxide$-H_2O$ flows in the pipe and previous experimental formulas have been employed to find its properties [5]. The length of duct was 1.5 m and $L_{out} = L_{in} = 0.5$m. We utilized pipe with a diameter of 2 inch. Twisted tapes can be introduced with five geometric parameters (y, t, l, b, p) which were depicted in Fig. 4.23, and we considered b as variable in this chapter. Simulations have been done for various $b = 5$, 10, and 15 mm. In addition, Reynolds ranges indicate the turbulent flow (Re = 5000, 10,000, 15,000). In current research, we utilized $k-\varepsilon$ RNG. Several papers were published to investigate nanofluid laminar and turbulent flow [6–60].

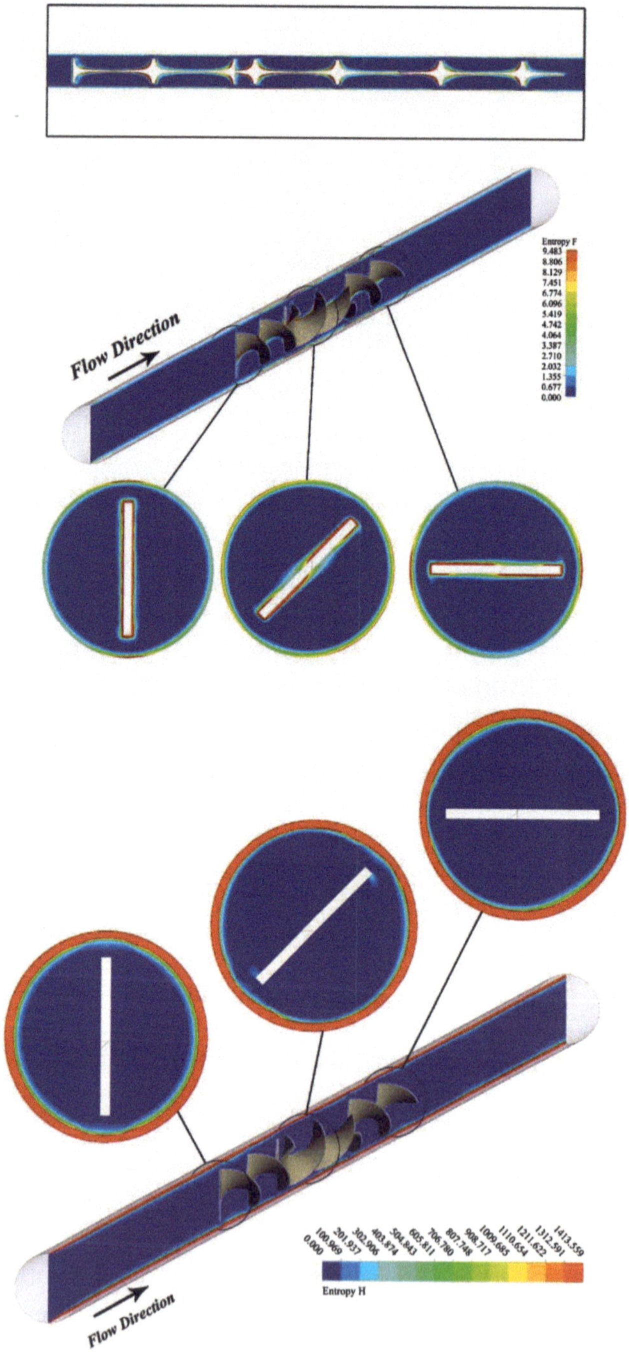

FIGURE 4.11

Frictional and thermal entropy generation contours at $\beta = 45°$, Re $= 5000$.

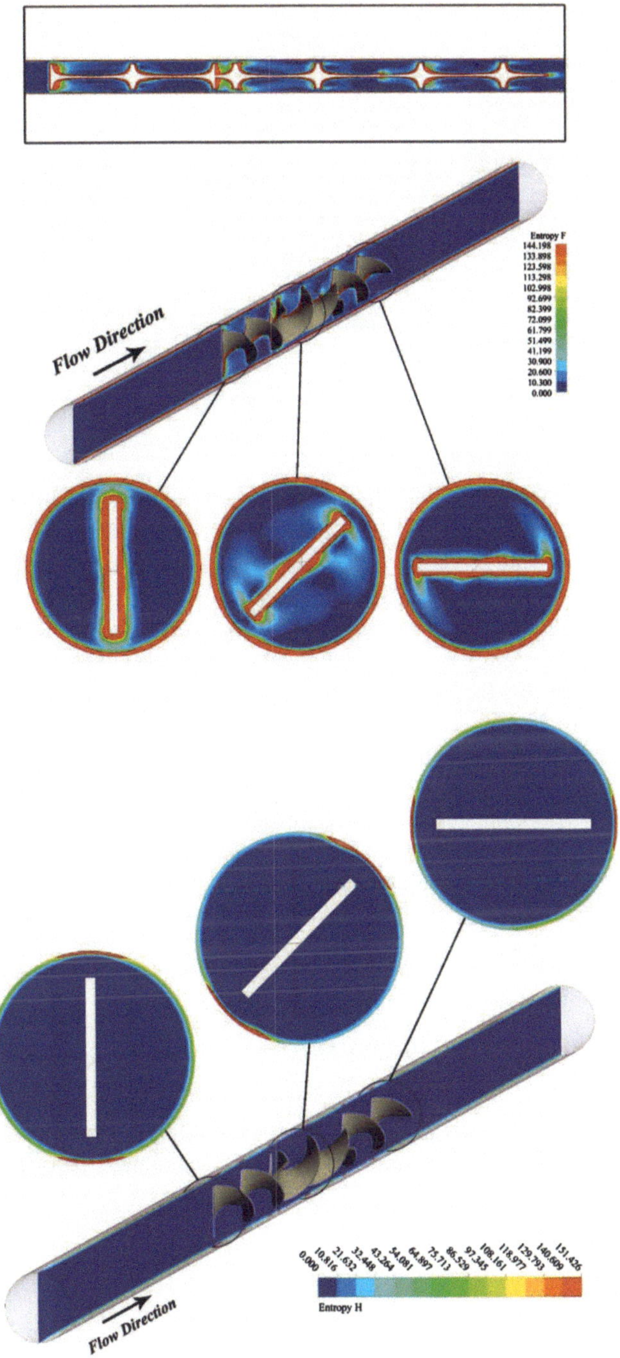

FIGURE 4.12

Frictional, thermal entropy generation, and velocity contours at $\beta = 45^\circ$, Re $= 20000$.

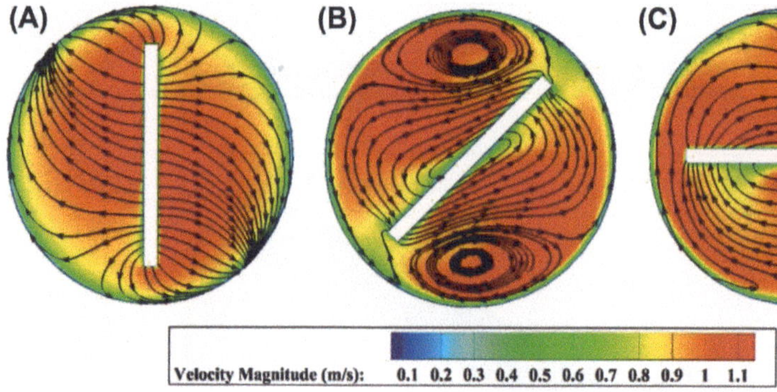

FIGURE 4.12 cont'd.

4.4.2 Effects of active parameters

Thermal and frictional irreversibility in a tube with double twisted tapes involving CuO—water nanofluid under turbulent flow have been simulated. FVM give us opportunity to model this problem. In the below part, velocity, generation of entropy, and temperature were scrutinized, and these subjects were involved with b and Re.

Figs. 4.24—4.26 discussed the variation of S_{gen}, temperature, and velocity. The strength of secondary eddy rises with augment of inlet velocity. More circulation occurs when b enhances. $S_{gen,th}$ concerned with temperature difference between bulk fluid and wall. So, as pumping power augments, this component of S_{gen} declines. In addition, greater viscous dissipation can be found with augmentation of inlet velocity which results in higher $S_{gen,f}$. Flow disturbance augments with rising b, which clarifies higher frictional resistance. Therefore, $S_{gen,f}$ has direct relationship with b. $S_{gen,th}$ depicts decrease tendency with augmenting of b. Entropy generation has two components in current problem in which their variations were summarized as follows:

$$S_{gen,f} = 6.8 \times 10^{-3} + 5.7 \times 10^{-3}\mathrm{Re}^* + 2.16 \times 10^{-3} b + 1.68 \times 10^{-3} b\,\mathrm{Re}^* + 6.67 \times 10^{-4}\, b^2 \tag{4.32}$$

$$S_{gen,th} = 20.74 - 19.83\mathrm{Re}^* - 1.46b + 1.44b\,\mathrm{Re}^* + 0.13b^2 \tag{4.33}$$

Corresponding to above equations, Fig. 4.27 has been drawn. $S_{gen,th}$ was inversely proportional to Re and b, while opposite relationship can be reported for $S_{gen,f}$. Also as expected, more contribution to S_{gen}, have been reported for $S_{gen,th}$.

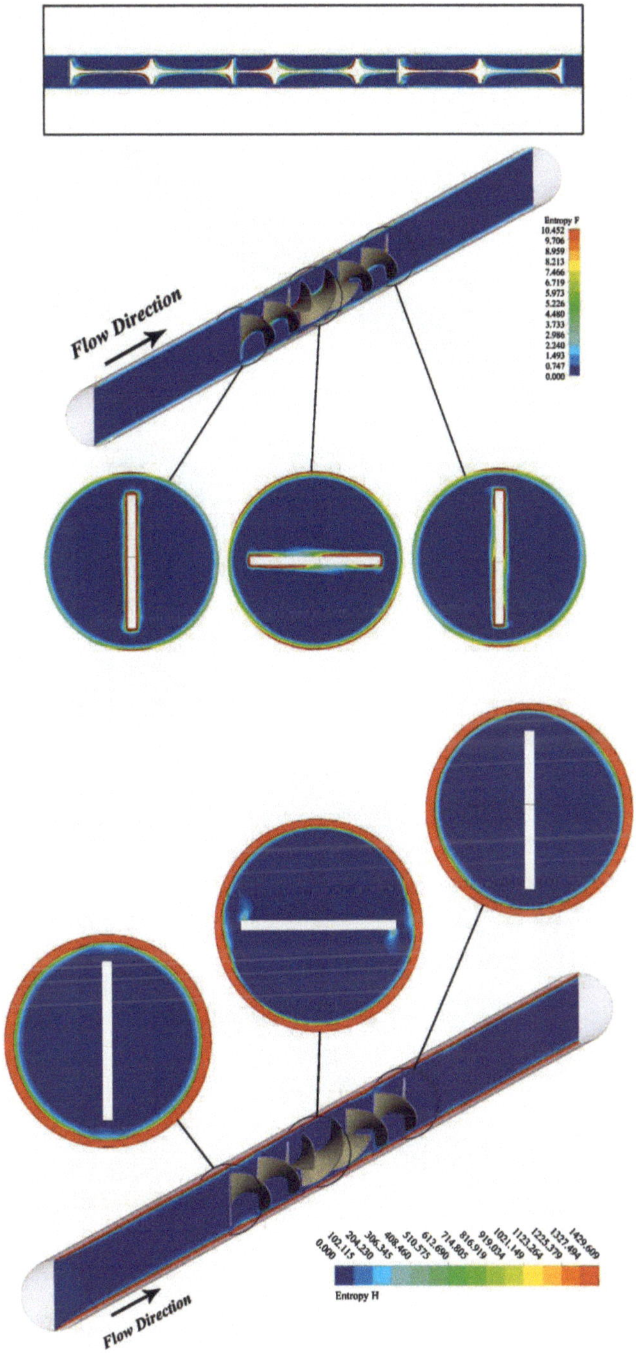

FIGURE 4.13

Frictional and thermal entropy generation contours at $\beta = 90°$, Re $= 5000$.

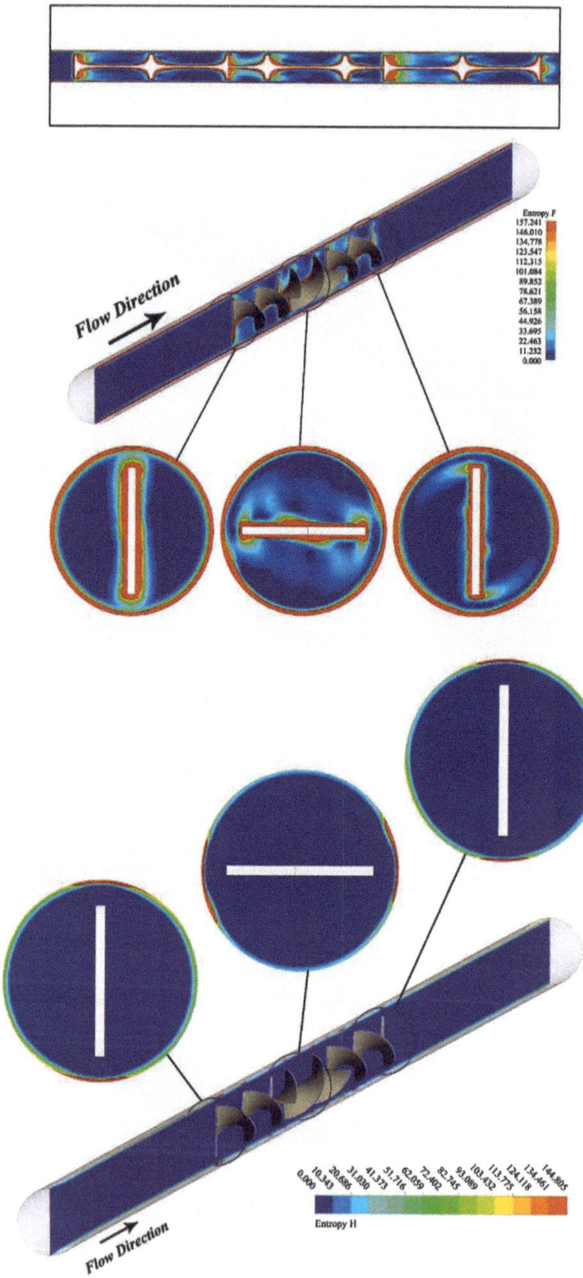

FIGURE 4.14

Frictional, thermal entropy generation and velocity contours at $\beta = 90°$, Re $= 20000$.

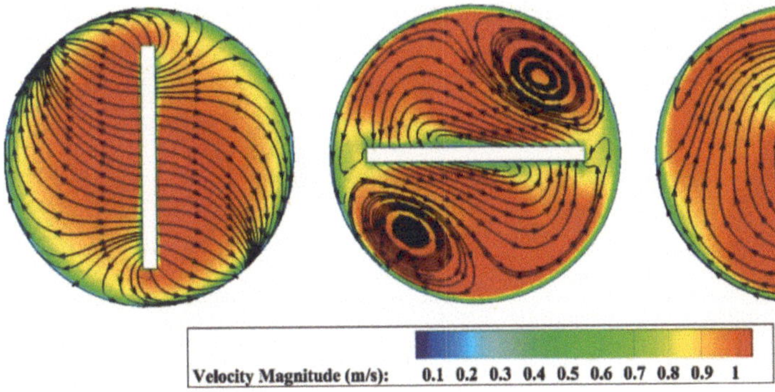

FIGURE 4.14 cont'd.

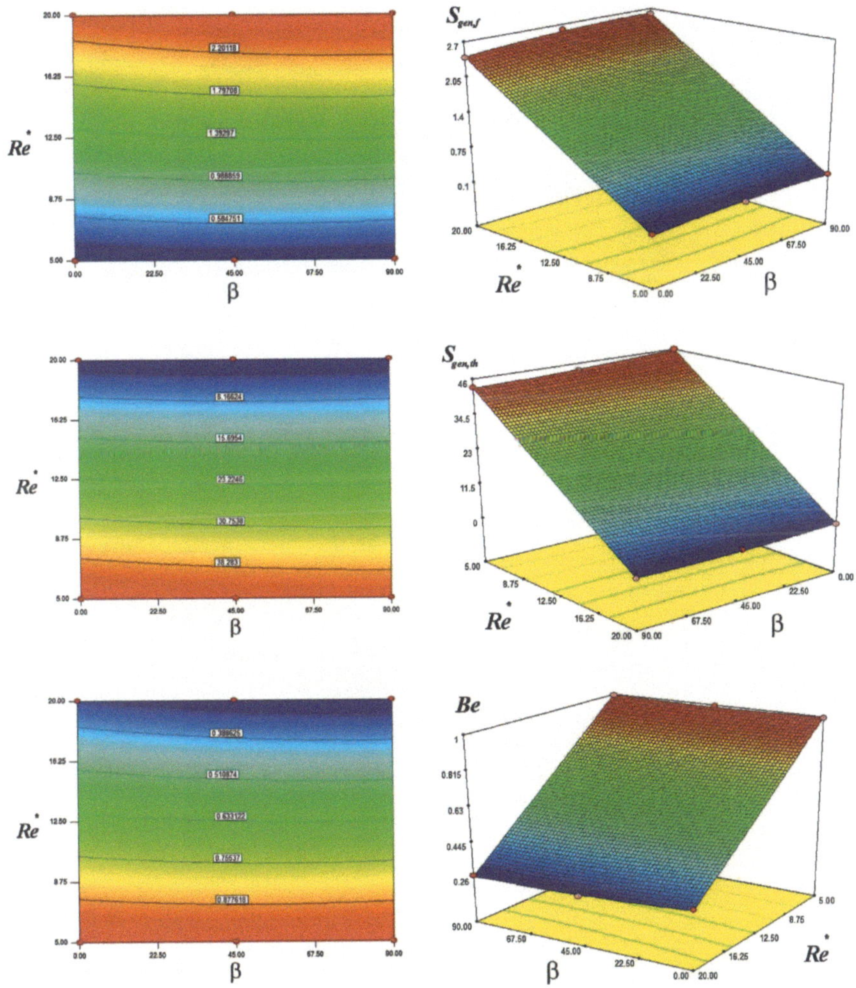

FIGURE 4.15

Effects of Re and β on $S_{gen,f}$, $S_{gen,th}$, Be.

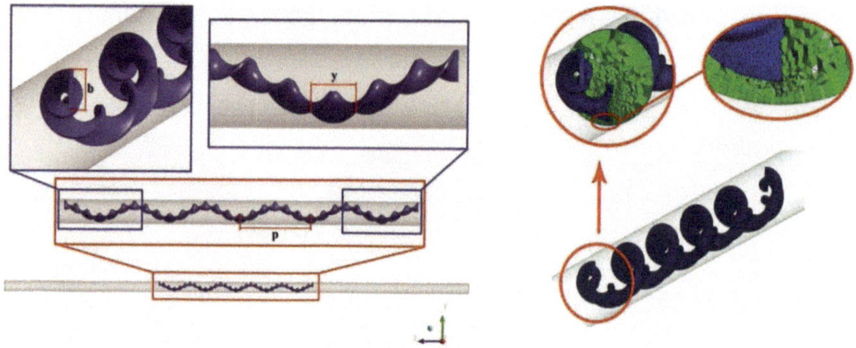

FIGURE 4.16

Geometry of the problem and sample mesh.

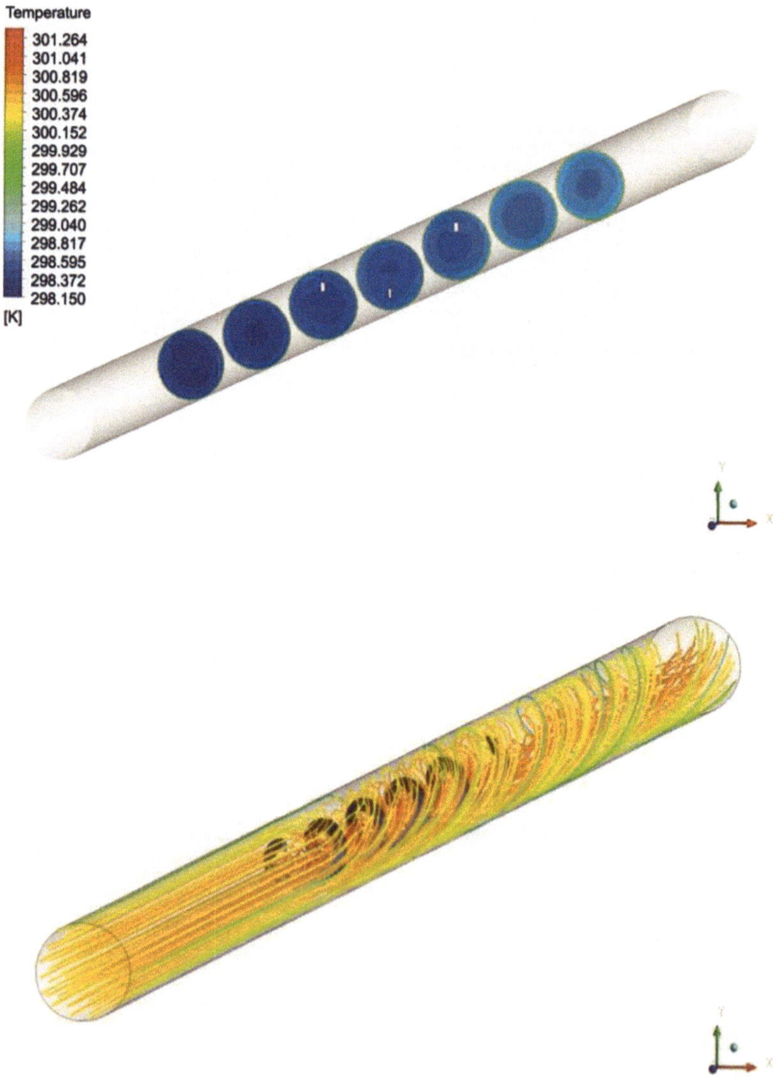

FIGURE 4.17

Stream tracer, temperature, entropy generation contours at $PR = 3$, $BR = 0.25$, $\mathrm{Re} = 15{,}000$.

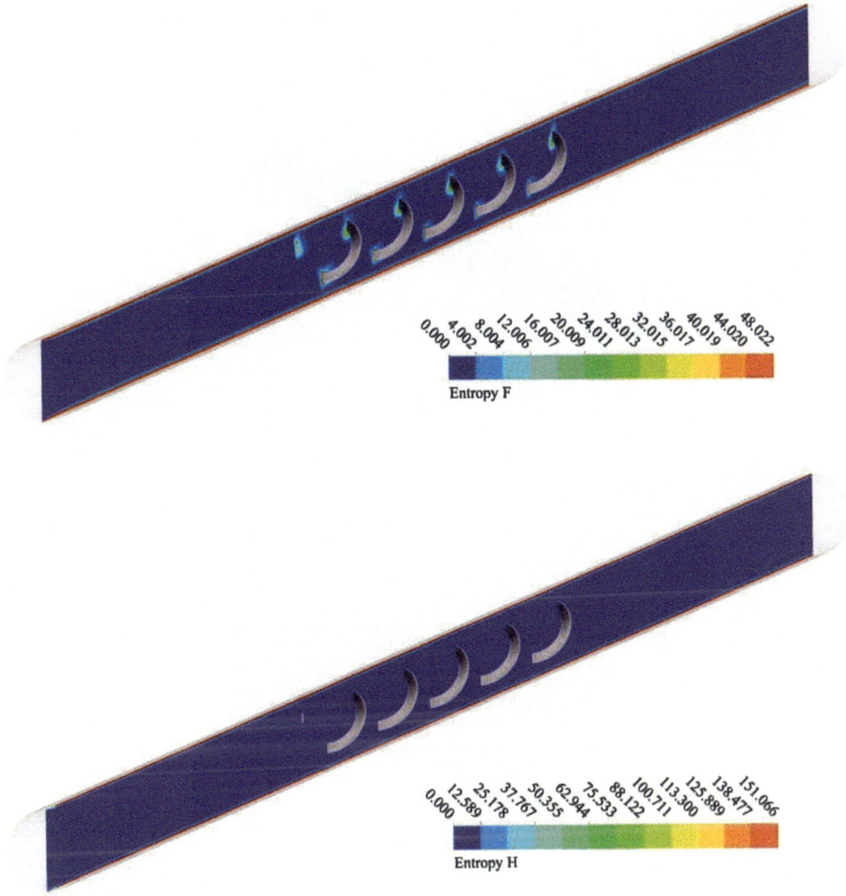

FIGURE 4.17 cont'd.

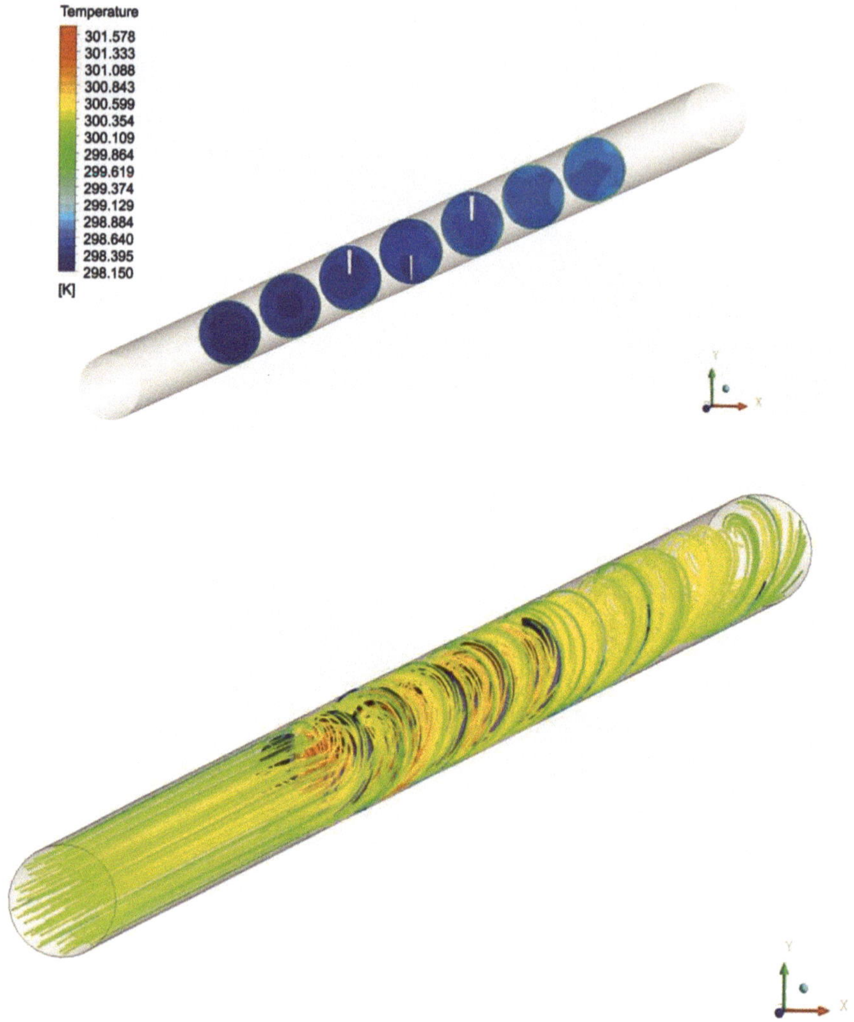

FIGURE 4.18

Stream tracer, temperature, entropy generation contours at $PR = 3$, $BR = 0.75$, $\mathrm{Re} = 15{,}000$.

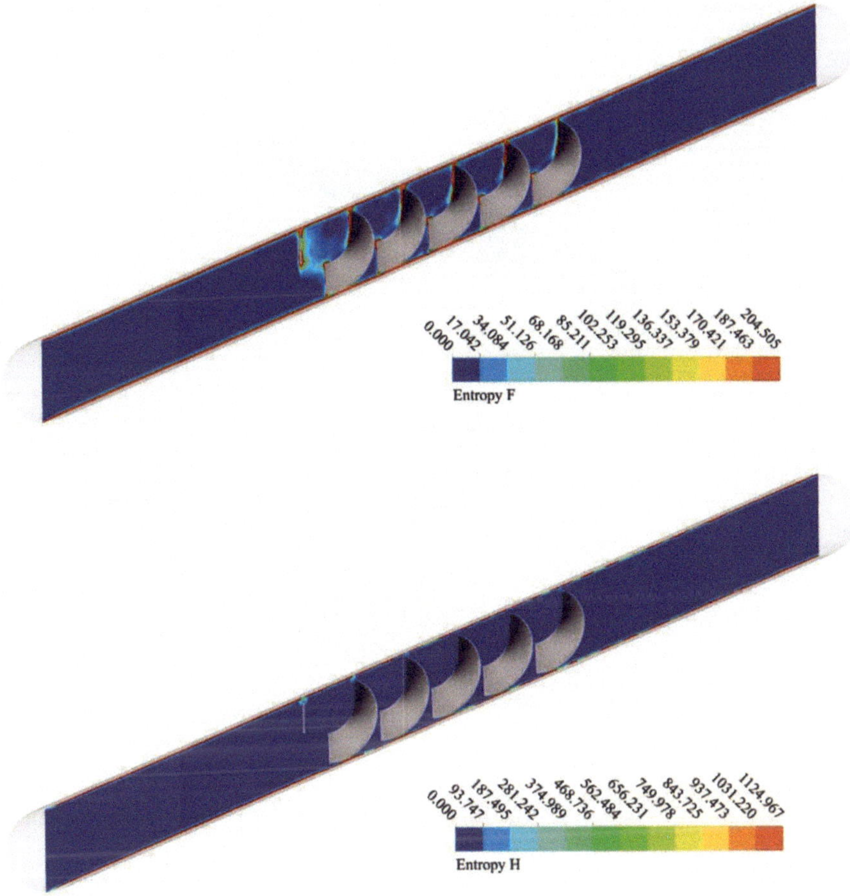

FIGURE 4.18 cont'd.

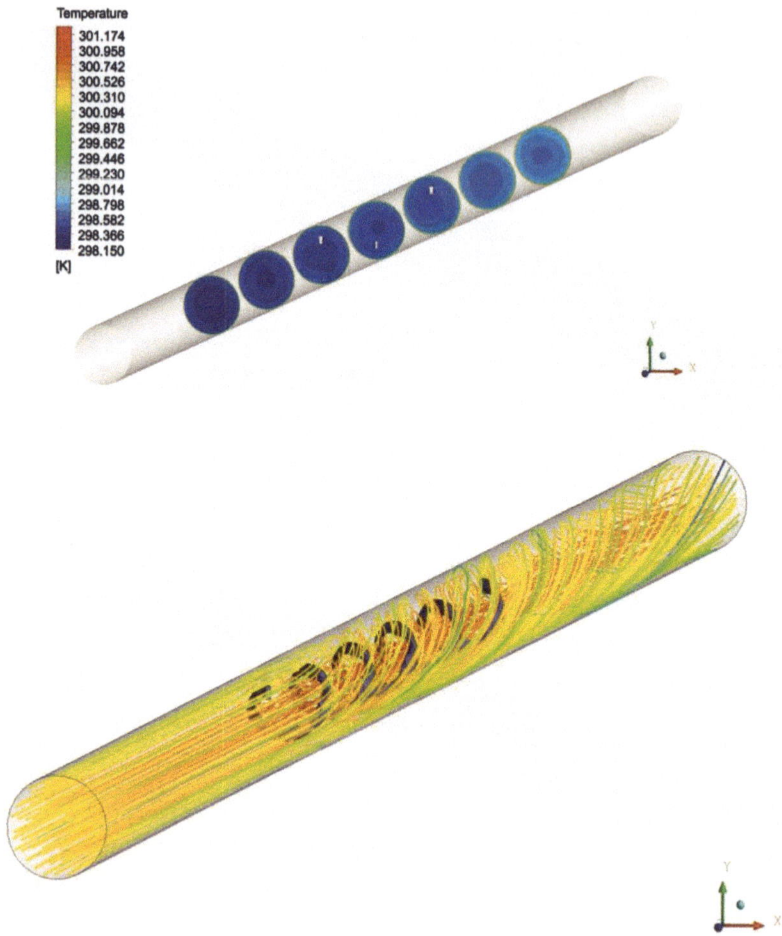

FIGURE 4.19

Stream tracer, temperature, entropy generation contours at $PR = 1.5$, $BR = 0.25$, $\mathrm{Re} = 15{,}000$.

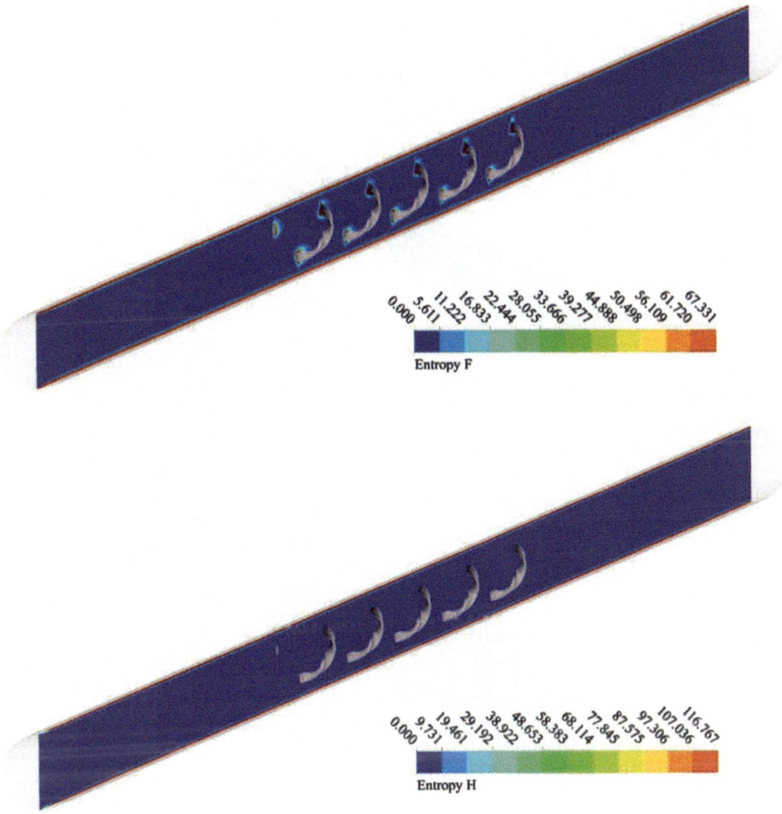

FIGURE 4.19 cont'd.

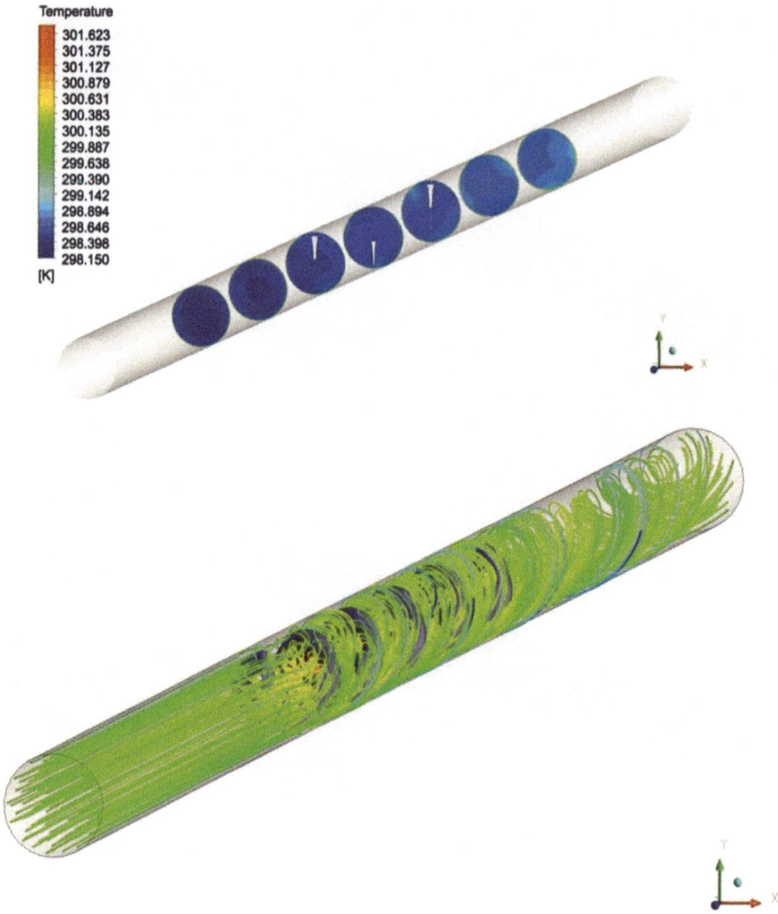

FIGURE 4.20

Stream tracer, temperature, entropy generation contours at $PR = 1.5$, $BR = 0.75$, Re = 15,000.

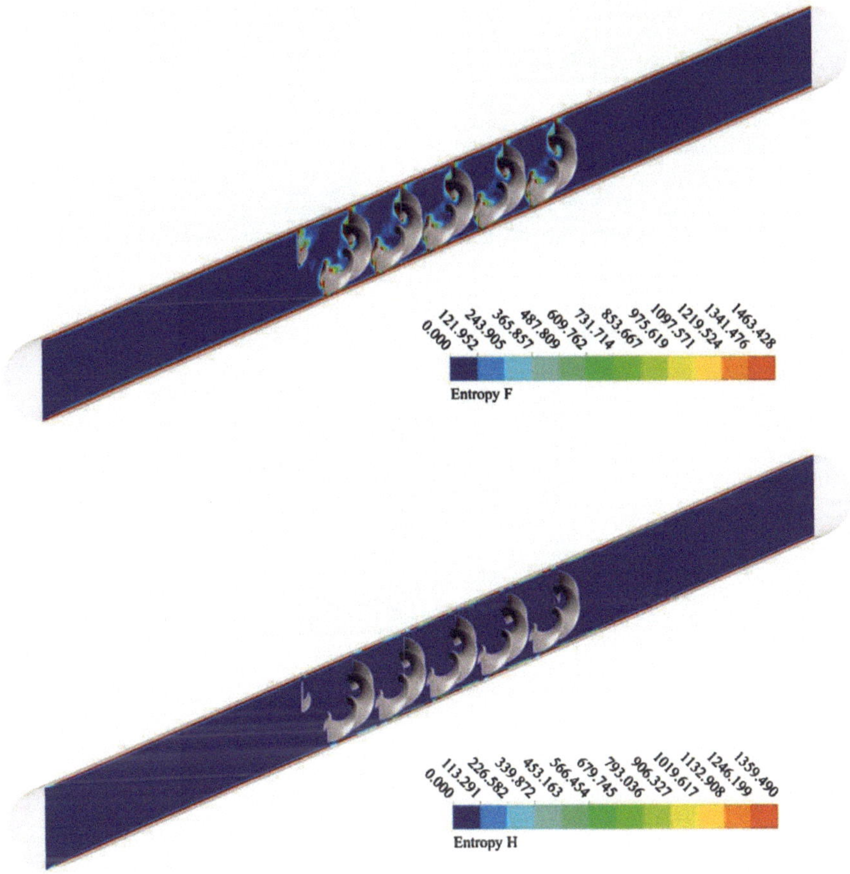

FIGURE 4.20 cont'd.

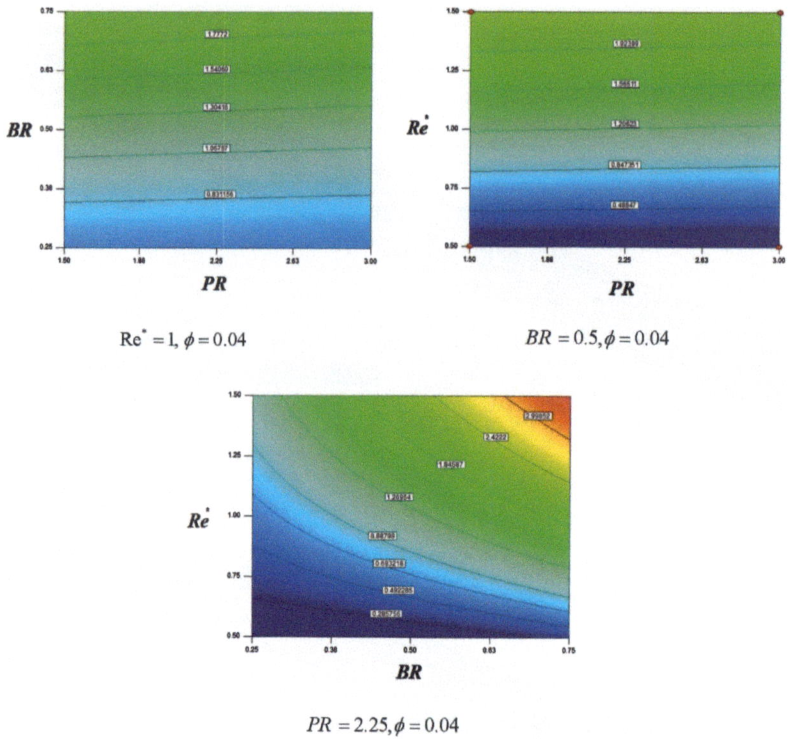

FIGURE 4.21

Effects of Re*, *PR*, and *BR* on $S_{gen,f}$.

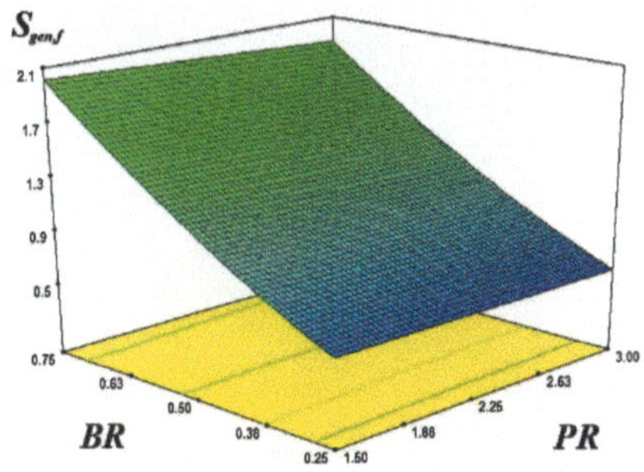

$Re^* = 1, \phi = 0.04$

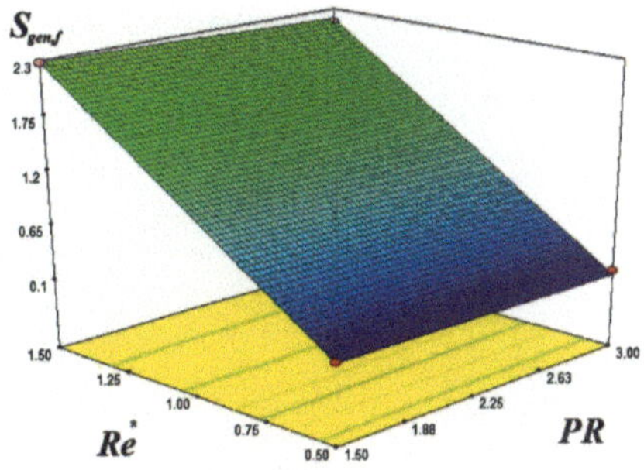

$BR = 0.5, \phi = 0.04$

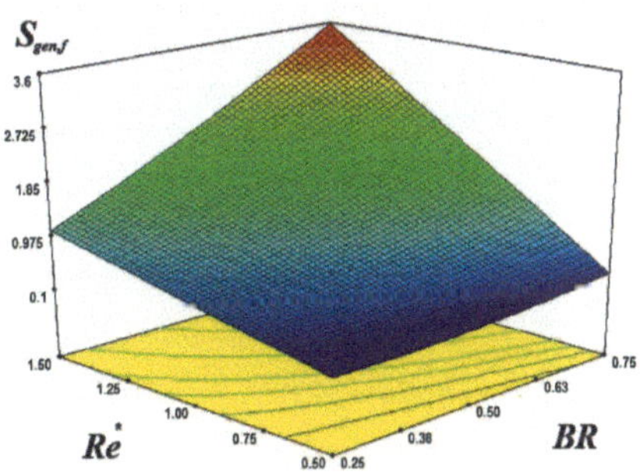

$PR = 2.25, \phi = 0.04$

FIGURE 4.21 cont'd.

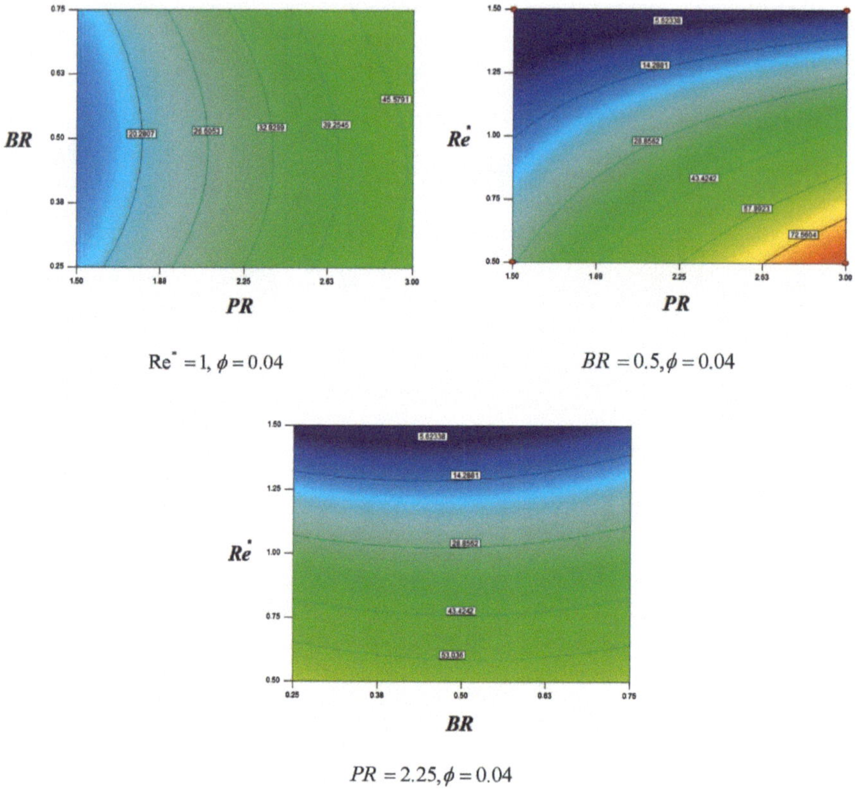

FIGURE 4.22

Effects of Re*, PR, and BR on $S_{gen,th}$.

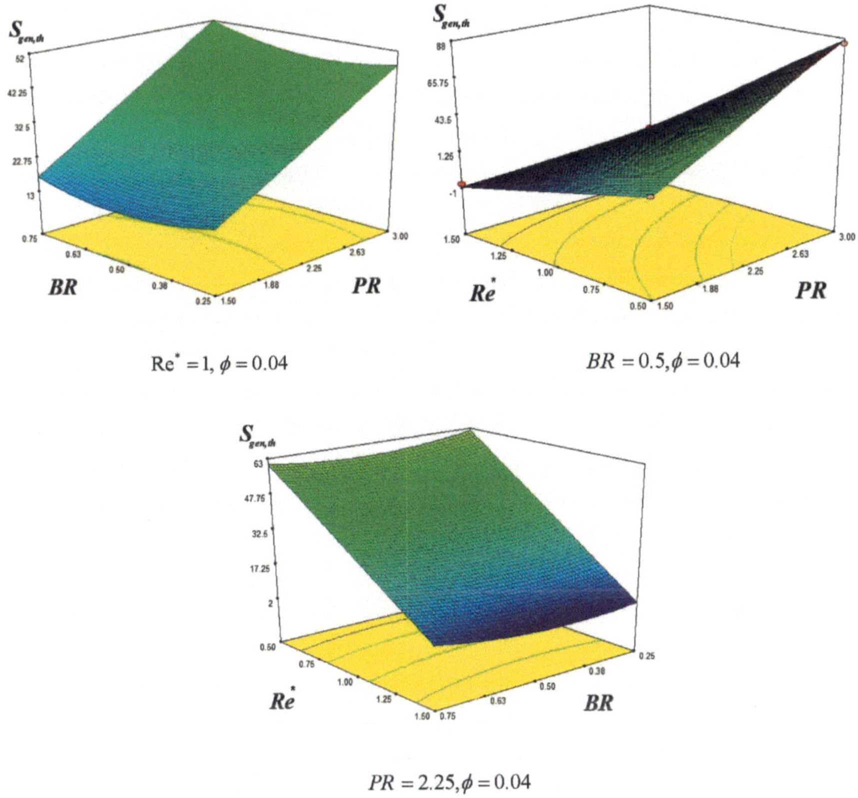

FIGURE 4.22 cont'd.

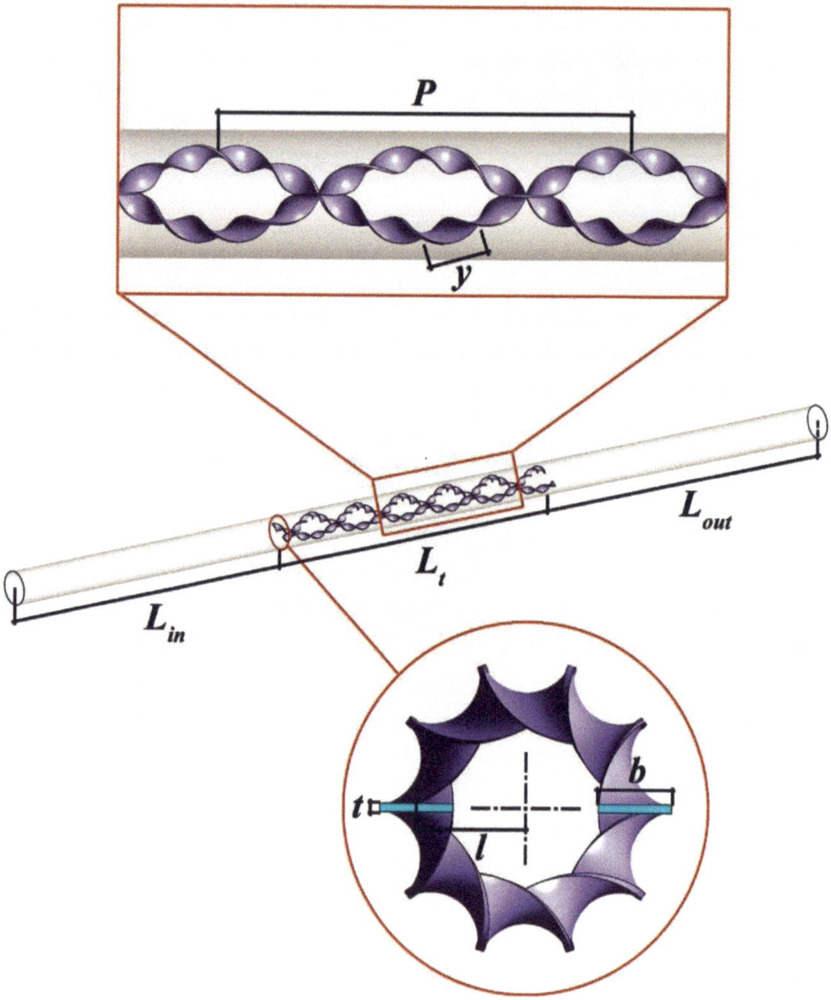

D	0.0508 m
L_{in} , L_{out} , L_t	0.5 m
P	166.667 mm
y	23.93292 mm
t	1 mm
l	15 mm

FIGURE 4.23

Configuration of heat exchanger and mesh.

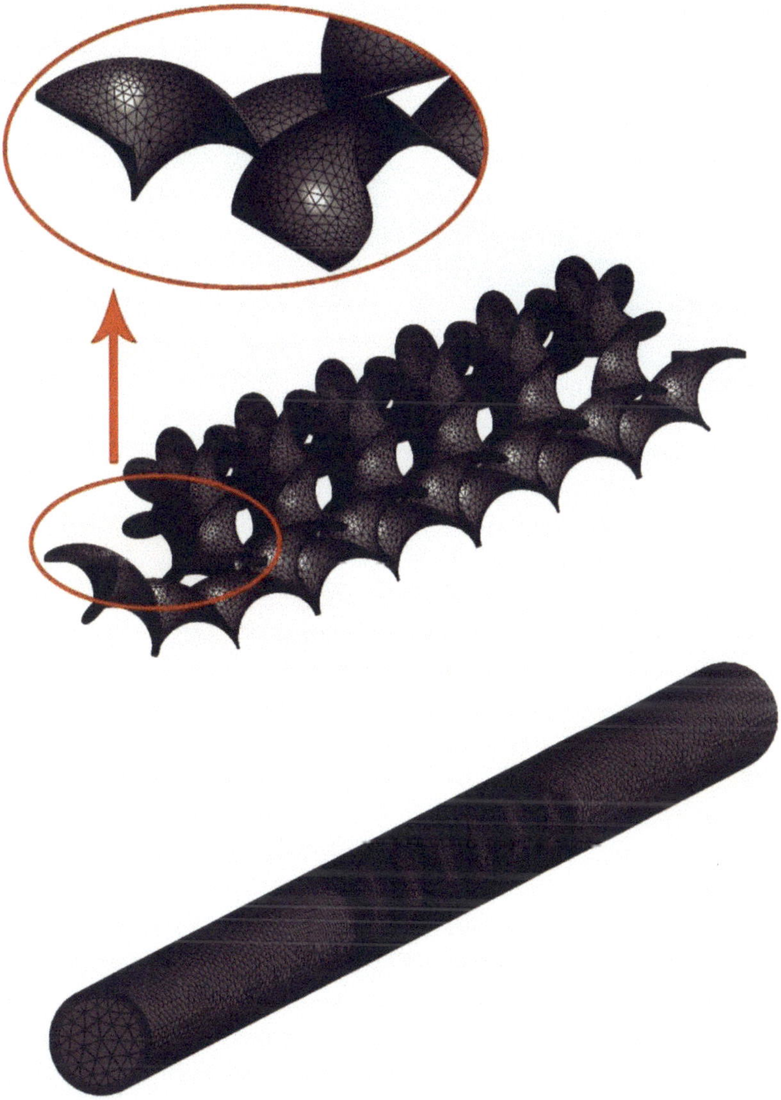

FIGURE 4.23 cont'd.

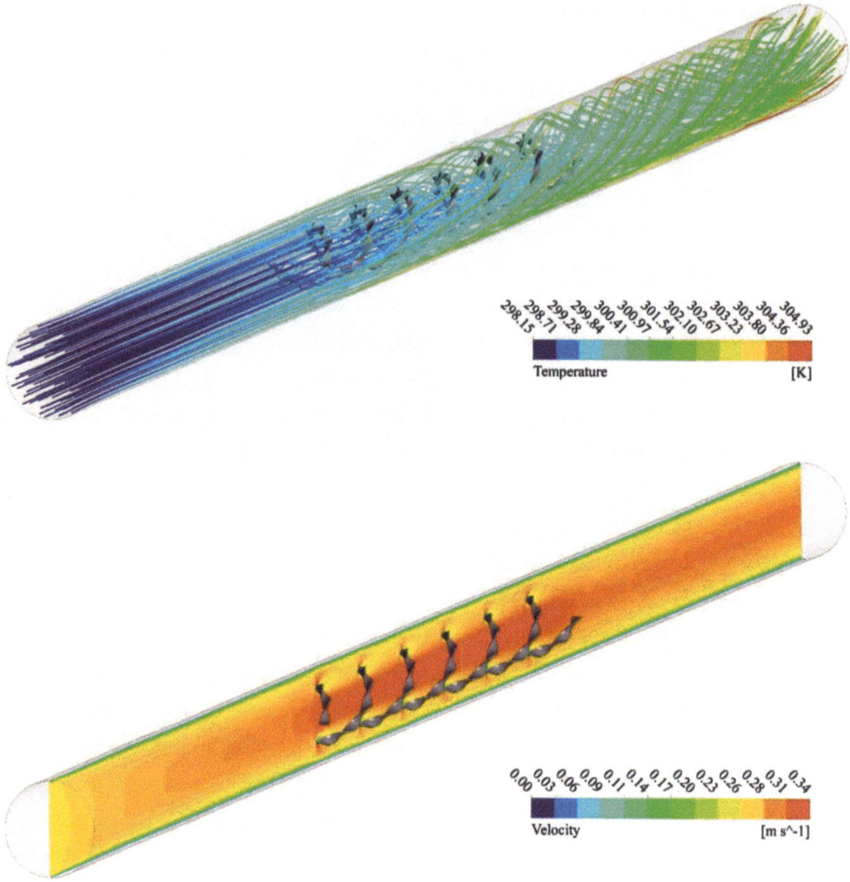

FIGURE 4.24

Velocity, T, S_{gen} contours for Re = 15,000, b = 5 mm.

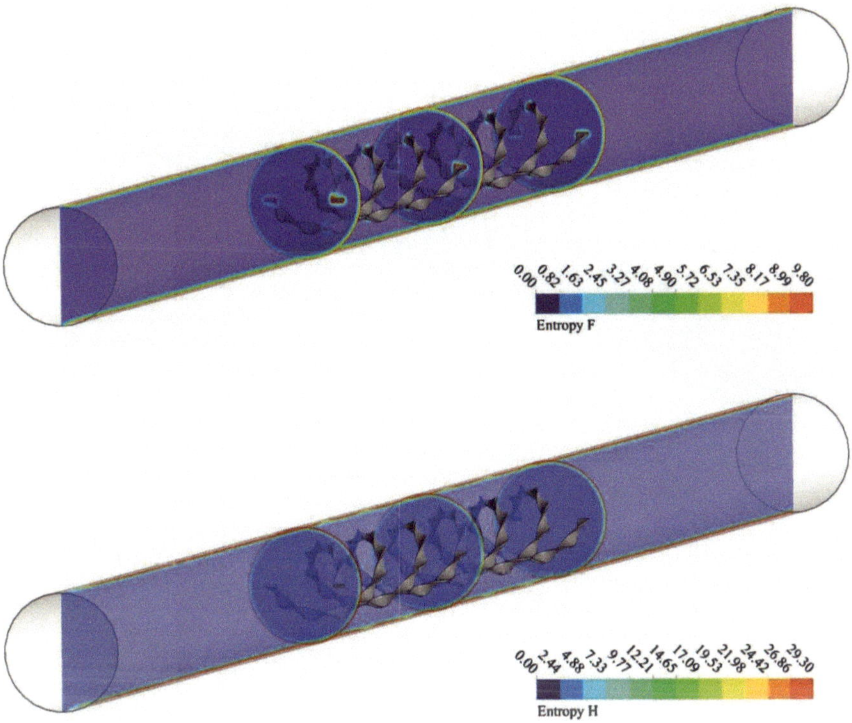

FIGURE 4.24 cont'd.

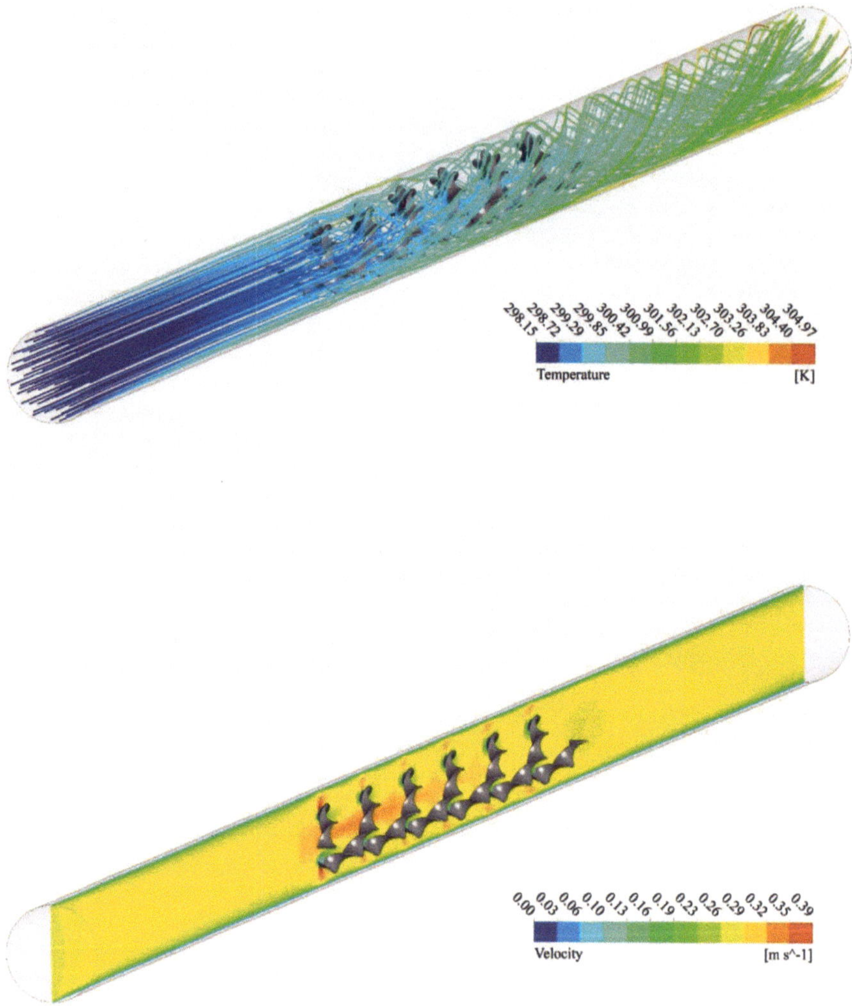

FIGURE 4.25

Velocity, T, S_{gen} contours for Re = 15,000, $b = 10$ mm.

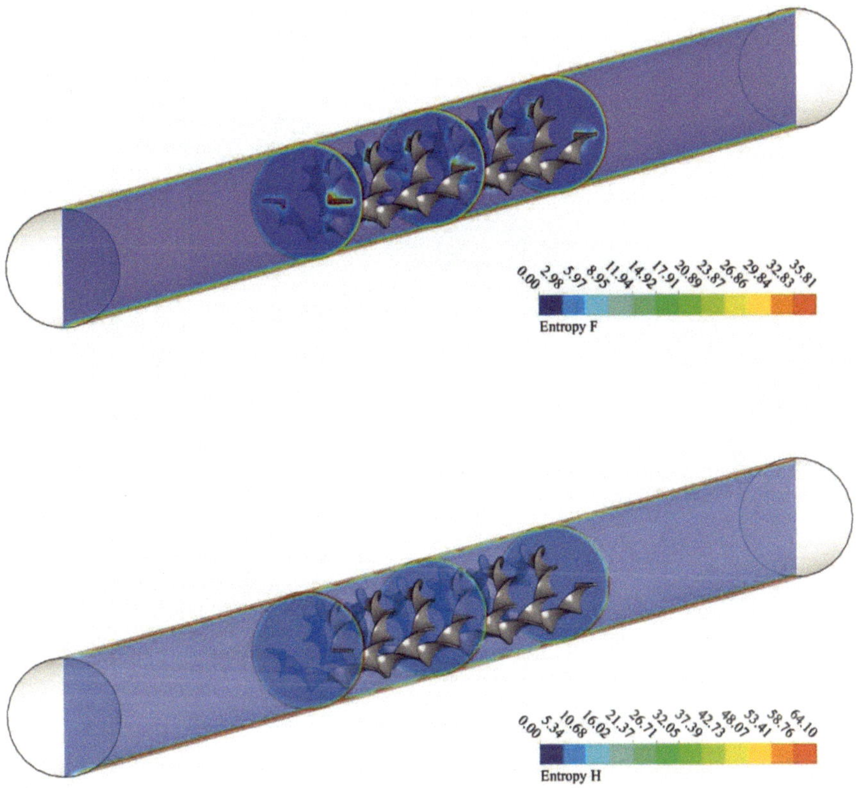

FIGURE 4.25 cont'd.

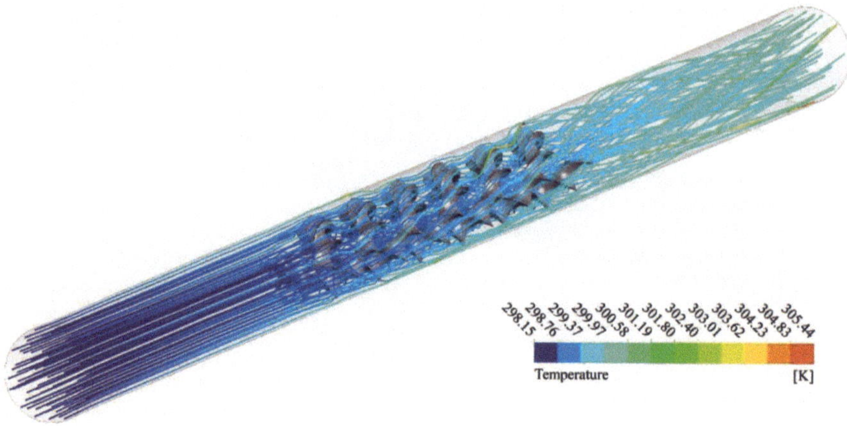

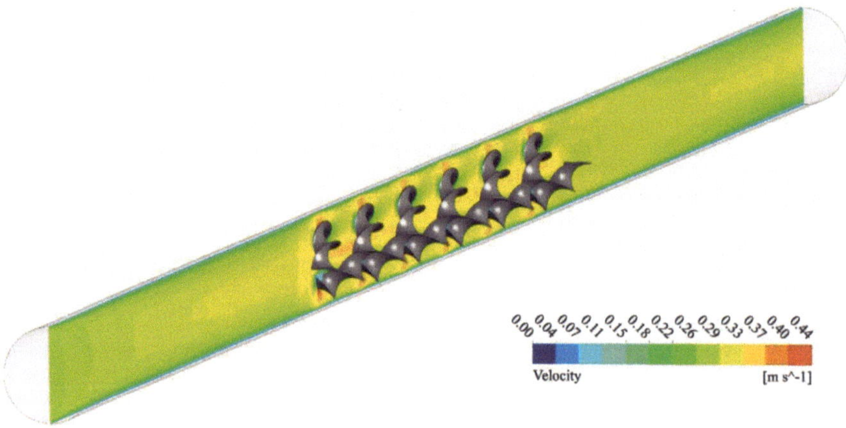

FIGURE 4.26

Velocity, T, S_{gen} contours for Re = 15,000, b = 15 mm.

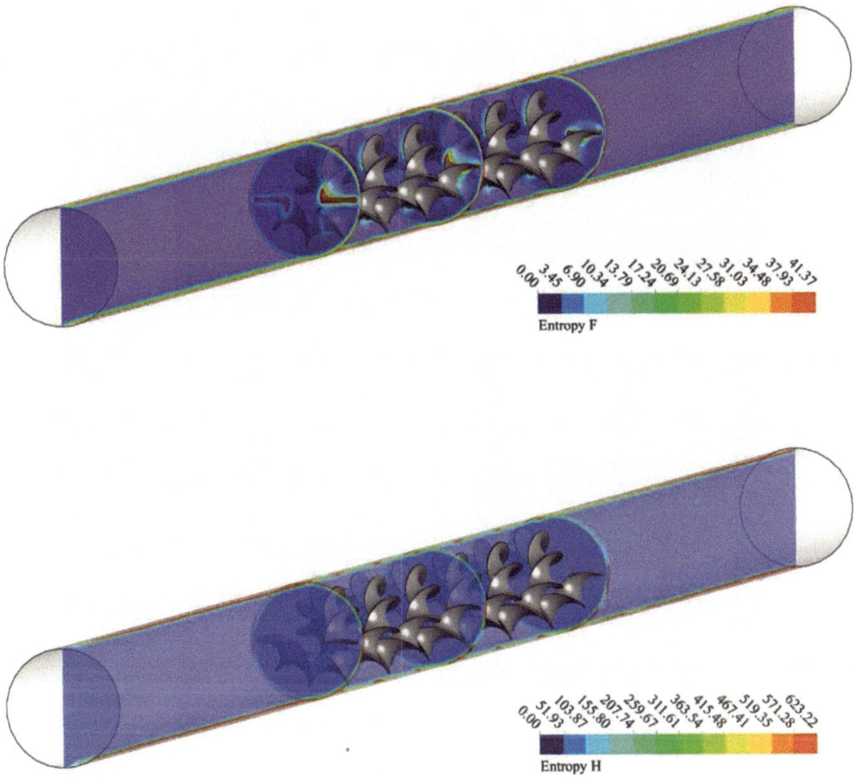

FIGURE 4.26 cont'd.

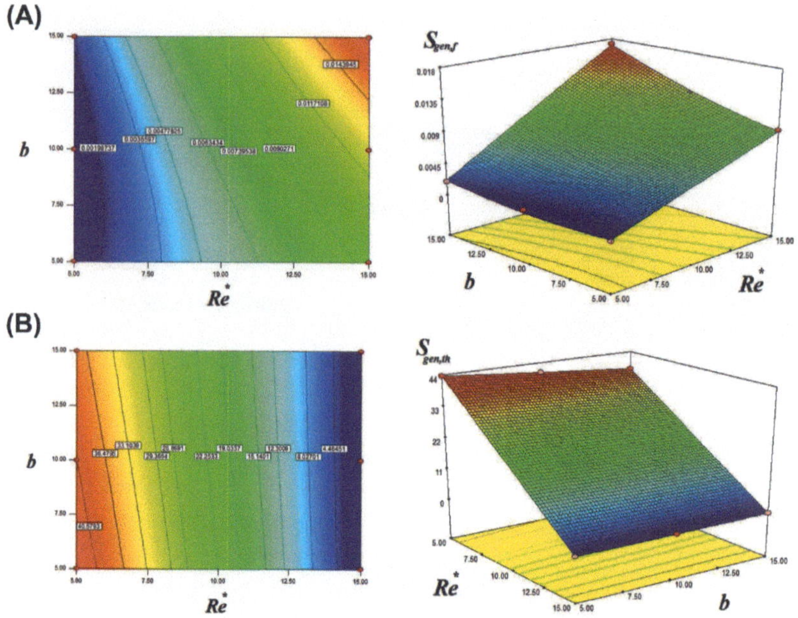

FIGURE 4.27

$S_{gen,f}$ and $S_{gen,th}$ for various Re and *b*.

References

[1] M. Sheikholeslami, M. Jafaryar, Z. Li, Second law analysis for nanofluid turbulent flow inside a circular duct in presence of twisted tape turbulators, Journal of Molecular Liquids 263 (2018) 489–500.

[2] M. Sheikholeslami, S.A. Shehzad, Z. Li, S. Ahmad, Numerical modeling for Alumina nanofluid magnetohydrodynamic convective heat transfer in a permeable medium using Darcy law, International Journal of Heat and Mass Transfer 127 (2018) 614–622.

[3] M. Sheikholeslami, M. Jafaryar, S. Ahmad, Z. Li, Nanofluid heat transfer and entropy generation through a heat exchanger considering a new turbulator and CuO nanoparticles, Journal of Thermal Analysis and Calorimetry 134 (3) (2018) 2295–2303, https://doi.org/10.1007/s10973-018-7866-7.

[4] Z. Li, M. Sheikholeslami, M. Jafaryar, S. Ahmad, A.J. Chamkha, Investigation of nanofluid entropy generation in a heat exchanger with helical twisted tapes, Journal of Molecular Liquids 266 (2018) 797–805.

[5] M. Sheikholeslami, M. Jafaryar, S. Ahmad, Z. Li, H. Rizwan-ul, Heat transfer of nanoparticles employing innovative turbulator considering entropy generation, International Journal of Heat and Mass Transfer 136 (2019) 1233–1240.

[6] M. Sheikholeslami, M.M. Bhatti, Forced convection of nanofluid in presence of constant magnetic field considering shape effects of nanoparticles, International Journal of Heat and Mass Transfer 111 (2017) 1039–1049.

[7] M. Sheikholeslami, M.M. Bhatti, Active method for nanofluid heat transfer enhancement by means of EHD, International Journal of Heat and Mass Transfer 109 (2017) 115–122.

[8] M. Sheikholeslami, S.A. Shehzad, Thermal radiation of ferrofluid in existence of Lorentz forces considering variable viscosity, International Journal of Heat and Mass Transfer 109 (2017) 82–92.

[9] M. Sheikholeslami, H.B. Rokni, Nanofluid two phase model analysis in existence of induced magnetic field, International Journal of Heat and Mass Transfer 107 (2017) 288–299.

[10] M. Sheikholeslami, S.A. Shehzad, Magnetohydrodynamic nanofluid convection in a porous enclosure considering heat flux boundary condition, International Journal of Heat and Mass Transfer 106 (2017) 1261–1269.

[11] M. Sheikholeslami, R. Ellahi, Three dimensional mesoscopic simulation of magnetic field effect on natural convection of nanofluid, International Journal of Heat and Mass Transfer 89 (2015) 799–808.

[12] Y. Qin, J. Liang, K. Tan, F. Li, A side by side comparison of the cooling effect of building blocks with retro-reflective and diffuse-reflective walls, Solar Energy 133 (2016) 172–179.

[13] Y. Qin, J. Liang, H. Yang, Z. Deng, Gas permeability of pervious concrete and its implications on the application of pervious pavements, Measurement 78 (2016) 104–110.

[14] M. Sheikholeslami, Numerical investigation of nanofluid free convection under the influence of electric field in a porous enclosure, Journal of Molecular Liquids 249 (2018) 1212–1221.

[15] M. Sheikholeslami, CuO-water nanofluid flow due to magnetic field inside a porous media considering Brownian motion, Journal of Molecular Liquids 249 (2018) 921–929.

[16] M. Sheikholeslami, Numerical investigation for CuO-H_2O nanofluid flow in a porous channel with magnetic field using mesoscopic method, Journal of Molecular Liquids 249 (2018) 739–746.

[17] M. Sheikholeslami, H.B. Rokni, Numerical modeling of nanofluid natural convection in a semi annulus in existence of Lorentz force, Computer Methods in Applied Mechanics and Engineering 317 (2017) 419–430.

[18] M. Sheikholeslami, A. Zeeshan, Analysis of flow and heat transfer in water based nanofluid due to magnetic field in a porous enclosure with constant heat flux using CVFEM, Computer Methods in Applied Mechanics and Engineering 320 (2017) 68–81.

[19] Y. Qin, J.E. Hiller, Understanding pavement-surface energy balance and its implications on cool pavement development, Energy and Buildings 85 (2014) 389–399.

[20] Y. Qin, M. Zhang, G. Mei, A new simplified method for measuring the permeability characteristics of highly porous media, Journal of Hydrology 562 (2018) 725–732.

[21] Y. Qin, Urban canyon albedo and its implication on the use of reflective cool pavements, Energy and Buildings 96 (2015) 86–94.

[22] M.M. Sheikholeslami, M.A. Sheremet, A. Shafee, I. Tlili, Simulation of nanoliquid thermogravitational convection within a porous chamber imposing magnetic and radiation impacts, Physica A: Statistical Mechanics and its Applications (2020), https://doi.org/10.1016/j.physa.2019.124058.

[23] M. Sheikholeslami, A.N. Keshteli, H. Babazadeh, Nanoparticles favorable effects on performance of thermal storage units, Journal of Molecular Liquids 300 (2020) 112329.

[24] M. Sheikholeslami, Effect of uniform suction on nanofluid flow and heat transfer over a cylinder, Journal of the Brazilian Society of Mechanical Sciences and Engineering 37 (2015) 1623–1633.
[25] M. Sheikholeslami, Lattice Boltzmann Method simulation of MHD non-Darcy nanofluid free convection, Physica B: Condensed Matter 516 (2017) 55–71.
[26] M. Sheikholeslami, Magnetic field influence on CuO -H_2O nanofluid convective flow in a permeable cavity considering various shapes for nanoparticles, International Journal of Hydrogen Energy 42 (2017) 19611–19621.
[27] M. Sheikholeslami, Numerical simulation for solidification in a LHTESS by means of Nano-enhanced PCM, Journal of the Taiwan Institute of Chemical Engineers 86 (2018) 25–41.
[28] M. Sheikholeslami, Numerical modeling of Nano enhanced PCM solidification in an enclosure with metallic fin, Journal of Molecular Liquids 259 (2018) 424–438.
[29] M. Sheikholeslami, Solidification of NEPCM under the effect of magnetic field in a porous thermal energy storage enclosure using CuO nanoparticles, Journal of Molecular Liquids 263 (2018) 303–315.
[30] M. Sheikholeslami, Influence of magnetic field on Al_2O_3-H_2O nanofluid forced convection heat transfer in a porous lid driven cavity with hot sphere obstacle by means of LBM, Journal of Molecular Liquids 263 (2018) 472–488.
[31] M. Sheikholeslami, New computational approach for exergy and entropy analysis of nanofluid under the impact of Lorentz force through a porous media, Computer Methods in Applied Mechanics and Engineering 344 (2019) 319–333.
[32] M. Sheikholeslami, Numerical approach for MHD Al_2O_3-water nanofluid transportation inside a permeable medium using innovative computer method, Computer Methods in Applied Mechanics and Engineering 344 (2019) 306–318.
[33] X. Ma, M. Sheikholeslami, M. Jafaryar, S. Ahmad, T. Nguyen-Thoi, Z. Li, Solidification inside a clean energy storage unit utilizing phase change material with copper oxide nanoparticles, Journal of Cleaner Production (2019), https://doi.org/10.1016/j.jclepro.2019.118888.
[34] M. Sheikholeslami, Omid Mahian, Enhancement of PCM solidification using inorganic nanoparticles and an external magnetic field with application in energy storage systems, Journal of Cleaner Production 215 (2019) 963–977.
[35] M. Sheikholeslami, B. Rezaeianjouybari, M. Darzi, S. Ahmad, Z. Li, T.K. Nguyen, Application of nano-refrigerant for boiling heat transfer enhancement employing an experimental study, International Journal of Heat and Mass Transfer 141 (2019) 974–980.
[36] M. Sheikholeslami, M. Jafaryar, M. Hedayat, S. Ahmad, Z. Li, T.K. Nguyen, M. Bakouri, Heat transfer and turbulent simulation of nanomaterial due to compound turbulator including irreversibility analysis, International Journal of Heat and Mass Transfer 137 (2019) 1290–1300.
[37] M. Sheikholeslami, Application of Darcy law for nanofluid flow in a porous cavity under the impact of Lorentz forces, Journal of Molecular Liquids 266 (2018) 495–503.
[38] M. Sheikholeslami, Finite element method for PCM solidification in existence of CuO nanoparticles, Journal of Molecular Liquids 265 (2018) 347–355.
[39] M. Sheikholeslami, R.-ul Haq, S. Ahmad, Z. Li, Y.G. Elaraki, I. Tlili, Heat transfer simulation of heat storage unit with nanoparticles and fins through a heat exchanger, International Journal of Heat and Mass Transfer 135 (2019) 470–478.

[40] M. Sheikholeslami, R.-ul Haq, S. Ahmad, Z. Li, Heat transfer behavior of Nanoparticle enhanced PCM solidification through an enclosure with V shaped fins, International Journal of Heat and Mass Transfer 130 (2019) 1322–1342.

[41] M. Sheikholeslami, Z. Li, S. Ahmad, Lorentz forces effect on NEPCM heat transfer during solidification in a porous energy storage system, International Journal of Heat and Mass Transfer 127 (2018) 665–674.

[42] M. Sheikholeslami, M. Jafaryar, S. Saleem, Z. Li, S. Ahmad, Y. Jiang, Nanofluid heat transfer augmentation and exergy loss inside a pipe equipped with innovative turbulators, International Journal of Heat and Mass Transfer 126 (2018) 156–163.

[43] M. Sheikholeslami, A. Ghasemi, Z. Li, S. Ahmad, S. Saleem, Influence of CuO nanoparticles on heat transfer behavior of PCM in solidification process considering radiative source term, International Journal of Heat and Mass Transfer 126 (2018) 1252–1264.

[44] M. Sheikholeslami, M. Darzi, Z. Li, Experimental investigation for entropy generation and exergy loss of nano-refrigerant condensation process, International Journal of Heat and Mass Transfer 125 (2018) 1087–1095.

[45] M. Sheikholeslami, S.A. Shehzad, Z. Li, Water based nanofluid free convection heat transfer in a three dimensional porous cavity with hot sphere obstacle in existence of Lorenz forces, International Journal of Heat and Mass Transfer 125 (2018) 375–386.

[46] M. Sheikholeslami, M. Jafaryar, Z. Li, Nanofluid turbulent convective flow in a circular duct with helical turbulators considering CuO nanoparticles, International Journal of Heat and Mass Transfer 124 (2018) 980–989.

[47] M. Sheikholeslami, A. Ghasemi, Solidification heat transfer of nanofluid in existence of thermal radiation by means of FEM, International Journal of Heat and Mass Transfer 123 (2018) 418–431.

[48] M. Sheikholeslami, S.A. Shehzad, CVFEM simulation for nanofluid migration in a porous medium using Darcy model, International Journal of Heat and Mass Transfer 122 (2018) 1264–1271.

[49] M. Sheikholeslami, M. Darzi, M.K. Sadoughi, Heat transfer improvement and pressure drop during condensation of refrigerant-based nanofluid; an experimental procedure, International Journal of Heat and Mass Transfer 122 (2018) 643–650.

[50] M. Sheikholeslami, S.A. Shehzad, Simulation of water based nanofluid convective flow inside a porous enclosure via Non-equilibrium model, International Journal of Heat and Mass Transfer 120 (2018) 1200–1212.

[51] M. Sheikholeslami, M. Seyednezhad, Simulation of nanofluid flow and natural convection in a porous media under the influence of electric field using CVFEM, International Journal of Heat and Mass Transfer 120 (2018) 772–781.

[52] M. Sheikholeslami, H.B. Rokni, Numerical simulation for impact of Coulomb force on nanofluid heat transfer in a porous enclosure in presence of thermal radiation, International Journal of Heat and Mass Transfer 118 (2018) 823–831.

[53] M. Sheikholeslami, S.A. Shehzad, Numerical analysis of Fe_3O_4 $-H_2O$ nanofluid flow in permeable media under the effect of external magnetic source, International Journal of Heat and Mass Transfer 118 (2018) 182–192.

[54] M. Sheikholeslami, M.K. Sadoughi, Simulation of CuO- water nanofluid heat transfer enhancement in presence of melting surface, International Journal of Heat and Mass Transfer 116 (2018) 909–919.

[55] M. Sheikholeslami, H.B. Rokni, Simulation of nanofluid heat transfer in presence of magnetic field: a review, International Journal of Heat and Mass Transfer 115 (2017) 1203–1233.
[56] A. Shafee, M. Sheikholeslami, M. Jafaryar, H. Babazadeh, Utilizing copper oxide nanoparticles for expedition of solidification within a storage system, Journal of Molecular Liquids 302 (2020) 112371, https://doi.org/10.1016/j.molliq.2019.112371.
[57] M. Sheikholeslami, S.A. Shehzad, CVFEM for influence of external magnetic source on $Fe_3O_4 - H_2O$ nanofluid behavior in a permeable cavity considering shape effect, International Journal of Heat and Mass Transfer 115 (2017) 180–191.
[58] M. Sheikholeslami, M. Seyednezhad, Nanofluid heat transfer in a permeable enclosure in presence of variable magnetic field by means of CVFEM, International Journal of Heat and Mass Transfer 114 (2017) 1169–1180.
[59] M. Sheikholeslami, B. Houman, Rokni, Melting heat transfer influence on nanofluid flow inside a cavity in existence of magnetic field, International Journal of Heat and Mass Transfer 114 (2017) 517–526.
[60] M. Sheikholeslami, M. Sadoughi, Mesoscopic method for MHD nanofluid flow inside a porous cavity considering various shapes of nanoparticles, International Journal of Heat and Mass Transfer 113 (2017) 106–114.

CHAPTER

Exergy behavior for heat exchanger in a pipe with modified turbulators

5

5.1 Nanofluid exergy loss within a pipe equipped with turbulators

5.1.1 Definition of the problem

In this chapter, a new turbolentor is utilized to improve thermal efficiency (Figs. 5.1 and 5.2). Three parts (inlet, outlet, and test sections) are taken into account. $CuO-H_2O$ nanofluid moves in z direction. In this chapter, $d = 5$ mm, $D = 20$ mm, and $L = 900$ mm. Turbulator inserts in middle part (test section) [1]. According to first and third parts, we can ensure about fully developed flow. Tables 5.1 and 5.2 provide the needed information of working fluid. Forced convection of nanofluid with high Reynolds number must be presented as

$$\frac{\partial(u_i)}{\partial x_i} = 0 \tag{5.1}$$

$$\frac{\partial}{\partial x_j}(u_j \rho_{nf} u_i) = -\frac{\partial p}{\partial x_i} + \frac{\partial}{\partial x_j}\left(\mu_{nf}\left(\frac{\partial u_i}{\partial x_j} + \frac{\partial u_j}{\partial x_i}\right)\right) + \frac{\partial}{\partial x_j}\left(-\rho_{nf}\overline{u_j' u_i'}\right) \tag{5.2}$$

$$\frac{\partial}{\partial x_i}(\rho_{nf} T u_i) = \frac{\partial}{\partial x_i}\left((\mu_t / Pr_t + \mu_{nf} / Pr_{nf})\frac{\partial T}{\partial x_i}\right), \tag{5.3}$$

$\rho_{nf}\overline{u_j' u_i'}$ and μ_t are:

$$\mu_t\left(\frac{\partial u_i}{\partial x_j} + \frac{\partial u_j}{\partial x_i}\right) - \frac{2}{3}\rho_{nf} k \delta_{ij} - \frac{2}{3}\mu_t \frac{\partial u_k}{\partial x_k}\delta_{ij} = -\overline{u_j' u_i'}\rho_{nf} \tag{5.4}$$

$$\mu_t = \rho_{nf} C_\mu k^2 / \varepsilon \tag{5.5}$$

Formals of k, ε are

Nanofluid in Heat Exchangers for Mechanical Systems. https://doi.org/10.1016/B978-0-12-821923-2.00005-1

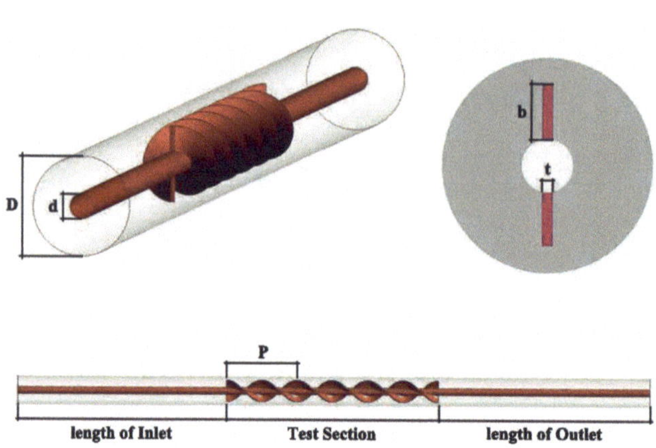

FIGURE 5.1

Geometry of the problem.

$$\frac{\partial}{\partial x_j}\left(\left(\frac{\mu_t}{\sigma_k}+\mu_{nf}\right)\frac{\partial k}{\partial x_j}\right)-\rho_{nf}\varepsilon+G_k=\frac{\partial}{\partial x_i}(u_i\rho_{nf}k), G_k=-\frac{\partial u_j}{\partial x_i}\rho_{nf}\overline{u'_j u'_i} \quad (5.6)$$

$$\frac{\partial}{\partial x_i}(u_i\rho_{nf}\varepsilon)=\frac{\partial}{\partial x_j}\left(\left(\frac{\mu_t}{\sigma_\varepsilon}+\mu_{nf}\right)\frac{\partial \varepsilon}{\partial x_j}\right)+\frac{\varepsilon}{k}G_k C_{1\varepsilon}-\rho_{nf}\frac{\varepsilon^2}{k}C_{2\varepsilon} \quad (5.7)$$

$$C_{1\varepsilon}=1.42, C_\mu=0.0845, C_{2\varepsilon}=1.68, Pr_t=0.85, \sigma_k=1, \sigma_\varepsilon=1.3 \quad (5.8)$$

The software which we used is ANSYS Fluent 14. Our selected turbulent model is k−ε (RNG). More details were mentioned in previous work [2].

ρ_{nf}, $(\rho C_p)_{nf}$, k_{nf}, and μ_{nf} are [3]

$$\rho_{nf}=\rho_f(1-\phi)+\rho_s\phi \quad (5.9)$$

$$(\rho C_p)_{nf}=-(\phi-1)(\rho C_p)_f+(\rho C_p)_s\phi \quad (5.10)$$

$$\frac{k_{nf}}{k_f}=1+3\frac{\phi(k_p/k_f-1)}{-\phi(k_p/k_f-1)+(2+k_p/k_f)}+5\times10^4 g'(\phi,T,d_p)\phi\rho_f c_{p,f}\sqrt{\frac{\kappa_b T}{\rho_p d_p}}$$

$$g'(\phi,T,d_p)=\Big(a_1+a_2Ln(d_p)+a_5Ln(d_p)^2+a_4Ln(d_p)Ln(\phi)+a_3Ln(\phi)\Big)Ln(T)$$
$$+\Big(a_6+a_7Ln(d_p)+a_{10}Ln(d_p)^2+a_8Ln(\phi)+a_9Ln(d_p)Ln(\phi)\Big),\frac{4\times10^{-8}}{d_p}$$
$$=-1/k_p+1/k_{p,eff}$$

(5.11)

$$\frac{\mu_{nf}}{\mu_f}=(1-\phi)^{-2.5}+k_{Brownian}(\mathrm{Pr}k_f)^{-1} \quad (5.12)$$

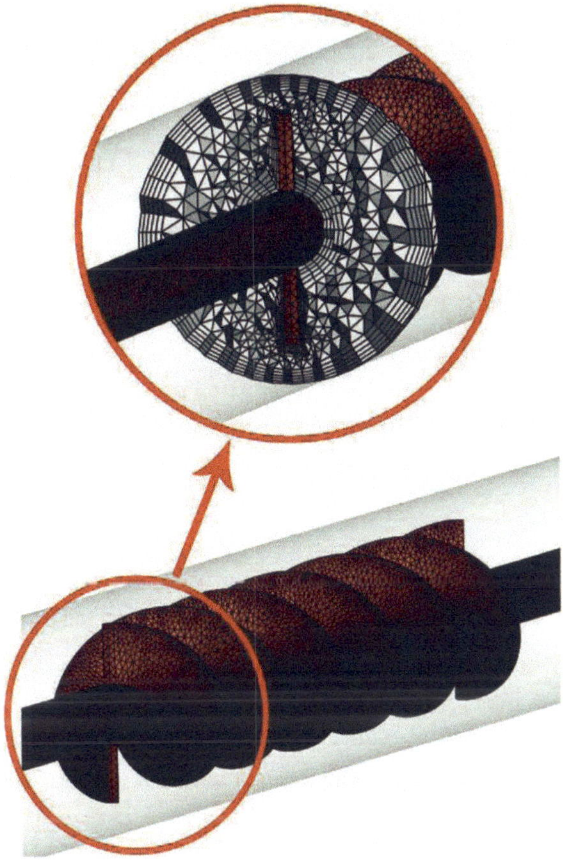

FIGURE 5.2

Sample of mesh.

Table 5.1 The coefficient values of *CuO* – Water nanofluid.

Coefficient values	CuO – Water
a_1	−26.5933108
a_2	−0.403818333
a_3	−33.3516805
a_4	−1.915825591
a_5	6.421858E-02
a_6	48.40336955
a_7	−9.787756683
a_8	190.245610009
a_9	10.9285386565
a_{10}	−0.72009983664

Table 5.2 Thermophysical properties of water and nanoparticles.

	ρ(kg/m³)	Cp(j/kgk)	k(W/m.k)	μ(Pa.s)
Pure water	997.1	4179	0.613	0.0010003
CuO	6500	540	18	–

At z = 0 and z = 900 mm, we have

$$I = 0.16(\mathrm{Re})^{\frac{-1}{8}}, u_i = 0, v_i = 0, T_i = cte, w_i = cte, \tag{5.13}$$

$$\frac{\partial \mathrm{v}}{\partial \mathrm{z}} = \frac{\partial \mathrm{u}}{\partial \mathrm{z}} = \frac{\partial \mathrm{w}}{\partial \mathrm{z}} = 0, \frac{\partial \mathrm{T}}{\partial \mathrm{z}} = 0 \tag{5.14}$$

Second law efficiency and exergy drop can be calculated as

$$\eta_{II} = 1 - \frac{X_d}{X_q} \tag{5.15}$$

$$\begin{aligned} X_q &= A\left(1 - \frac{T_0}{T_s}\right)q'' \\ X_d &= \dot{m}C_p(T_{in} - T_{out}) + \dot{m}T_0C_p \ln(T_{out}/T_{in}) + X_q \end{aligned} \tag{5.16}$$

5.1.2 Effects of active parameters

In this chapter, exergy drop and second law efficiency for turbulent flow of nanofluid are studied. Swirling device is used as passive technique. Properties of working fluid are estimated by considering Brownian motion effect.

Fig. 5.3 illustrates the temperature and velocity contours. Rotating flow becomes stronger for lower values of pitch ratio. Also, similar trend is reported for height ratio and Reynolds number. Pressure loss augments with increase of Reynolds number. Besides, thermal boundary layer thickness detracts with augment of Re. Higher height ratio and lower pitch ratio make temperature gradient to enhance. Temperature gradient improves with augment of height ratio. This is because of better fluid mixing and greater turbulent intensity.

Two formulas can be offered according to data:

$$\begin{aligned} X_d = {} & 5.17 + 0.51PR - 0.39BR - 1.76\mathrm{Re}^* + 0.28(PR)(BR) \\ & + 0.86\mathrm{Re}^{*2} + 0.037\mathrm{Re}^*(BR) - 0.19(PR)(\mathrm{Re}^*) \end{aligned} \tag{5.17}$$

$$\begin{aligned} \eta_{II} = {} & 0.84 - 0.014PR + 0.012BR + 0.04\mathrm{Re}^* - 5.6 \times 10^{-4}(BR)\mathrm{Re}^* - 0.015\mathrm{Re}^{*2} \\ & - 9.5 \times 10^{-3}(PR)(BR) + 3.84 \times 10^{-3}(PR)(\mathrm{Re}^*) \end{aligned} \tag{5.18}$$

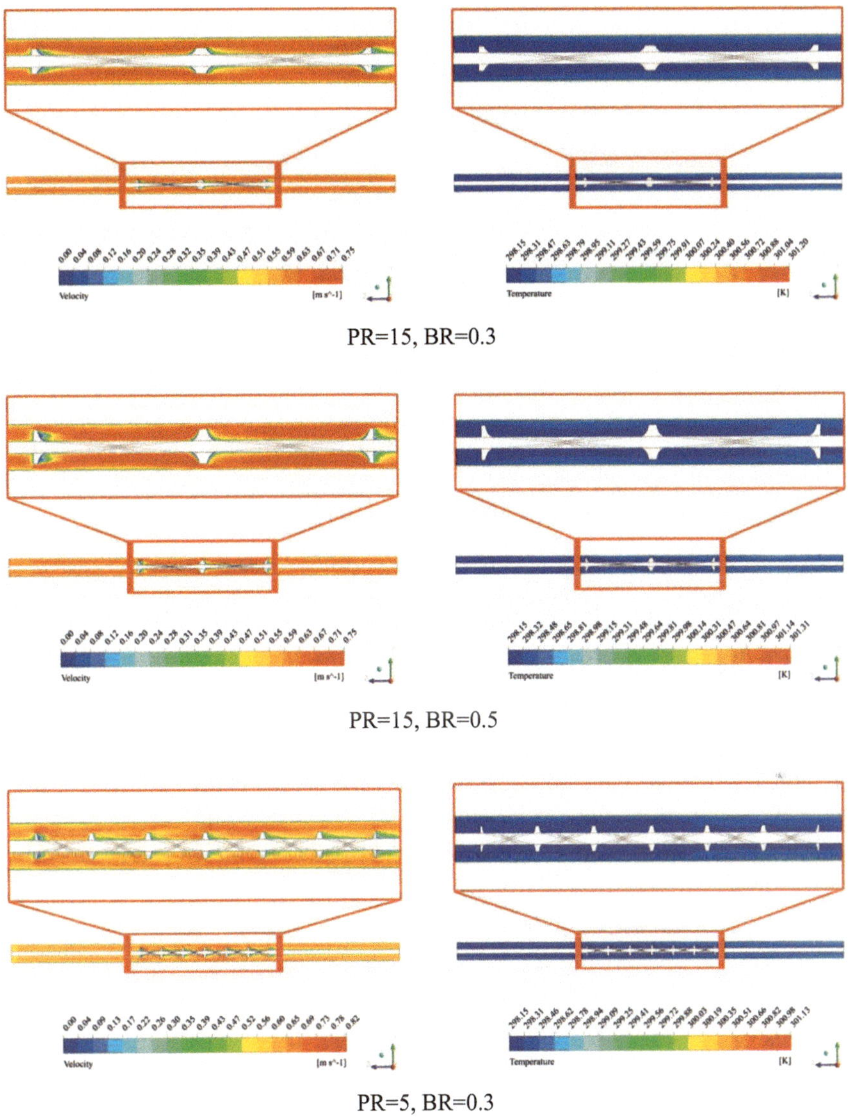

FIGURE 5.3

Velocity (right) and temperature (left) contours when Re = 20,000.

where $\text{Re}^* = \text{Re}/1000$. Figs. 5.4 and 5.5 depict the effects of *PR*, Re, and *BR* on exergy drop and second law efficiency. X_d is a reducing function of inlet velocity and height ratio, while it enhances with rise of *PR*. Effects of active parameters on second law efficiency are opposite of exergy reduction. Second law efficiency

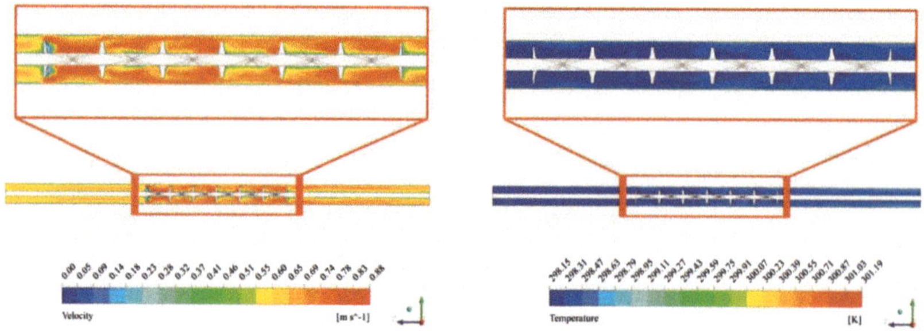

PR=5, BR=0.5

FIGURE 5.3 cont'd.

enhances with rise of height ratio and Reynolds number. Because the secondary vortexes becomes stronger with augment of these parameters and in turn more reduction in thickness of thermal boundary layer can be found.

5.2 Exergy and hydrothermal behavior of nanofluid through a tube using passive methods

5.2.1 Definition of the problem

Helical swirling device has been used to enhance heat transfer of nanofluid (see Fig. 5.6). To eliminate back flow, test section is considered in the middle part. Total length of the pipe is 900 mm and diameters are 20 and 5 mm [3].

Total entropy generation, Bejan number, and exergy loss are

$$S_{gen,total} = \underbrace{\frac{1}{T^2}k_{nf}\left[\left(\frac{\partial T}{\partial z}\right)^2 + \left(\frac{\partial T}{\partial x}\right)^2 + \left(\frac{\partial T}{\partial y}\right)^2\right]}_{S_{gen,th}} + \frac{\mu_{nf}}{T}\underbrace{\left\{\begin{array}{c}\left(\frac{\partial v_y}{\partial x} + \frac{\partial v_x}{\partial y}\right)^2 + 2\left[\left(\frac{\partial v_z}{\partial z}\right)^2 + \left(\frac{\partial v_x}{\partial x}\right)^2 + \left(\frac{\partial v_y}{\partial y}\right)^2\right] \\ + \left(\frac{\partial v_z}{\partial y} + \frac{\partial v_y}{\partial z}\right)^2 + \left(\frac{\partial v_z}{\partial x} + \frac{\partial v_x}{\partial z}\right)^2\end{array}\right\}}_{S_{gen,f}} \tag{5.19}$$

$$Be = \frac{S_{gen,th}}{S_{gen,total}}, \tag{5.20}$$

$$X_d = \left(S_{gen,f} + S_{gen,th}\right)T_0 \tag{5.21}$$

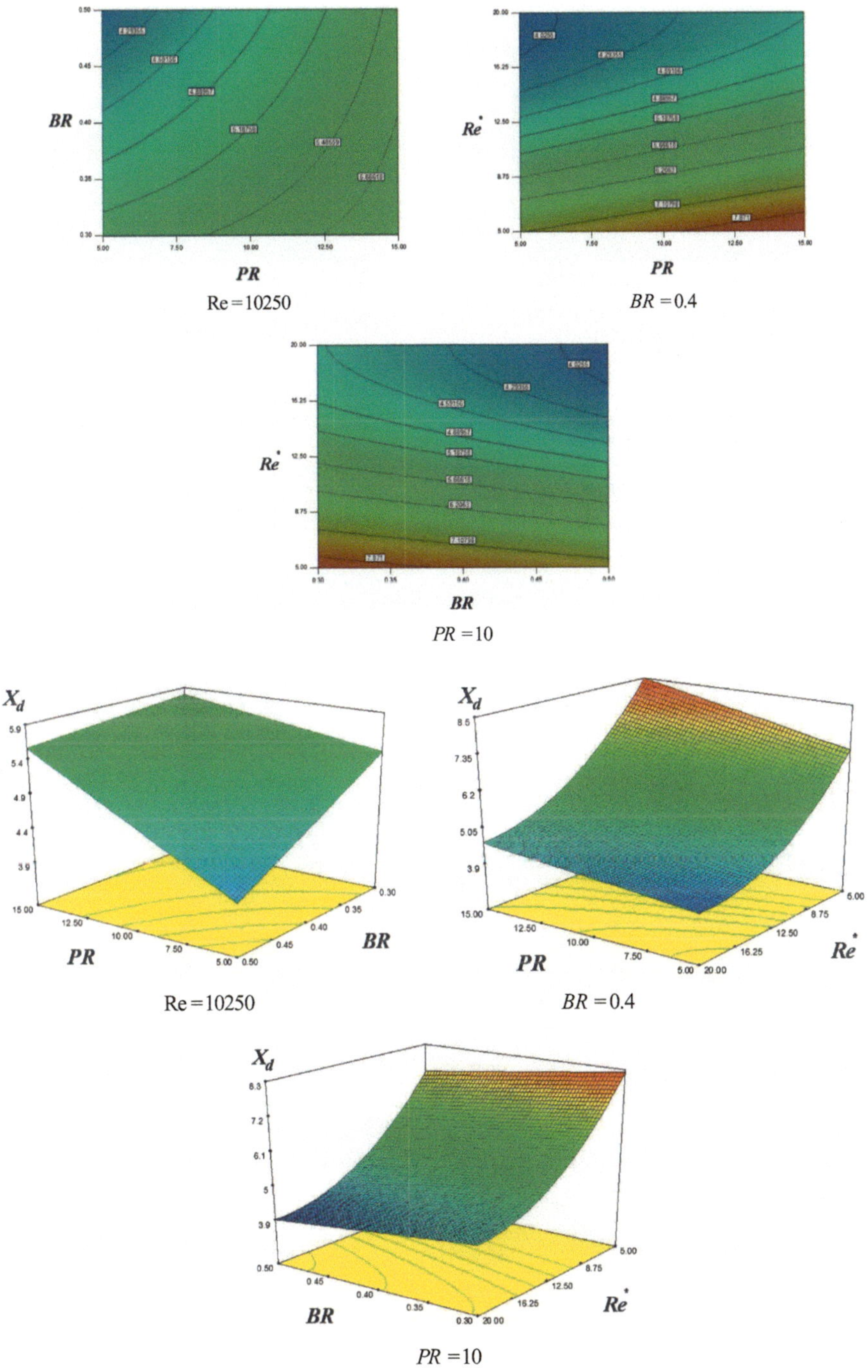

FIGURE 5.4

Effect of Re, *PR*, and *BR* on exergy destruction.

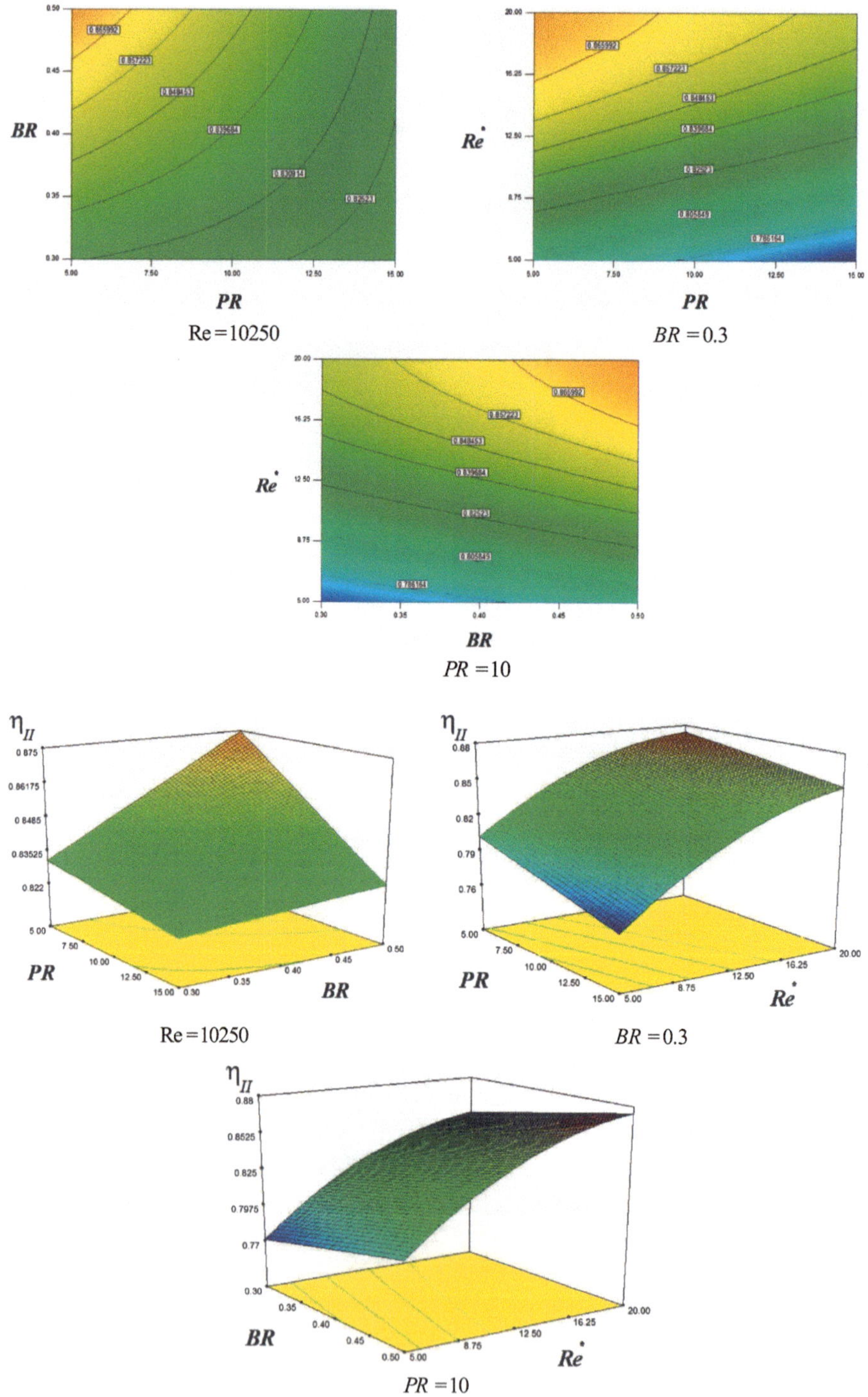

FIGURE 5.5

Effect of Re, *PR*, and *BR* on second law efficiency.

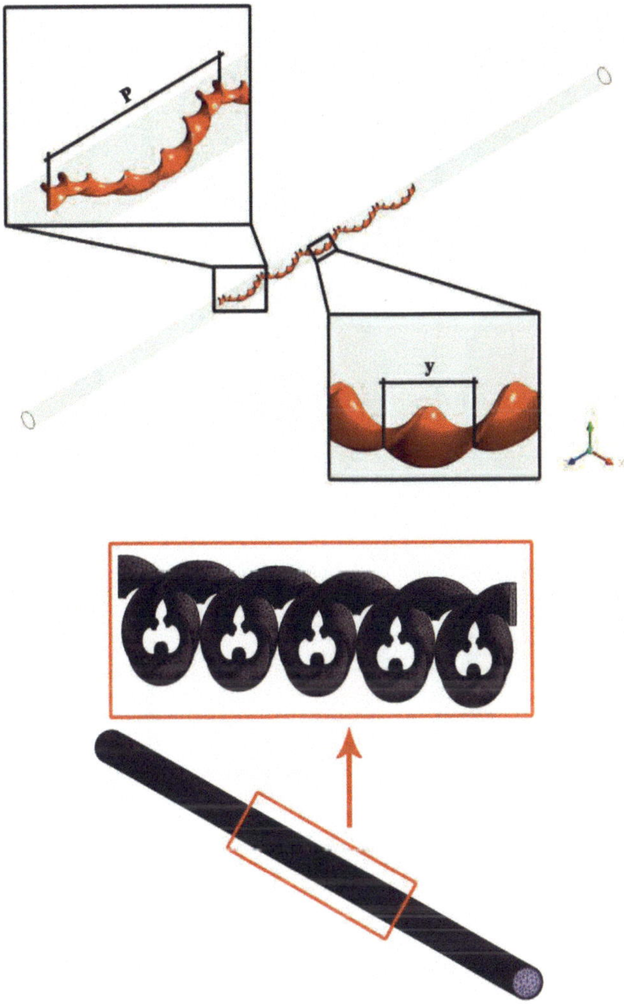

FIGURE 5.6

Current heat exchanger and grid.

5.2.2 Effects of active parameters

In this chapter, nanofluid migration and exergy loss through a circular tube with helical swirling device are portrayed by means of finite volume method. Properties of working fluid are calculated considering homogenous model. Outputs are shown as exergy loss, temperature, velocity, and streamline contours for various cases.

Fig. 5.7 illustrates the streamline and velocity contours in three sections (at start, middle, and end of test sections). Velocity augments when going along the

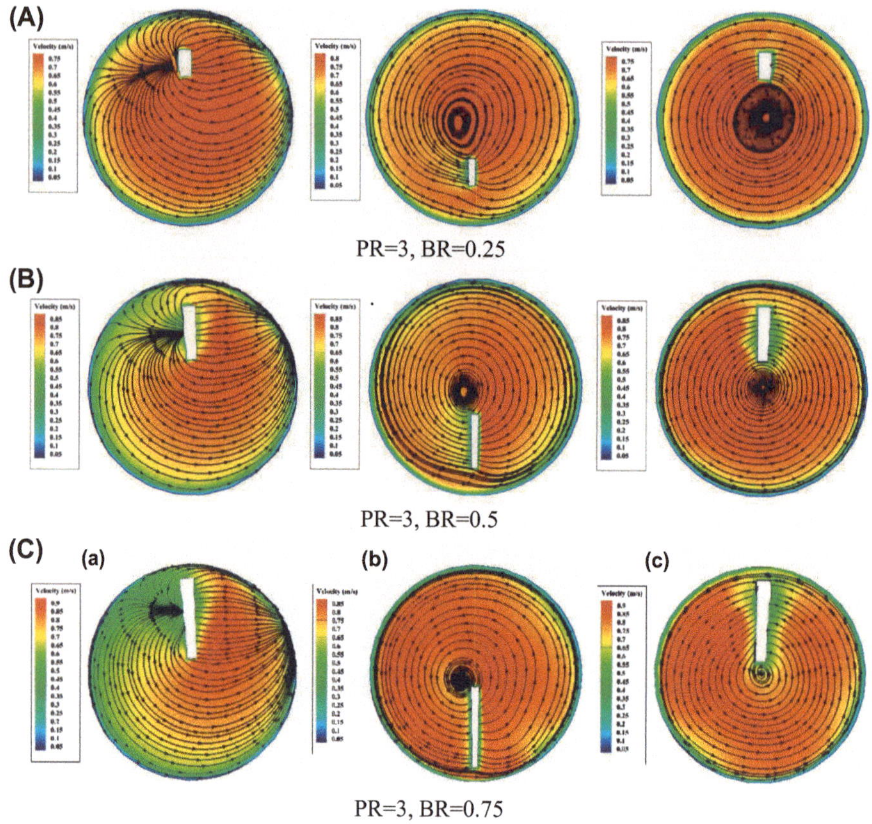

FIGURE 5.7

Velocity and streamline contours at Re = 15,000. (A) *PR*=3, *BR*=0.25; (B) *PR*=3, *BR*=0.5; (C) *PR*=3, *BR*=0.75; (D) *PR*=1.5, *BR*=0.25; (E) *PR*=1.5, *BR*=0.5; (F) *PR*=1.5, *BR*=0.75. (a) start of test section; (b) middle of test section; (c) end of test section.

helical twisted tape and velocity magnitude enhances at third section. As *PR* reduces, working fluid rotates more and velocity enhances. Also, similar impact is shown for height ratio. Therefore, swirling flow enhances with rise of *BR*. Although rate of heat transfer augments in higher *BR* and lower *PR*, pressure loss shows the same behavior.

Fig. 5.8 demonstrates the temperature contours. As it progresses along the pipe, temperature of nanofluid enhances due to more interaction with hot surface. Higher temperature gradient can be seen with rise of *BR*. Decreasing *PR* leads to enhancement in convective flow due to greater secondary flows. Exergy loss contours are illustrated in Fig. 5.9. As convective flow enhances, exergy loss reduces. So,

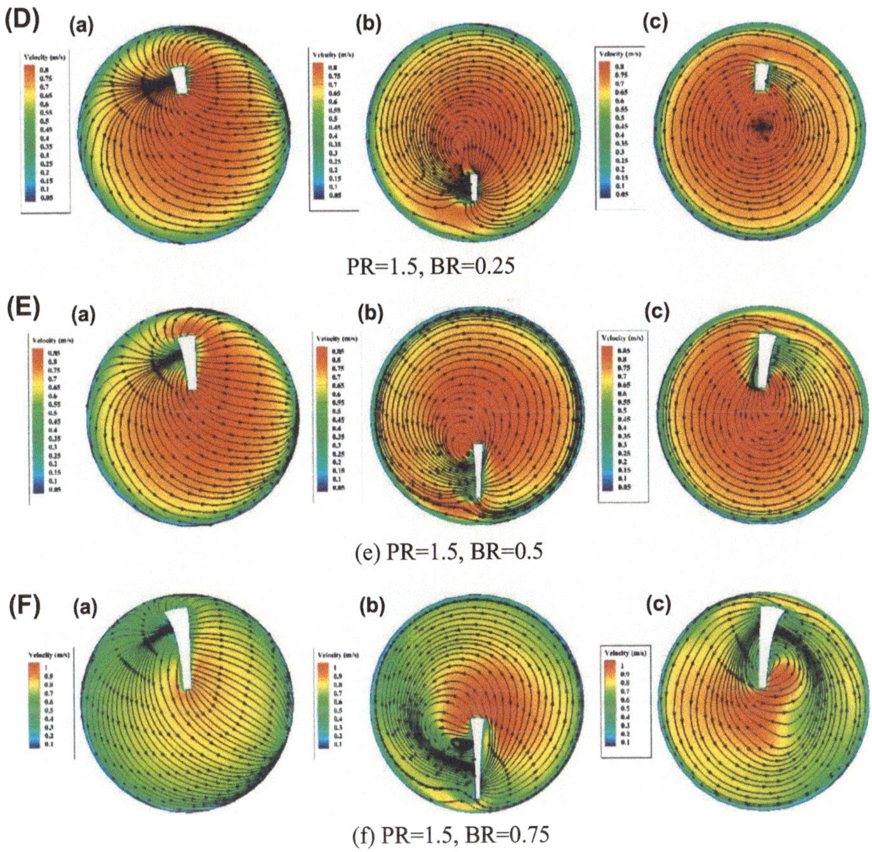

FIGURE 5.7 cont'd.

decreasing *PR* leads to reduction in X_d. As height of turbolentor augments, swirling flow improves and exergy loss reduces.

Following formulas can be considered for *Be* and X_d:

$$Be = 0.86 + 0.041PR - 0.069BR - 0.14\text{Re}^* + 0.038(\text{Re}^*)(PR) - (\text{Re}^*)(0.066)(BR) + (BR)(0.02)(PR) \quad (5.22)$$

$$X_d = 10120.3 + 4544.2PR - 411.67BR - 7863.8\text{Re}^* - 142.5(PR)(BR) - 3876.16(PR)(\text{Re}^*) + 280.34(BR)(\text{Re}^*) \quad (5.23)$$

where $\text{Re}^* = \text{Re}/10000$. Influences of BR, PR, and Re on Bejan number and exergy reduction are depicted in Figs. 5.10 and 5.11. Bejan number is influenced by total entropy generation, and similar behavior is found for exergy loss.

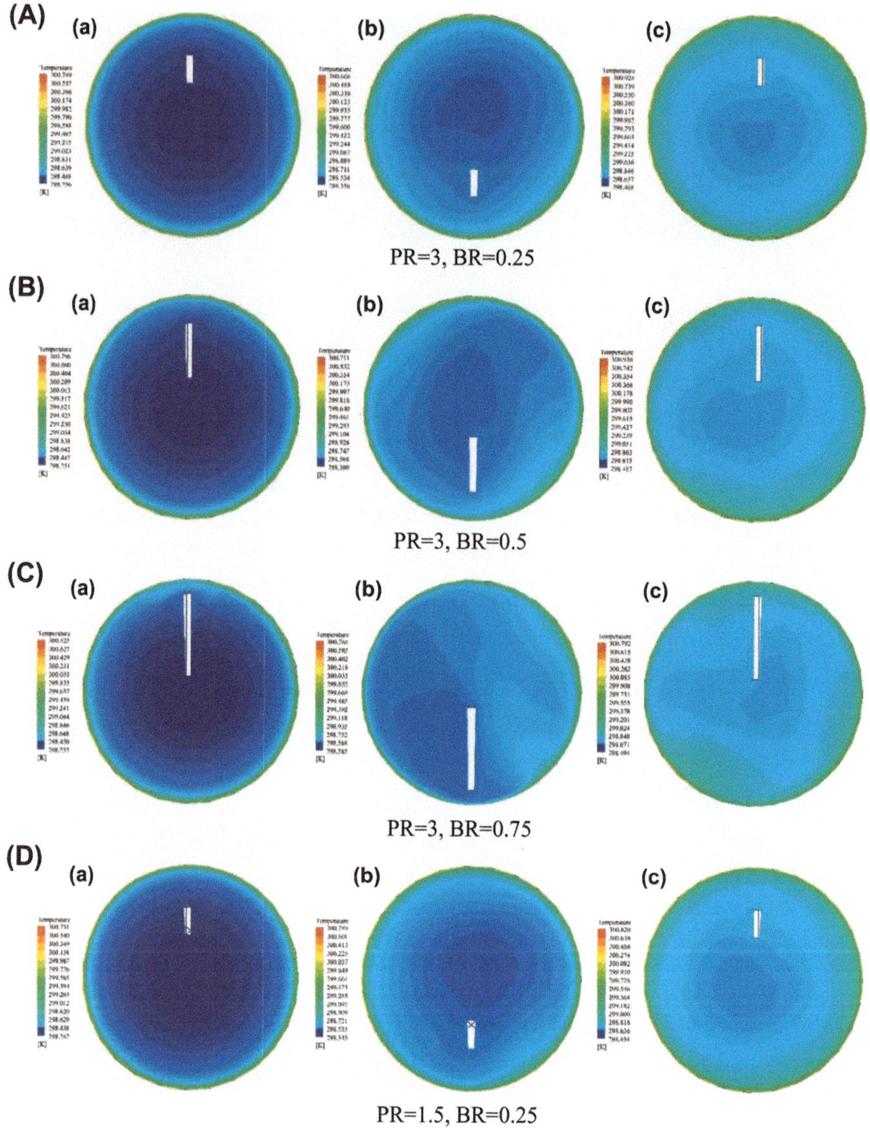

FIGURE 5.8

Temperature contours at Re = 15,000. (A) *PR*=3, *BR*=0.25; (B) *PR*=3, *BR*=0.5; (C) *PR*=3, *BR*=0.75; (D) *PR*=1.5, *BR*=0.25; (E) *PR*=1.5, *BR*=0.5; (F) *PR*=1.5, *BR*=0.75. (a) start of test section; (b) middle of test section; (c) end of test section.

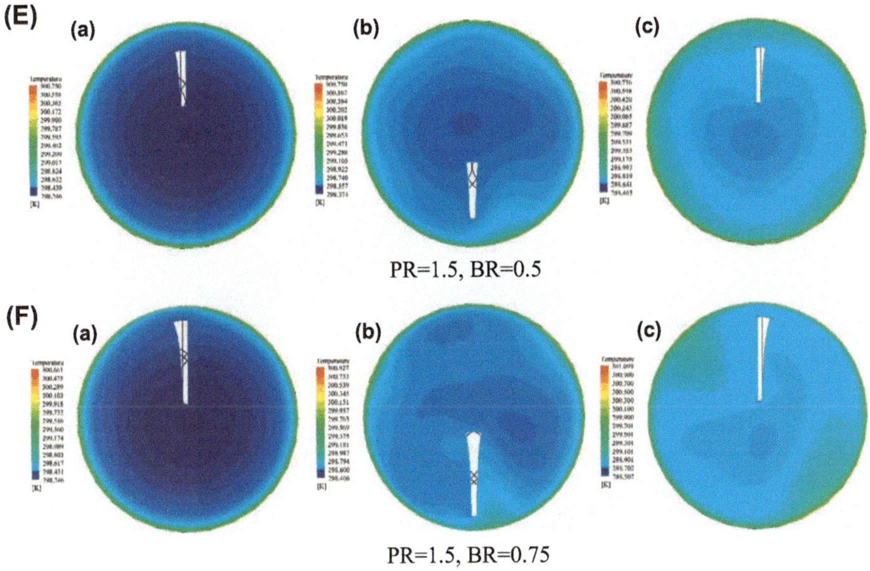

FIGURE 5.8 cont'd.

As inlet velocity augments, X_d reduces. Exergy loss has reverse relationship with *BR*. Increasing number of revolutions can reduce the Bejan number. Influence of *BR* can be neglected in comparison with effects of two other parameters.

5.3 Numerical simulation for turbulent flow in a tube with combined swirl flow device

5.3.1 Definition of the problem

The geometry of turbulator and employed boundary conditions have been illustrated in Fig. 5.12. We selected a pipe with diameter of 2 inch which has 1.5 m length. Constant geometric parameters were mentioned in this graph, too. In addition, sample grid was depicted which indicates the denser mesh near the solid walls to satisfy the limitation of turbulent models. Mixture of water and copper oxide is taken into account as the testing fluid, which is assumed to be homogeneous fluid. All detail of properties of current nanomaterial was mentioned in Ref. [2]. The Reynolds number (Re) and widths of turbulator (b) vary from 5000 to 15,000 and 5–15 mm, respectively. Steady, laminar turbulent flow in three dimensional has been assumed. Free convection is ignored. We proposed k−ε model in this simulation. This model is very accurate in appearance of turbulators. No slip conditions have been imposed on all walls, and constant heat flux was concerned for outer wall.

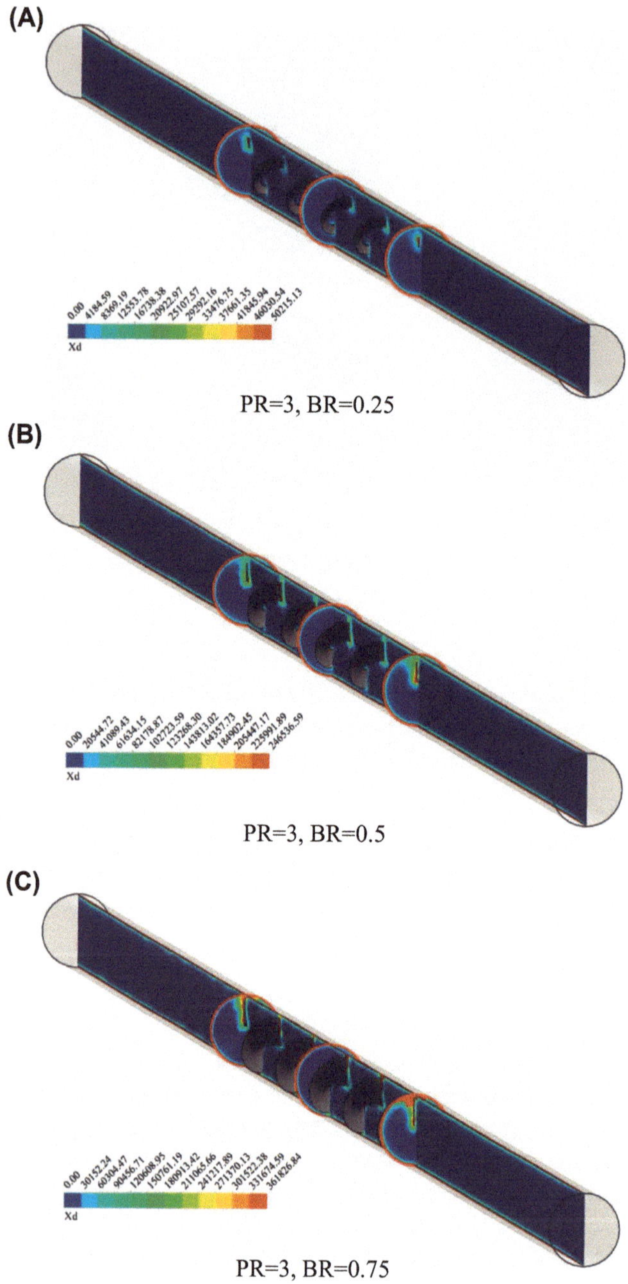

FIGURE 5.9

Exergy loss contours at Re = 15,000. (A) *PR*=3, *BR*=0.25; (B) *PR*=3, *BR*=0.5; (C) *PR*=3, *BR*=0.75; (D) *PR*=1.5, *BR*=0.25; (E) *PR*=1.5, *BR*=0.5; (F) *PR*=1.5, *BR*=0.75.

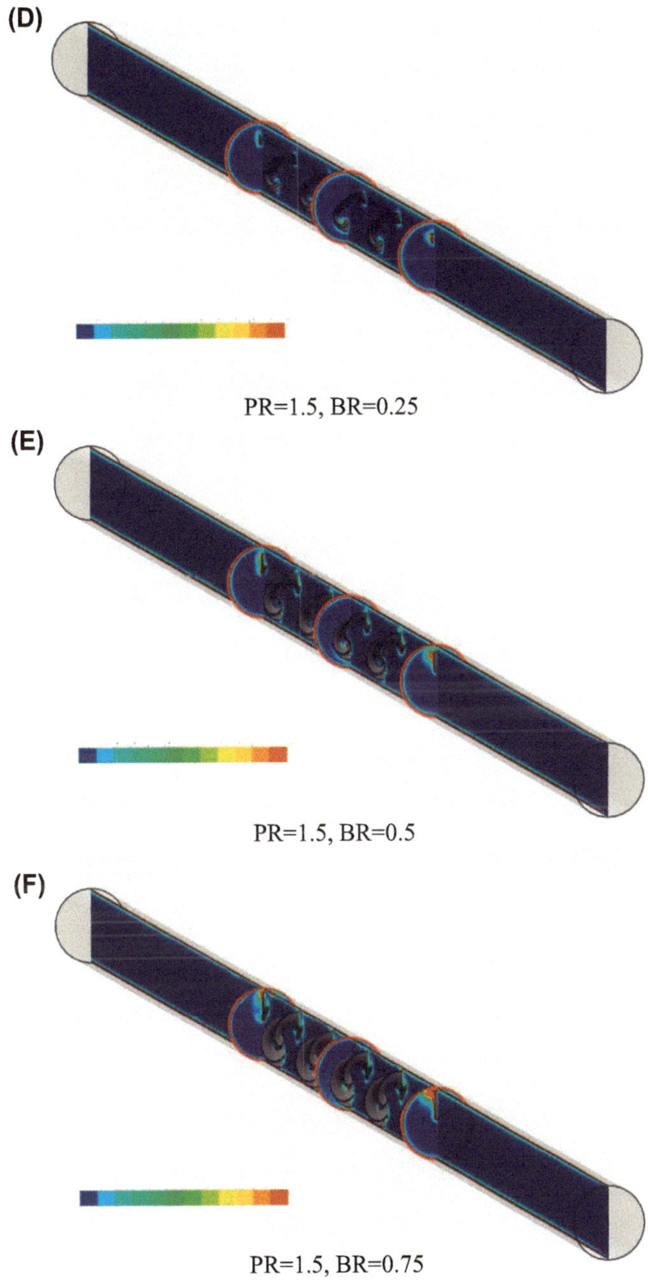

FIGURE 5.9 cont'd.

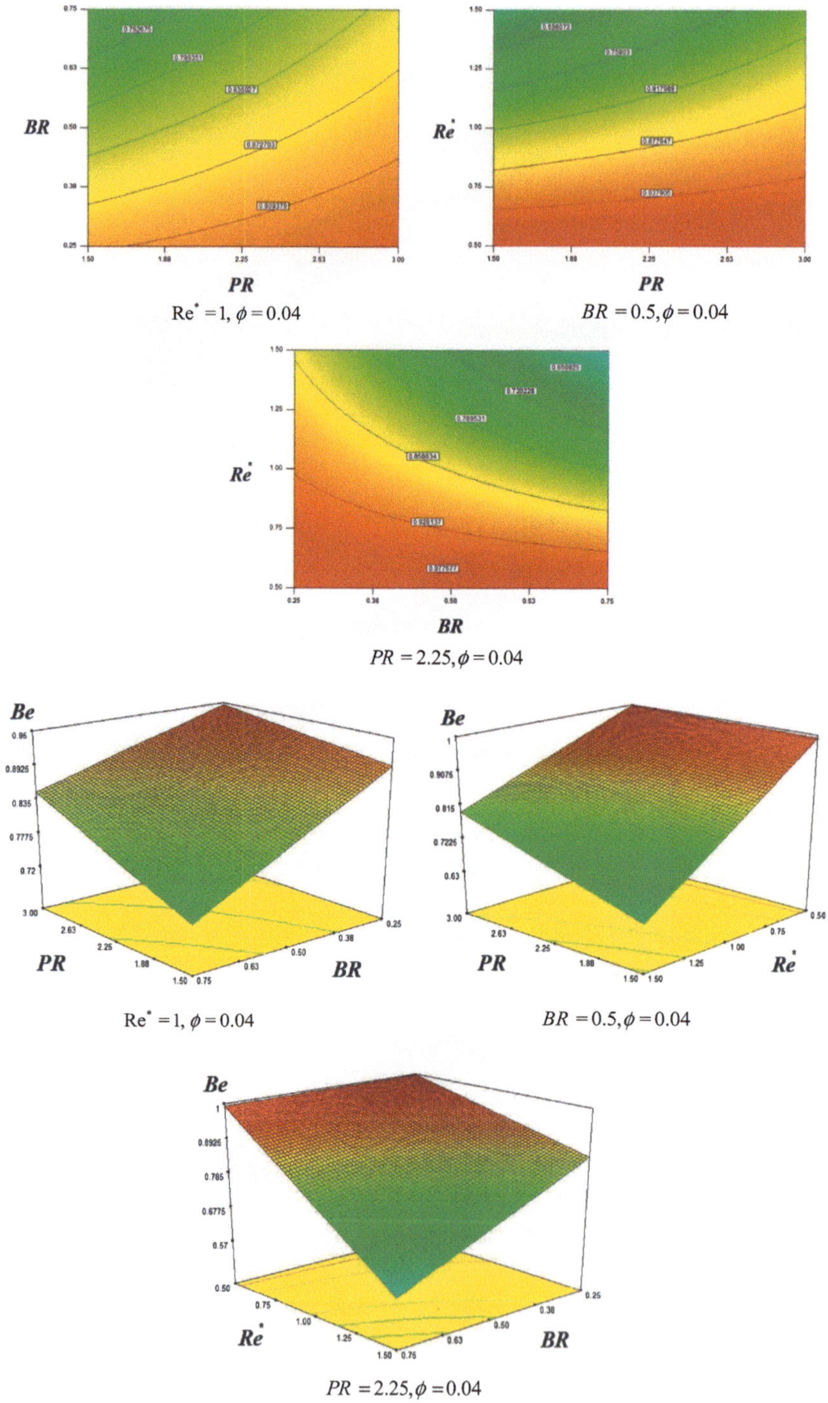

FIGURE 5.10

Impacts of Re*, *PR*, and *BR* on *Be*.

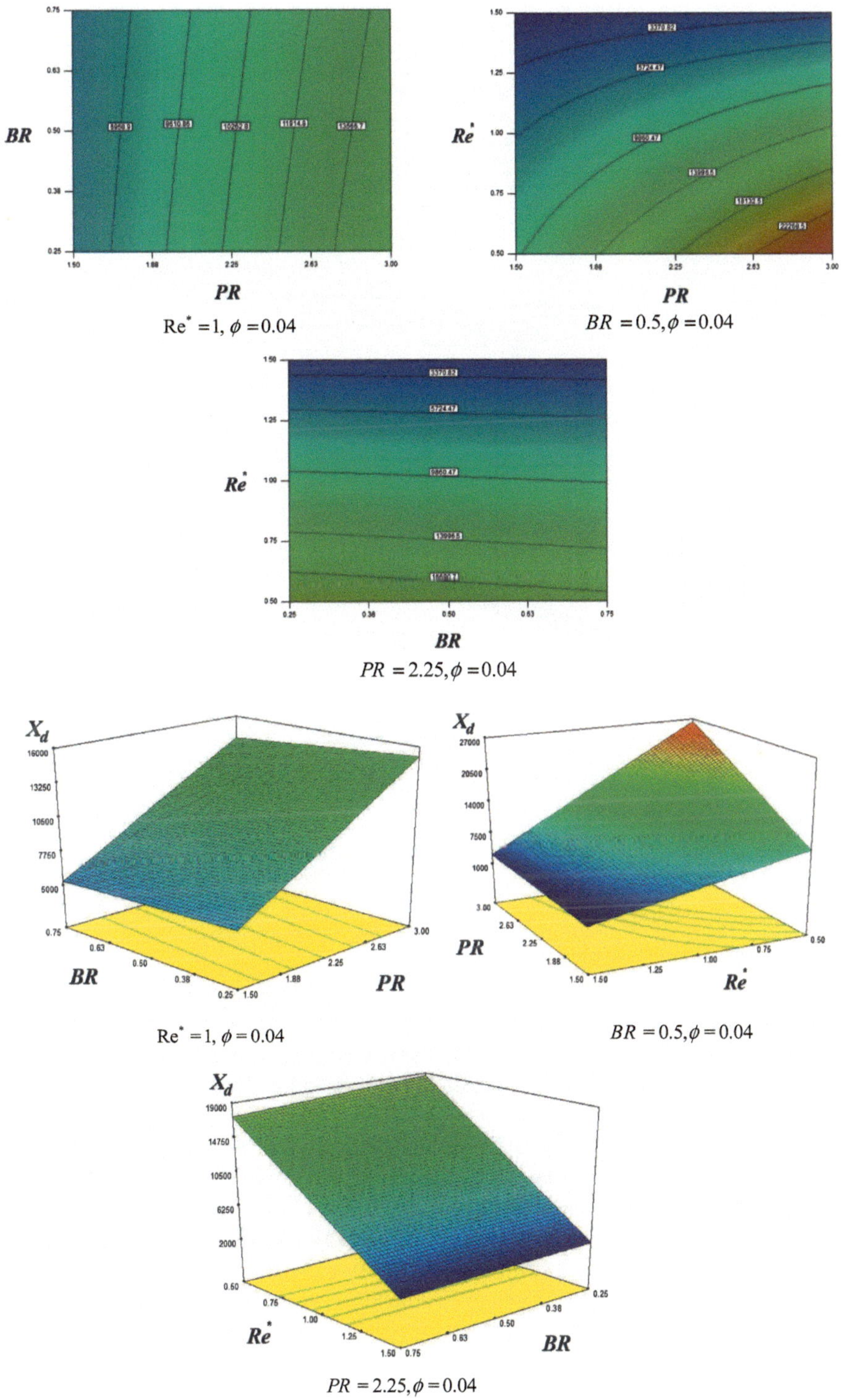

FIGURE 5.11

Influences of Re*, *PR*, and *BR* on X_d.

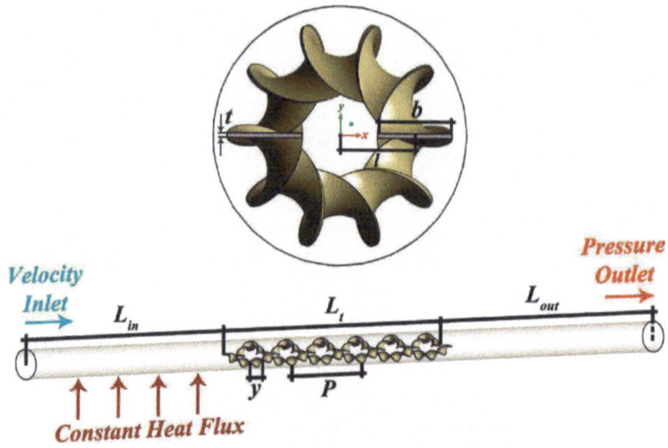

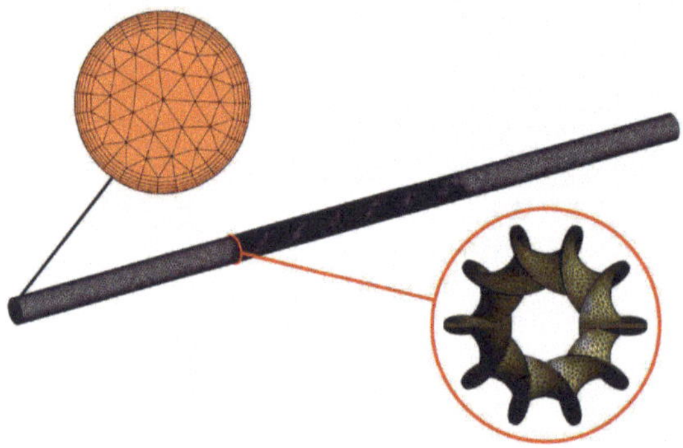

D (in)	2 in
L_{in}, L_{out}, L_t	0.5 m
P	166.667 mm
y	23.93292 mm
b	5, 10 and 15 mm
t	1 mm
l	15 mm
Re	5000, 10000, 15000

FIGURE 5.12

Tube fitted with turbulator and mesh.

5.3.2 Effects of active parameters

To decline the flow resistance and augmentation of heat transfer inside a pipe, complex shape of turbulator was proposed. In addition, exergy loss of carrier fluid, which is copper oxide-H_2O nanomaterial, has been scrutinized. Numerical modeling has been conducted to obtain an understanding of nanofluid hydrothermal treatment and its exergy drop.

Testing fluid is a mixture of Al_2O_3 and H_2O. Small deviation of present outputs and previous data indicates the nice accuracy of such modeling. Contour of X_d, temperature, pressure, velocity, and streamline have been presented in Figs. 5.13–5.15. Inserting turbulator leads to separation of thermal boundary layer. So, temperature of outer wall declines which results in greater convective flow. In addition, utilizing such device enhances the flow resistance. In lower pumping

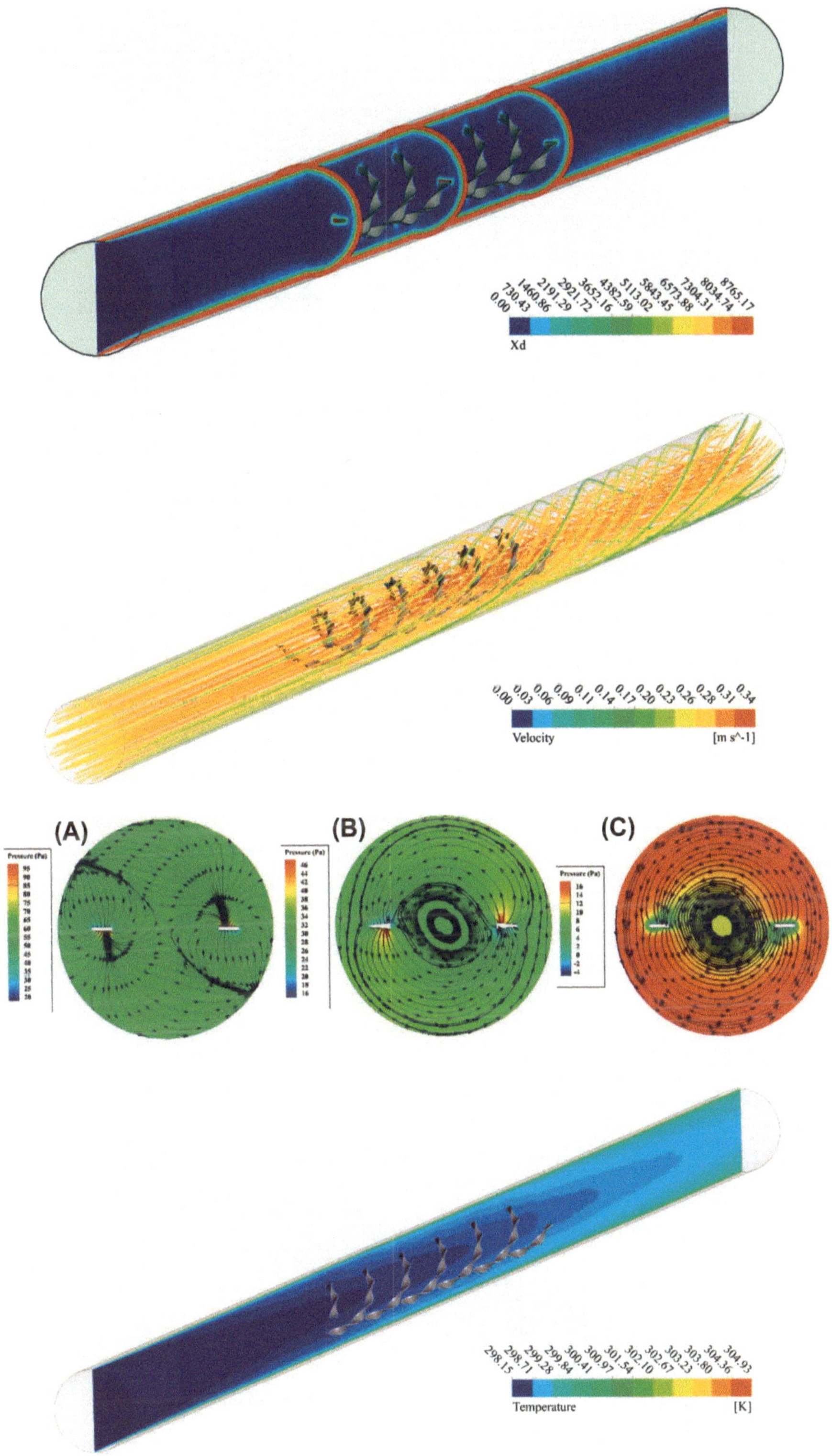

FIGURE 5.13

Contours of X_d, temperature, pressure, velocity, streamline at Re = 15,000, b = 5 mm. (A) start of test section; (B) middle of test section; (C) end of test section.

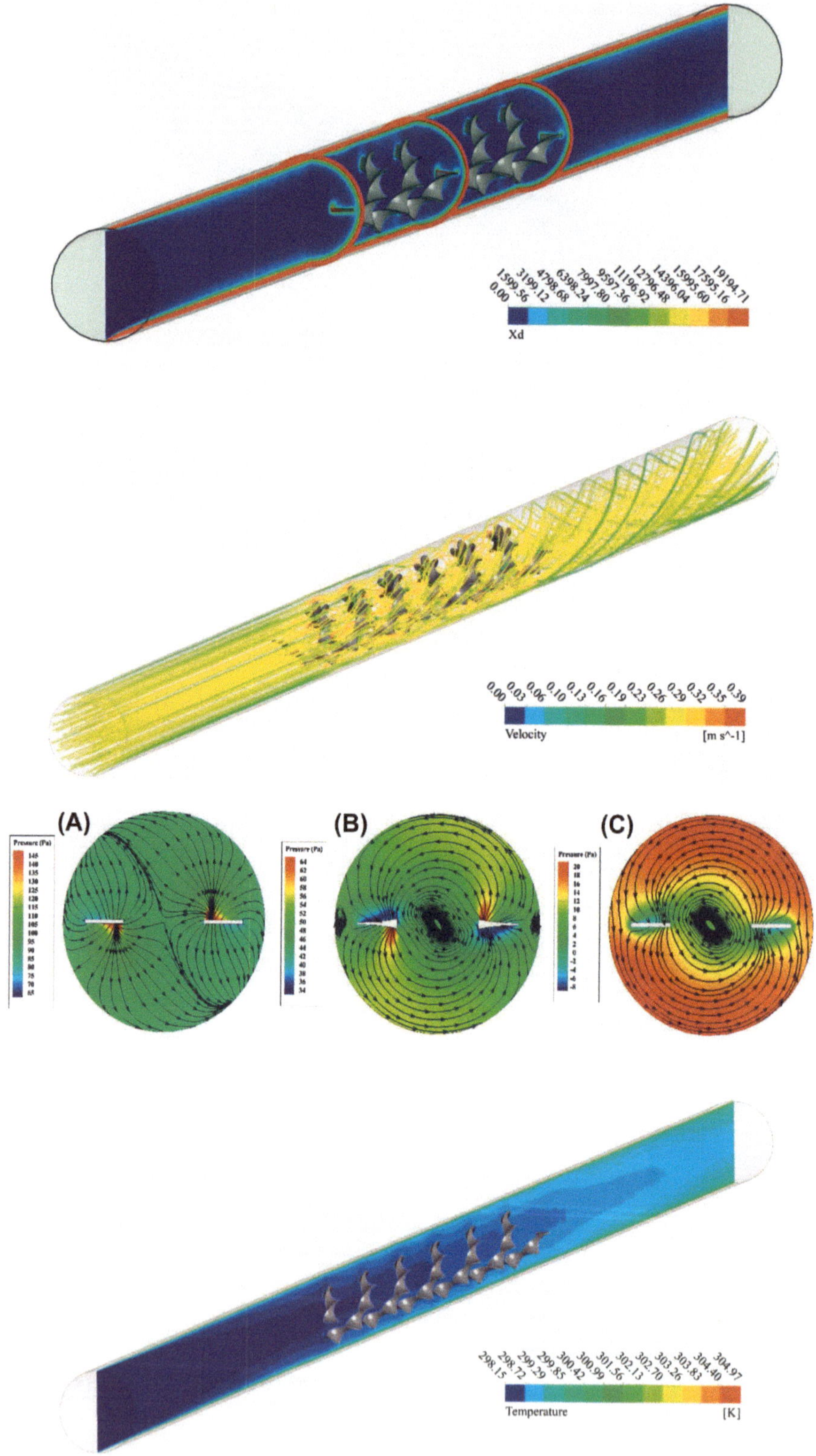

FIGURE 5.14

Contours of X_d, temperature, pressure, velocity, streamline at Re = 15,000, b = 10 mm. (A) start of test section; (B) middle of test section; (C) end of test section.

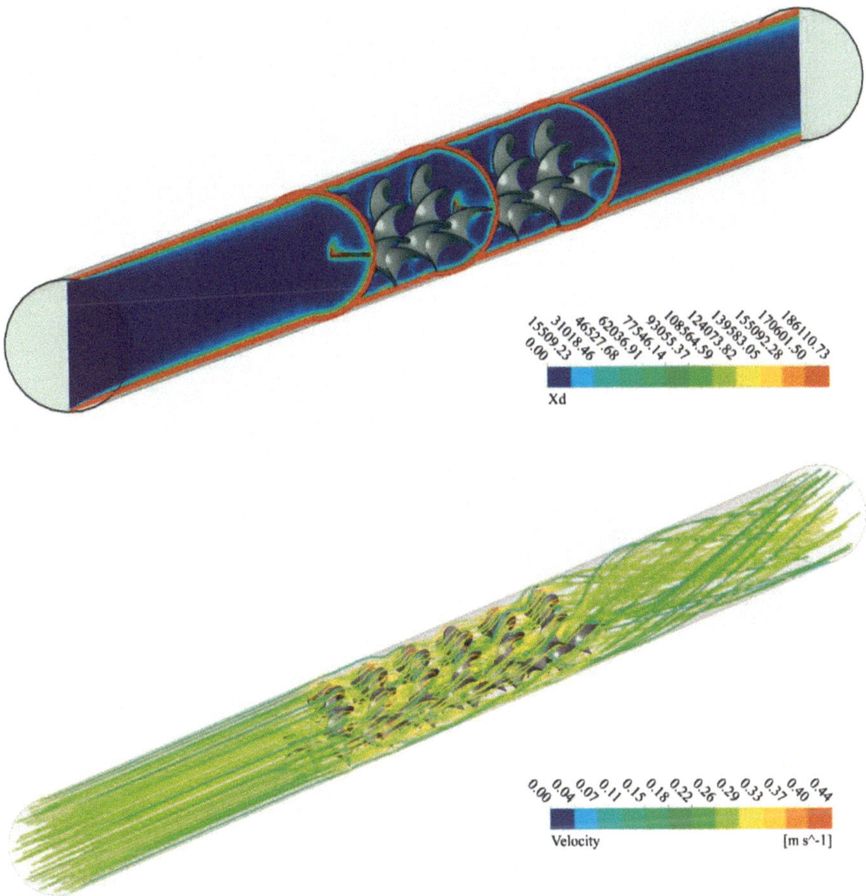

FIGURE 5.15

Contours of X_d, temperature, pressure, velocity, streamline at Re = 15,000, b = 15 mm. (A) start of test section; (B) middle of test section; (C) end of test section.

power, the effect of such device insertion is more observable. Swirl flow plays a critical role on disturbance of temperature boundary layer. As width of turbulator augments, temperature gradient near the outer wall enhances. The greater the b, the greater the friction factor. Stronger secondary vortex tends to be generated in greater values of inlet velocity. So, pressure drop enhances and stronger convective flow can be seen for higher Re. Unit with greater width of turbulator results in greater turbulent intensity. So, exergy loss declines with rise of b. X_d can be written as follows:

$$X_d = 6183.5 - 5907\text{Re}^* - 433.65b + 428.9b\text{Re}^* + 37.5b^2 \tag{5.24}$$

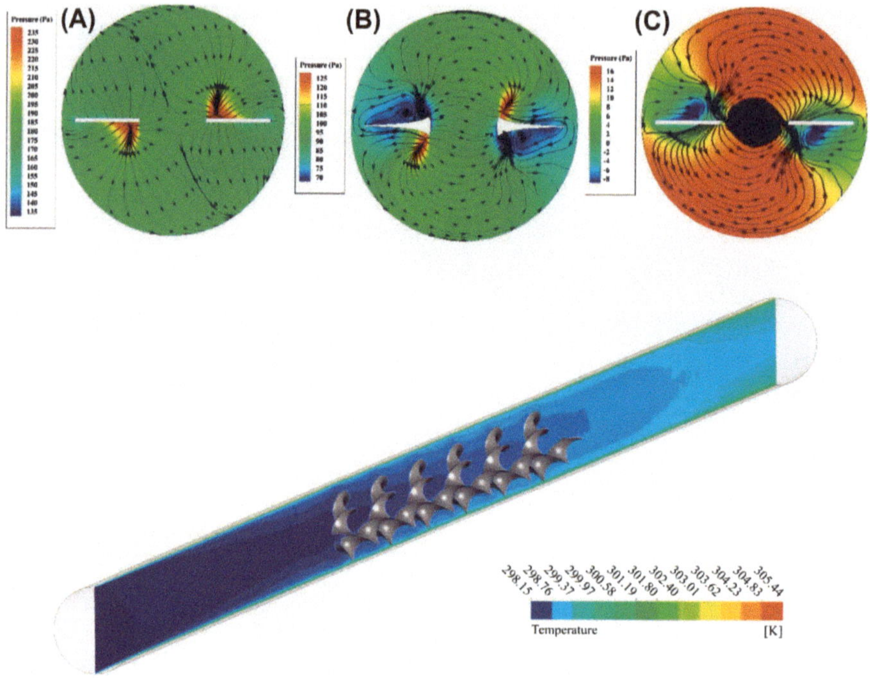

FIGURE 5.15 cont'd.

Exergy loss variation due to changing b and Re has been depicted in Fig. 5.16. More swirl flows due to increasing inlet velocity lead to thinner boundary layer which supports the greater convective flow. Weaker swirls generated by using turbulator with smaller b. Strength of convective flow act against the X_d. Exergy loss has reverse relationship with both parameters.

5.4 Simulation of nanomaterial turbulent modeling in appearance of compound swirl device

5.4.1 Definition of the problem

The unit of interest, as exhibited in Fig. 5.17, is a complex twisted tube inserted inside a round duct. The turbulator was installed in the middle part of domain. The length of tube is 1.5 m with two inch diameter. Carrier fluid in current work is copper oxide/H_2O with listed properties which exist in Ref. [4]. As exhibited in this graph, grids close to solid walls were denser and limitation of proposed model was satisfied in this simulation. As proved in Ref. [5], $k-\varepsilon$ model is a more

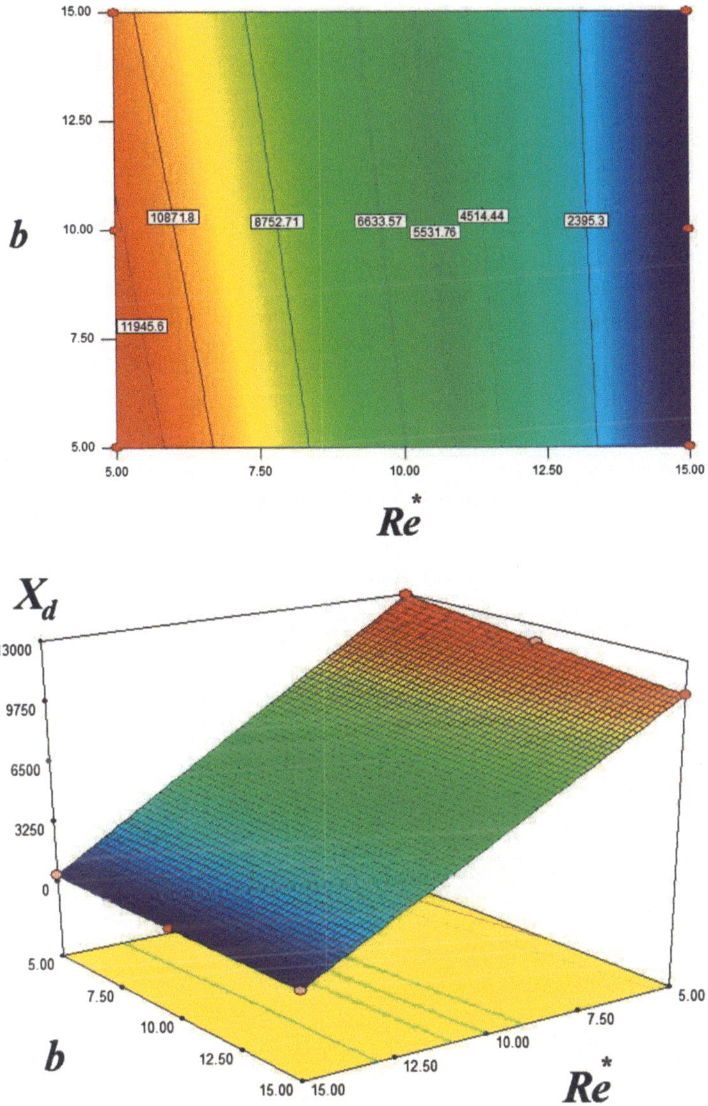

FIGURE 5.16

X_d for different Re and *b*.

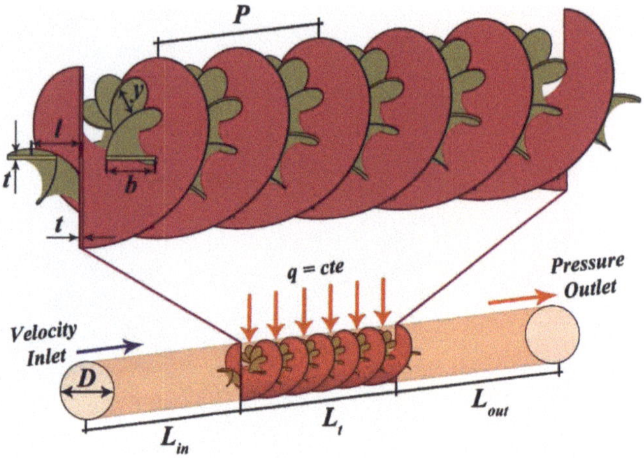

D	*2 in*
l	*15 mm*
$L_{in}=L_t=L_{out}$	*500 mm*
P	*166.667 mm*
t	*1 mm*
y	*23.933 mm*

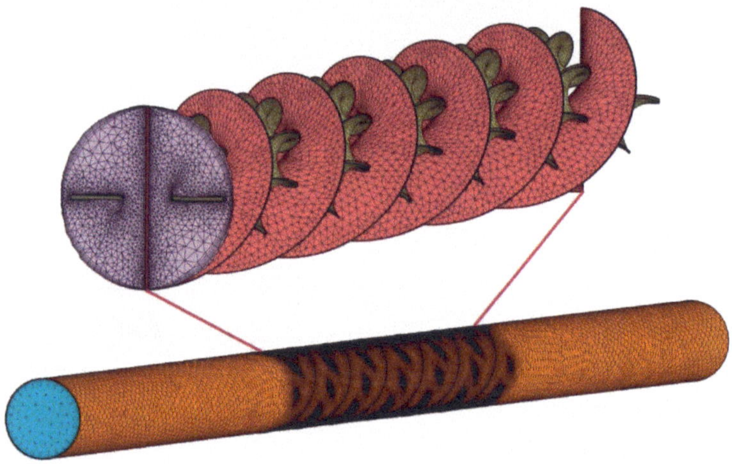

FIGURE 5.17

Present pipe with turbulator and sample grid.

accurate approach in tube with twisted tape. For cases with greatest values of scrutinized parameter, 1,055,747 nodes have been employed. Walls were considered impermeable with no slip condition. To demonstrate various parameters impacts, namely inlet velocity and tape width, finite volume method (ANSYS FLUENT 18.1) has been applied with same setup mentioned in Ref. [6–59].

5.4.2 Effects of active parameters

This chapter presents homogeneous nanomaterial behavior within a round duct improved with innovative turbulator considering copper oxide/H_2O nanofluid. The outputs were examined under the turbulent flow with Reynolds ranges of $5000 < \text{Re} < 15{,}000$. The outputs were exhibited in views of exergy drop, velocity, streamline, and temperature contours as well as distribution of X_d.

Contours of various scalars were shown in Figs. 5.18–5.23. Swirl flow induced with insertion of turbulator through the tube. Therefore, thorough mixing of carrier fluid happens. The greater convective flow of the turbulator with larger b is attributed to the greater interaction of nanofluid with surface. Turbulator involved with lower b and Re possesses the greater exergy destroyed. Intensity of swirl flow augments with rise of tape width. This reveals better fluid mixing and higher pressure loss. The lower inlet velocity yields the greatest exergy drop and attribute with weaker secondary flows. Disruption of thermal boundary layer enhances with enhance of Re. Relation of convention and conduction mode makes changes in heat transfer rate. Stronger convection with utilizing greater b, leaves to augment heat convection coefficient. Lower temperature difference (T_s-T_b) takes critical role for augmenting convective flow. Increasing inlet velocity leads to reduction in (T_s-T_b) and in turn Nusselt number enhances and exergy drop declines.

Fig. 5.24 demonstrates the variation of X_d with respect to two active parameters. As Re and b augment, convective flow intensifies due to thorough mixing of nanofluid. As tape width enhances, migration of nanoparticles disturbs the boundary layer and stronger secondary vortex generates. Therefore, X_d declines with augment of both scrutinized parameters. In addition, the below equation can show variation of X_d:

$$X_d = 6060.7 - 4886.25\text{Re}^* - 90.31b + 88.99b\text{Re}^* - 51.41b^2 \tag{5.25}$$

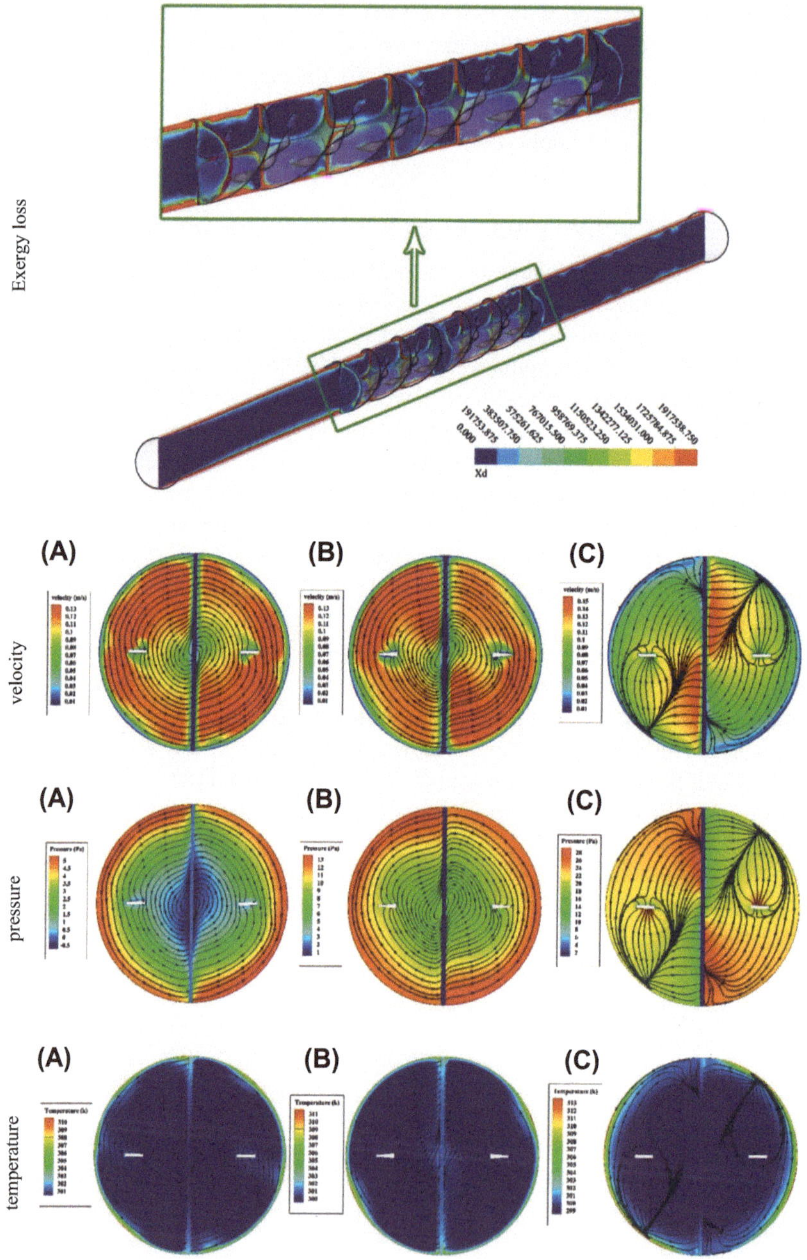

FIGURE 5.18

Contours of velocity, X_d, streamline, P, temperature when $b = 5$ mm and $Re = 5000$. (A) start of test section; (B) middle of test section; (C) end of test section.

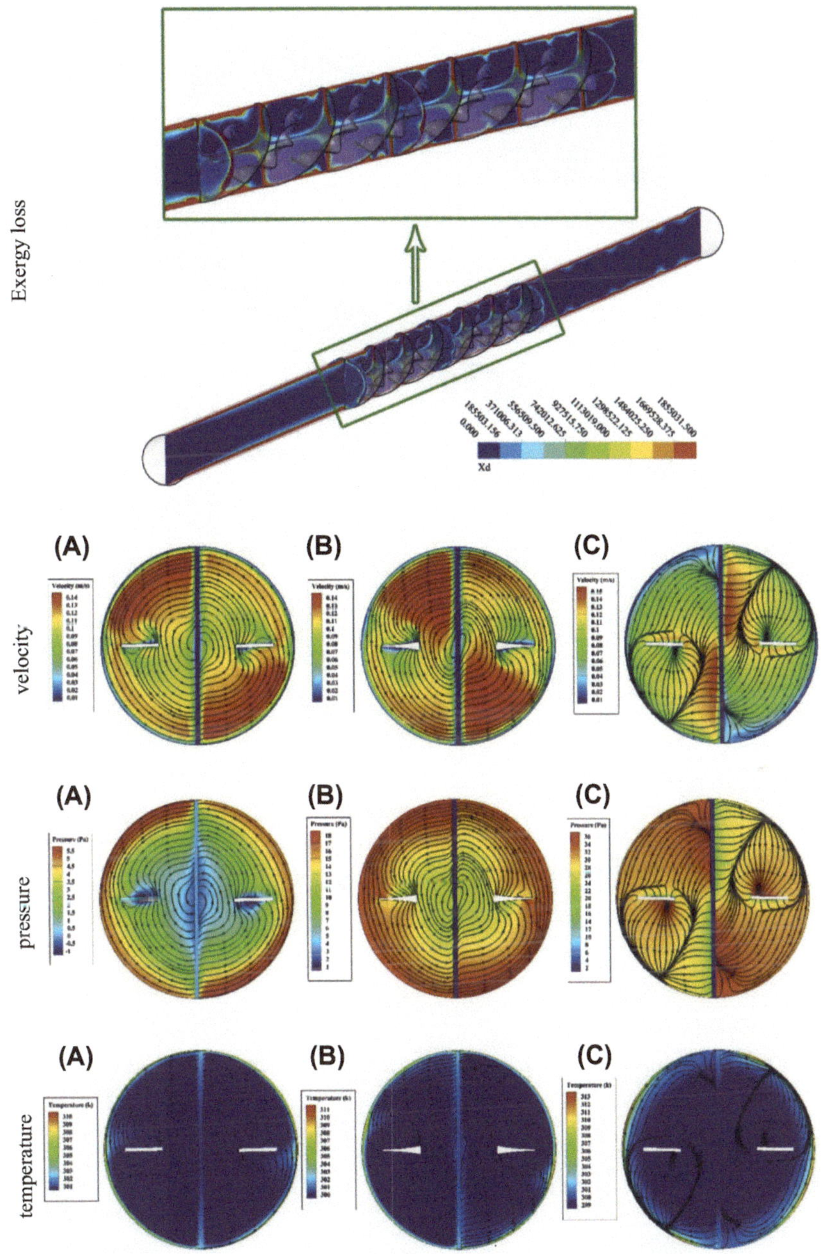

FIGURE 5.19

Contours of velocity, X_d, streamline, P, temperature when $b = 10$ mm and Re = 5000. (A) start of test section; (B) middle of test section; (C) end of test section.

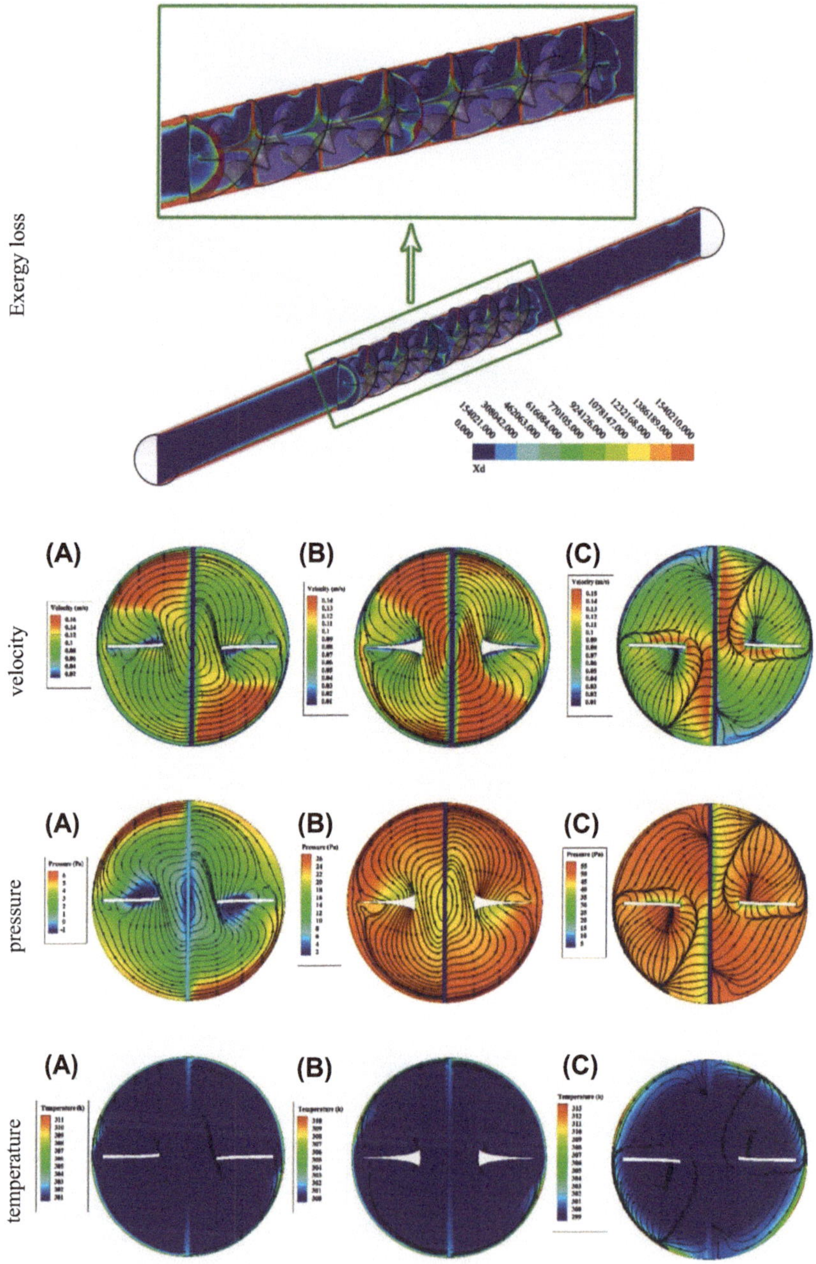

FIGURE 5.20

Contours of velocity, X_d, streamline, P, temperature when $b = 15$ mm and Re = 5000. (A) start of test section; (B) middle of test section; (C) end of test section.

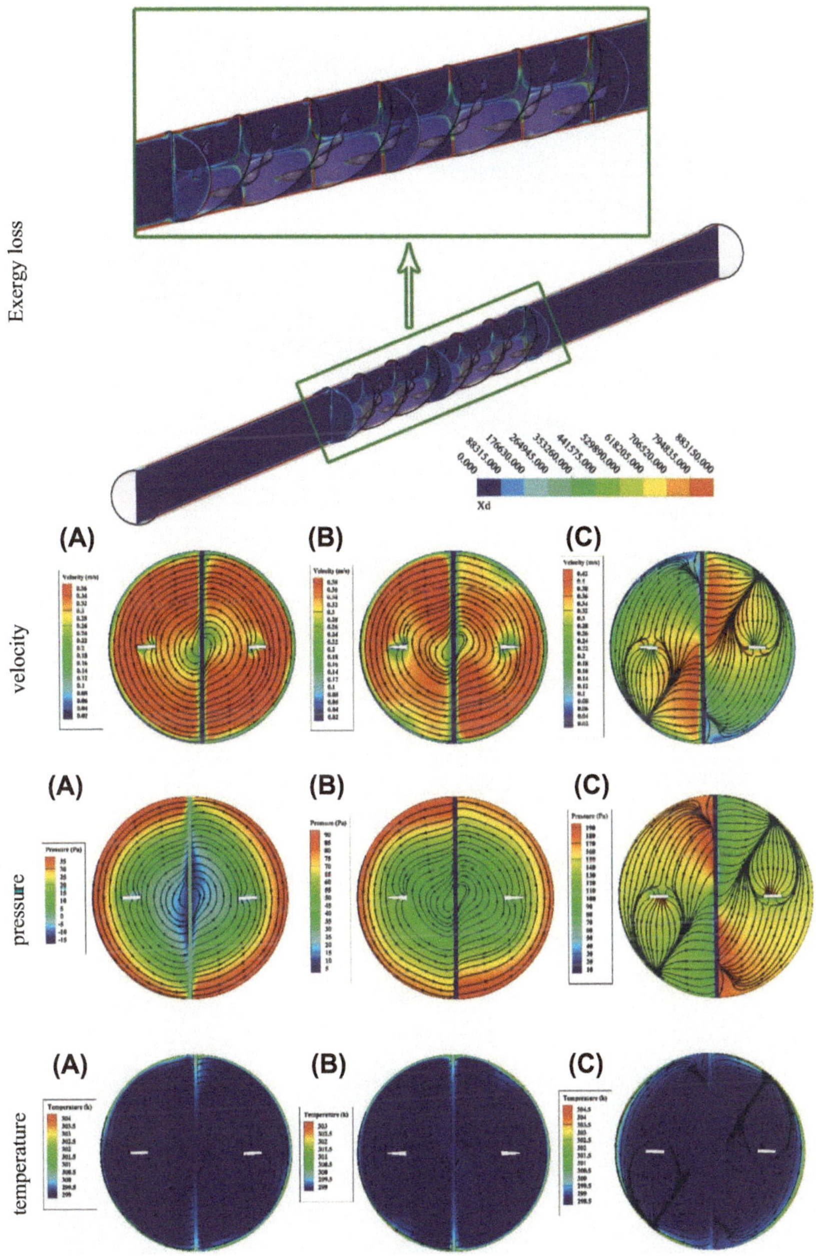

FIGURE 5.21

Contours of velocity, X_d, streamline, P, temperature when $b = 5$ mm and Re = 15,000.

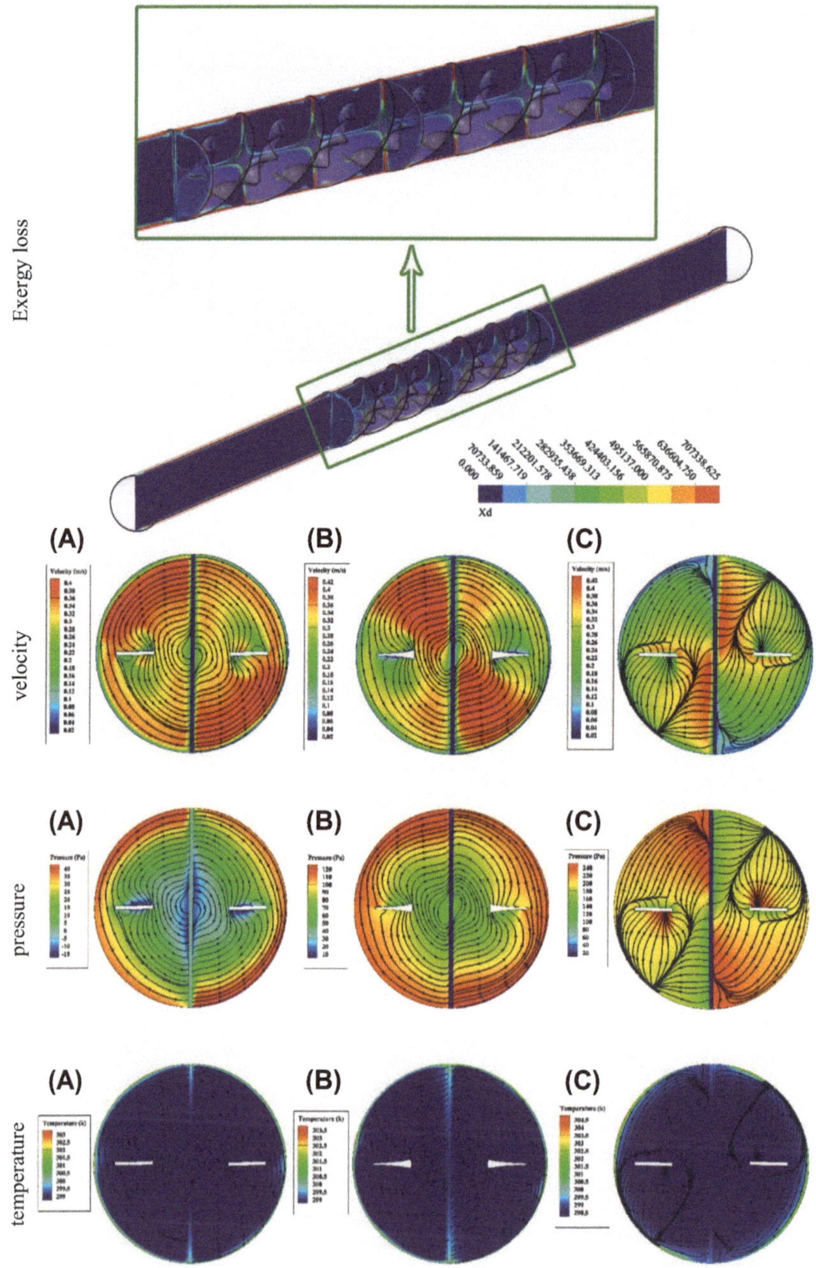

FIGURE 5.22

Contours of velocity, X_d, streamline, P, temperature when $b = 10$ mm and Re = 15,000. (A) start of test section; (B) middle of test section; (C) end of test section.

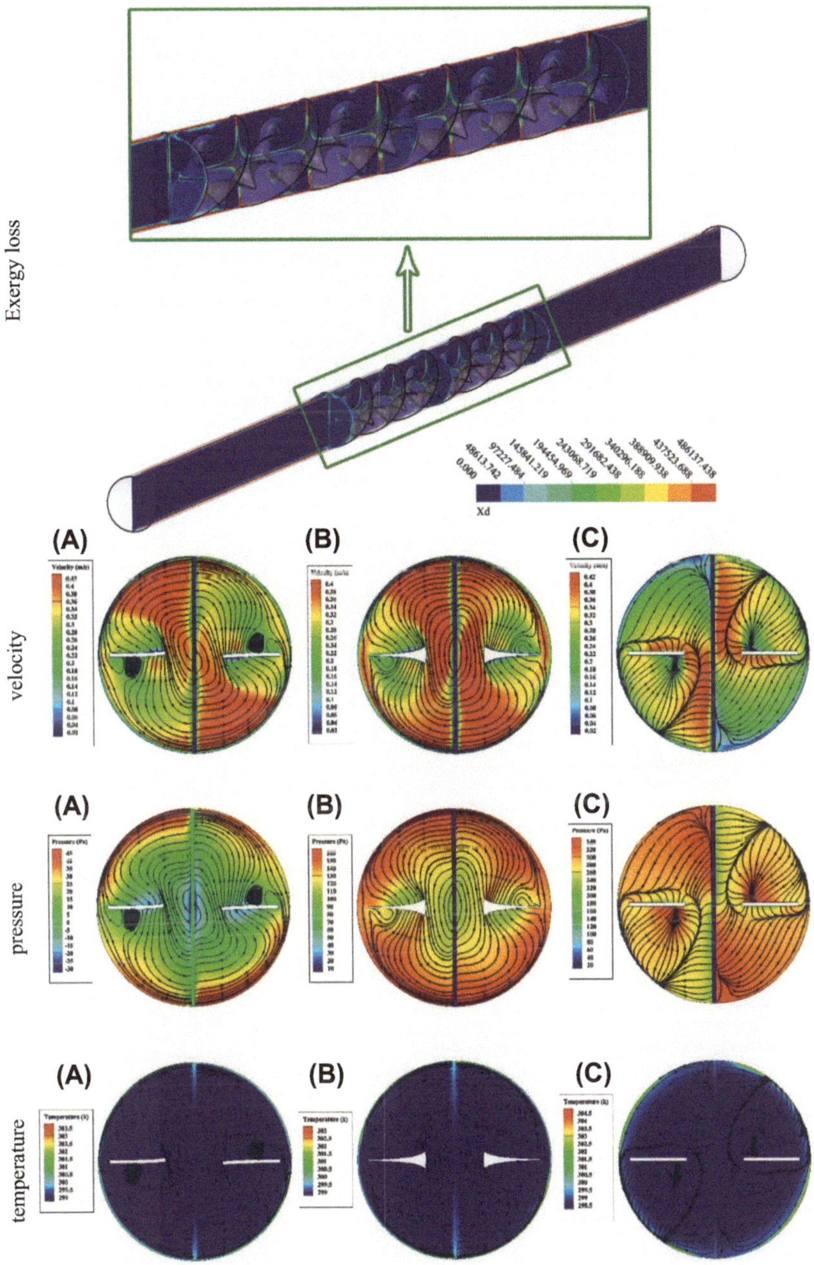

FIGURE 5.23

Contours of velocity, X_d, streamline, P, temperature when $b = 15$ mm and Re = 15,000. (A) start of test section; (B) middle of test section; (C) end of test section.

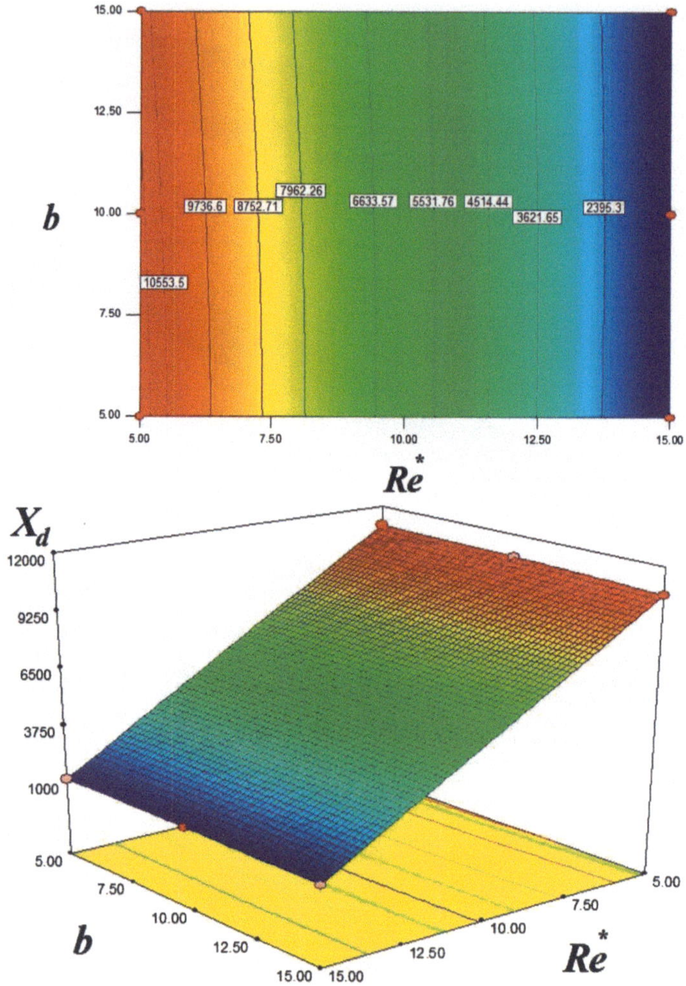

FIGURE 5.24

Variation of X_d with changing Re and b.

References

[1] M. Sheikholeslami, M. Jafaryar, S. Saleem, Z. Li, A. Shafee, Y. Jiang, Nanofluid heat transfer augmentation and exergy loss inside a pipe equipped with innovative turbulators, International Journal of Heat and Mass Transfer 126 (2018) 156–163.

[2] S. Ahmad, A. Arabkoohsar, M. Sheikholeslami, M. Jafaryar, M. Ayani, T. Nguyen-Thoi, Z. Li, D. Baba Basha, I. Tlili, Numerical simulation for turbulent flow in a tube with combined swirl flow device considering nanofluid exergy loss, Physica A: Statistical Mechanics and its Applications (2019).

[3] M. Sheikholeslami, M. Jafaryar, A. Shafee, Z. Li, Investigation of second law and hydrothermal behavior of nanofluid through a tube using passive methods, Journal of Molecular Liquids 269 (2018) 407–416.
[4] W. Sun, M. Jafaryar, M. Sheikholeslami, A. Shafee, T. Nguyen-Thoi, T.M. Yazdani, I. Tlili, Z. Li, Simulation of nanomaterial turbulent modeling in appearance of compound swirl device concerning exergy drop, Physica A: Statistical Mechanics and its Applications 534 (2019) 122121.
[5] M. Sheikholeslami, Lattice Boltzmann method simulation of MHD non-Darcy nanofluid free convection, Physica B 516 (2017) 55–71.
[6] M. Sheikholeslami, H.B. Rokni, Melting heat transfer influence on nanofluid flow inside a cavity in existence of magnetic field, International Journal of Heat and Mass Transfer 114 (2017) 517–526.
[7] M. Sheikholeslami, M. Sadoughi, Mesoscopic method for MHD nanofluid flow inside a porous cavity considering various shapes of nanoparticles, International Journal of Heat and Mass Transfer 113 (2017) 106–114.
[8] M. Sheikholeslami, M.M. Bhatti, Forced convection of nanofluid in presence of constant magnetic field considering shape effects of nanoparticles, International Journal of Heat and Mass Transfer 111 (2017) 1039–1049.
[9] M. Sheikholeslami, M.M. Bhatti, Active method for nanofluid heat transfer enhancement by means of EHD, International Journal of Heat and Mass Transfer 109 (2017) 115–122.
[10] M. Sheikholeslami, S.A. Shehzad, Thermal radiation of ferrofluid in existence of Lorentz forces considering variable viscosity, International Journal of Heat and Mass Transfer 109 (2017) 82–92.
[11] M. Sheikholeslami, H.B. Rokni, Nanofluid two phase model analysis in existence of induced magnetic field, International Journal of Heat and Mass Transfer 107 (2017) 288–299.
[12] M. Sheikholeslami, S.A. Shehzad, Magnetohydrodynamic nanofluid convection in a porous enclosure considering heat flux boundary condition, International Journal of Heat and Mass Transfer 106 (2017) 1261–1269.
[13] M. Sheikholeslami, R. Ellahi, Three dimensional mesoscopic simulation of magnetic field effect on natural convection of nanofluid, International Journal of Heat and Mass Transfer 89 (2015) 799–808.
[14] M. Sheikholeslami, Numerical investigation of nanofluid free convection under the influence of electric field in a porous enclosure, Journal of Molecular Liquids 249 (2018) 1212–1221.
[15] M. Sheikholeslami, CuO-water nanofluid flow due to magnetic field inside a porous media considering Brownian motion, Journal of Molecular Liquids 249 (2018) 921–929.
[16] M. Sheikholeslami, Numerical investigation for CuO-H_2O nanofluid flow in a porous channel with magnetic field using mesoscopic method, Journal of Molecular Liquids 249 (2018) 739–746.
[17] M. Sheikholeslami, A. Zeeshan, Analysis of flow and heat transfer in water based nanofluid due to magnetic field in a porous enclosure with constant heat flux using CVFEM, Computer Methods in Applied Mechanics and Engineering 320 (2017) 68–81.
[18] M. Sheikholeslami, H.B. Rokni, Numerical modeling of nanofluid natural convection in a semi annulus in existence of Lorentz force, Computer Methods in Applied Mechanics and Engineering 317 (2017) 419–430.

[19] Y. Qin, Pavement surface maximum temperature increases linearly with solar absorption and reciprocal thermal inertial, International Journal of Heat and Mass Transfer 97 (2016) 391–399.
[20] M. Sheikholeslami, Influence of magnetic field on Al_2O_3-H_2O nanofluid forced convection heat transfer in a porous lid driven cavity with hot sphere obstacle by means of LBM, Journal of Molecular Liquids 263 (2018) 472–488.
[21] M. Sheikholeslami, New computational approach for exergy and entropy analysis of nanofluid under the impact of Lorentz force through a porous media, Computer Methods in Applied Mechanics and Engineering 344 (2019) 319–333.
[22] Y. Qin, J. Liang, K. Tan, F. Li, A side by side comparison of the cooling effect of building blocks with retro-reflective and diffuse-reflective walls, Solar Energy 133 (2016) 172–179.
[23] M. Sheikholeslami, M.A. Sheremet, A. Shafee, I. Tlili, Simulation of nanoliquid thermogravitational convection within a porous chamber imposing magnetic and radiation impacts, Physica A: Statistical Mechanics and its Applications (2020), https://doi.org/10.1016/j.physa.2019.124058.
[24] M. Sheikholeslami, Magnetic field influence on CuO -H_2O nanofluid convective flow in a permeable cavity considering various shapes for nanoparticles, International Journal of Hydrogen Energy 42 (2017) 19611–19621.
[25] M. Sheikholeslami, Numerical simulation for solidification in a LHTESS by means of Nano-enhanced PCM, Journal of the Taiwan Institute of Chemical Engineers 86 (2018) 25–41.
[26] Y. Qin, Y. He, B. Wu, S. Ma, X. Zhang, Regulating top albedo and bottom emissivity of concrete roof tiles for reducing building heat gains, Energy and Buildings 156 (Suppl. C) (2017) 218–224.
[27] M. Sheikholeslami, A. Nematpour Keshteli, H. Babazadeh, Nanoparticles favorable effects on performance of thermal storage units, Journal of Molecular Liquids 300 (2020) 112329.
[28] M. Sheikholeslami, Effect of uniform suction on nanofluid flow and heat transfer over a cylinder, Journal of the Brazilian Society of Mechanical Sciences and Engineering 37 (2015) 1623–1633.
[29] A. Shafee, M. Sheikholeslami, M. Jafaryar, H. Babazadeh, Utilizing copper oxide nanoparticles for expedition of solidification within a storage system, Journal of Molecular Liquids 302 (2020) 112371, https://doi.org/10.1016/j.molliq.2019.112371.
[30] M. Sheikholeslami, Numerical modeling of Nano enhanced PCM solidification in an enclosure with metallic fin, Journal of Molecular Liquids 259 (2018) 424–438.
[31] M. Sheikholeslami, Solidification of NEPCM under the effect of magnetic field in a porous thermal energy storage enclosure using CuO nanoparticles, Journal of Molecular Liquids 263 (2018) 303–315.
[32] M. Sheikholeslami, O. Mahian, Enhancement of PCM solidification using inorganic nanoparticles and an external magnetic field with application in energy storage systems, Journal of Cleaner Production 215 (2019) 963–977.
[33] M. Sheikholeslami, B. Rezaeianjouybari, M. Darzi, A. Shafee, Z. Li, T.K. Nguyen, Application of nano-refrigerant for boiling heat transfer enhancement employing an experimental study, International Journal of Heat and Mass Transfer 141 (2019) 974–980.
[34] M. Sheikholeslami, M. Jafaryar, M. Hedayat, A. Shafee, Z. Li, T.K. Nguyen, M. Bakouri, Heat transfer and turbulent simulation of nanomaterial due to compound

turbulator including irreversibility analysis, International Journal of Heat and Mass Transfer 137 (2019) 1290–1300.
[35] M. Sheikholeslami, Application of Darcy law for nanofluid flow in a porous cavity under the impact of Lorentz forces, Journal of Molecular Liquids 266 (2018) 495–503.
[36] M. Sheikholeslami, Finite element method for PCM solidification in existence of CuO nanoparticles, Journal of Molecular Liquids 265 (2018) 347–355.
[37] M. Sheikholeslami, M. Jafaryar, A. Shafee, Z. Li, R-ul Haq, Heat transfer of nanoparticles employing innovative turbulator considering entropy generation, International Journal of Heat and Mass Transfer 136 (2019) 1233–1240.
[38] M. Sheikholeslami, R.-ul Haq, A. Shafee, Z. Li, Y.G. Elaraki, I. Tlili, Heat transfer simulation of heat storage unit with nanoparticles and fins through a heat exchanger, International Journal of Heat and Mass Transfer 135 (2019) 470–478.
[39] M. Sheikholeslami, R-ul Haq, A. Shafee, Z. Li, Heat transfer behavior of Nanoparticle enhanced PCM solidification through an enclosure with V shaped fins, International Journal of Heat and Mass Transfer 130 (2019) 1322–1342.
[40] M. Sheikholeslami, S.A. Shehzad, Z. Li, A. Shafee, Numerical modeling for Alumina nanofluid magnetohydrodynamic convective heat transfer in a permeable medium using Darcy law, International Journal of Heat and Mass Transfer 127 (2018) 614–622.
[41] M. Sheikholeslami, Z. Li, A. Shafee, Lorentz forces effect on NEPCM heat transfer during solidification in a porous energy storage system, International Journal of Heat and Mass Transfer 127 (2018) 665–674.
[42] M. Sheikholeslami, A. Ghasemi, Z. Li, A. Shafee, S. Saleem, Influence of CuO nanoparticles on heat transfer behavior of PCM in solidification process considering radiative source term, International Journal of Heat and Mass Transfer 126 (2018) 1252–1264.
[43] M. Sheikholeslami, M. Darzi, Z. Li, Experimental investigation for entropy generation and exergy loss of nano-refrigerant condensation process, International Journal of Heat and Mass Transfer 125 (2018) 1087–1095.
[44] M. Sheikholeslami, S.A. Shehzad, Z. Li, Water based nanofluid free convection heat transfer in a three dimensional porous cavity with hot sphere obstacle in existence of Lorenz forces, International Journal of Heat and Mass Transfer 125 (2018) 375–386.
[45] M. Sheikholeslami, M. Jafaryar, Z. Li, Nanofluid turbulent convective flow in a circular duct with helical turbulators considering CuO nanoparticles, International Journal of Heat and Mass Transfer 124 (2018) 980–989.
[46] M. Sheikholeslami, A. Ghasemi, Solidification heat transfer of nanofluid in existence of thermal radiation by means of FEM, International Journal of Heat and Mass Transfer 123 (2018) 418–431.
[47] M. Sheikholeslami, S.A. Shehzad, CVFEM simulation for nanofluid migration in a porous medium using Darcy model, International Journal of Heat and Mass Transfer 122 (2018) 1264–1271.
[48] M. Sheikholeslami, M. Darzi, M.K. Sadoughi, Heat transfer improvement and pressure drop during condensation of refrigerant-based nanofluid; an experimental procedure, International Journal of Heat and Mass Transfer 122 (2018) 643–650.
[49] M. Sheikholeslami, S.A. Shehzad, Simulation of water based nanofluid convective flow inside a porous enclosure via Non-equilibrium model, International Journal of Heat and Mass Transfer 120 (2018) 1200–1212.

[50] M. Sheikholeslami, M. Seyednezhad, Simulation of nanofluid flow and natural convection in a porous media under the influence of electric field using CVFEM, International Journal of Heat and Mass Transfer 120 (2018) 772–781.

[51] M. Sheikholeslami, H.B. Rokni, Numerical simulation for impact of Coulomb force on nanofluid heat transfer in a porous enclosure in presence of thermal radiation, International Journal of Heat and Mass Transfer 118 (2018) 823–831.

[52] M. Sheikholeslami, S.A. Shehzad, Numerical analysis of $Fe_3O_4-H_2O$ nanofluid flow in permeable media under the effect of external magnetic source, International Journal of Heat and Mass Transfer 118 (2018) 182–192.

[53] M. Sheikholeslami, M.K. Sadoughi, Simulation of CuO- water nanofluid heat transfer enhancement in presence of melting surface, International Journal of Heat and Mass Transfer 116 (2018) 909–919.

[54] M. Sheikholeslami, H.B. Rokni, Simulation of nanofluid heat transfer in presence of magnetic field: a review, International Journal of Heat and Mass Transfer 115 (2017) 1203–1233.

[55] M. Sheikholeslami, T. Hayat, A. Alsaedi, On simulation of nanofluid radiation and natural convection in an enclosure with elliptical cylinders, International Journal of Heat and Mass Transfer 115 (2017) 981–991.

[56] M. Sheikholeslami, S.A. Shehzad, CVFEM for influence of external magnetic source on $Fe_3O_4-H_2O$ nanofluid behavior in a permeable cavity considering shape effect, International Journal of Heat and Mass Transfer 115 (2017) 180–191.

[57] M. Sheikholeslami, M. Seyednezhad, Nanofluid heat transfer in a permeable enclosure in presence of variable magnetic field by means of CVFEM, International Journal of Heat and Mass Transfer 114 (2017) 1169–1180.

[58] M. Sheikholeslami, Numerical approach for MHD Al_2O_3-water nanofluid transportation inside a permeable medium using innovative computer method, Computer Methods in Applied Mechanics and Engineering 344 (2019) 306–318.

[59] X. Ma, M. Sheikholeslami, M. Jafaryar, A. Shafee, T. Nguyen-Thoi, Z. Li, Solidification inside a clean energy storage unit utilizing phase change material with copper oxide nanoparticles, Journal of Cleaner Production (2019), https://doi.org/10.1016/j.jclepro.2019.118888.

CHAPTER

Unsteady process in a heat exchanger during melting

6

6.1 Expediting melting of PCM inside a finned enclosure using nanoparticles

6.1.1 Definition of the problem

We proposed the two-dimensional geometry with three different configurations for fins which is depicted in Fig. 6.1. We considered L = 40 mm, D = 20 mm. Inside the inner cylinder, hot fluid is moved, and it functions as the heat source [1]. The square cylinder is adiabatic. CuO and n-octadecane paraffin are selected as nanoparticle and pure PCM. Finite volume method is involved for modeling of melting process. To model this phenomenon, below PDEs can be noticed:

$$\frac{\partial T}{\partial t}+\vec{V}\cdot\nabla T=-\frac{\partial(\rho L\lambda)_{nf}}{\partial t}+\frac{k_{nf}}{(\rho C_p)_{nf}}\nabla^2 T \tag{6.1}$$

$$\frac{\partial u}{\partial t}+\nabla u\cdot\vec{V}=\frac{1}{\rho_{nf}}\left(-\nabla P+\mu_{nf}\nabla^2 u\right)+Cu\frac{(1-\lambda)^2}{\lambda^3+\varepsilon} \tag{6.2}$$

$$\frac{\partial v}{\partial t}+\vec{V}\cdot\nabla v=\frac{1}{\rho_{nf}}\left(\mu_{nf}\nabla^2 v-\nabla P\right)+Cv\frac{(1-\lambda)^2}{\lambda^3+\varepsilon}+g(T-T_{ref})(\rho\beta)_{nf} \tag{6.3}$$

$$\nabla\cdot\vec{V}=0 \tag{6.4}$$

Boussinesq estimation is considered for buoyancy terms. $C=10^5, \varepsilon=10^{-3}$ are mushy zone constant and small number, respectively.

$(\rho L)_{nf}$, density, $(\rho\beta)_{nf}$, and $(\rho C_p)_{nf}$ have been calculated as

$$(\rho L)_{nf}=(1-\phi)(\rho L)_f \tag{6.5}$$

$$\rho_{nf}=\rho_f(1-\phi)+\phi\rho_s \tag{6.6}$$

$$(\rho\beta)_{nf}=\phi(\rho\beta)_s+(1-\phi)(\rho\beta)_f \tag{6.7}$$

Nanofluid in Heat Exchangers for Mechanical Systems. https://doi.org/10.1016/B978-0-12-821923-2.00006-3

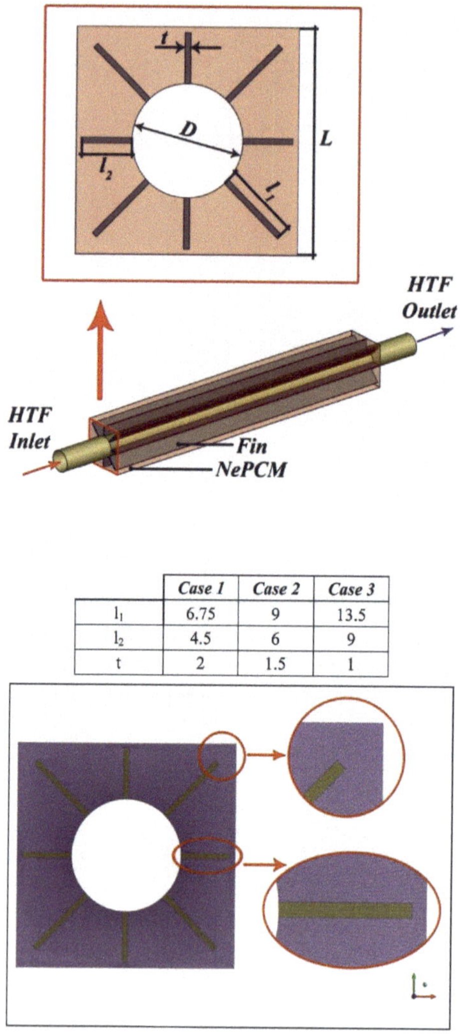

	Case 1	Case 2	Case 3
l_1	6.75	9	13.5
l_2	4.5	6	9
t	2	1.5	1

FIGURE 6.1

Finned heat exchanger for energy saving and sample grid.

$$(\rho C_p)_{nf} = \phi(\rho C_p)_s + (1-\phi)(\rho C_p)_f \tag{6.8}$$

k_{nf}, μ_{nf} have below formulas:

$$k_{nf} = k_f \frac{-2\phi(k_f - k_s) + 2k_f + k_p}{\phi(k_f - k_s) + k_p + 2k_f} \tag{6.9}$$

$$\mu_{nf} = \frac{\mu_f}{(1-\phi)^{2.5}} \tag{6.10}$$

The characteristics of copper oxide and pure PCM have been reported in Table 6.1.

The sensible and total enthalpies are

$$h = h_{ref} + \int_{T_{ret}}^{T} (C_p)_{nf} dT \tag{6.11}$$

$$H_e = h + \lambda L \tag{6.12}$$

where h_{ref} is the enthalpy at $T_{ref} = 273K$.

λ should be obtained from below formula:

$$\lambda = \begin{cases} \dfrac{T_s - T}{T_s - T_l} & T_s < T < T_l \\ 0 & T < T_s \\ 1 & T < T_l \end{cases} \tag{6.13}$$

We selected commercial software based on FVM to model current research. In order to pressure correction, we chose PRESTO. Advection terms were discretized with second-order upwind. Under-relaxation factor of pressure, temperature, and melt fraction are 0.3, 1, and 0.9, respectively. According to Ref. [1], we utilized porosity–enthalpy method. To show the accuracy of existence code, we recalculate previous melt fraction profile which was reported in Ref. [1] (see Fig. 6.2).

6.1.2 Effects of active parameters

Three arrangements of fins were tested to demonstrate the melting of nanoparticle enhanced phase change material (NEPCM). Numerical examinations have been employed to reach the outcomes which are streamline, isotherm, and liquid fraction. The results of case 1 during charging process (in three time

Table 6.1 Characteristics of n-octadecane paraffin and copper oxide.

Property	ρ[kg/m^3]	C_p[j/kg K]	k[w/mK]	μ×10^3[Pa.s]	β×10^5[K^{-1}]	L[j/kg]	Fusion point[°C]
N-octadecane	770	2196	0.148	3.85	91	243,500	28
Nanoparticles	6500	540	18	–	29	–	–

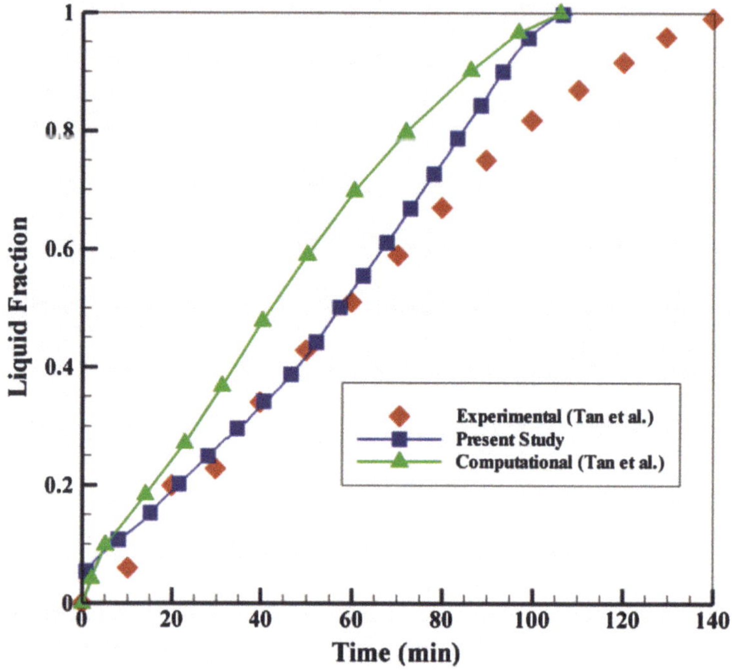

FIGURE 6.2

Verification of numerical code by comparing with Ref. [1].

steps) are reported in Figs. 6.3–6.5. At the beginning of charging, conduction mode affects the domain. So, there are only week eddies, and isotherms are parallel to each other. Gradually, buoyancy forces become stronger and main mode converts to convection. Thus, thermal plumes generate and shape melt front becomes more complex.

Figs. 6.6–6.8 illustrate the outputs of case 2. As length of the fins enhances, melting rate augments and hot surface affect more PCM. Liquid front reaches to adiabatic square cylinder with higher speed in comparison to the previous case. Thermal plume generates over the hot fins. Figs. 6.9–6.11 exhibit various stages of melting for case 3. In such configurations, highest melting efficiency and stronger eddies can be found. Liquid film occupies the top region sooner than case 1. Liquid fraction, heat transfer rate and heat storage for various cases have been reported in Figs. 6.12 and 6.13. Saving energy is changed considerably with variation of liquid fraction.

Greatest liquid fraction was examined for Case 3. One maximum point exists for heat capacity. At first, this factor augments then it reduces with progressing time. Due to greater heat penetration, the third case takes lowest time for charging.

FIGURE 6.3

Variation of Liquid fraction during melting for case 1.

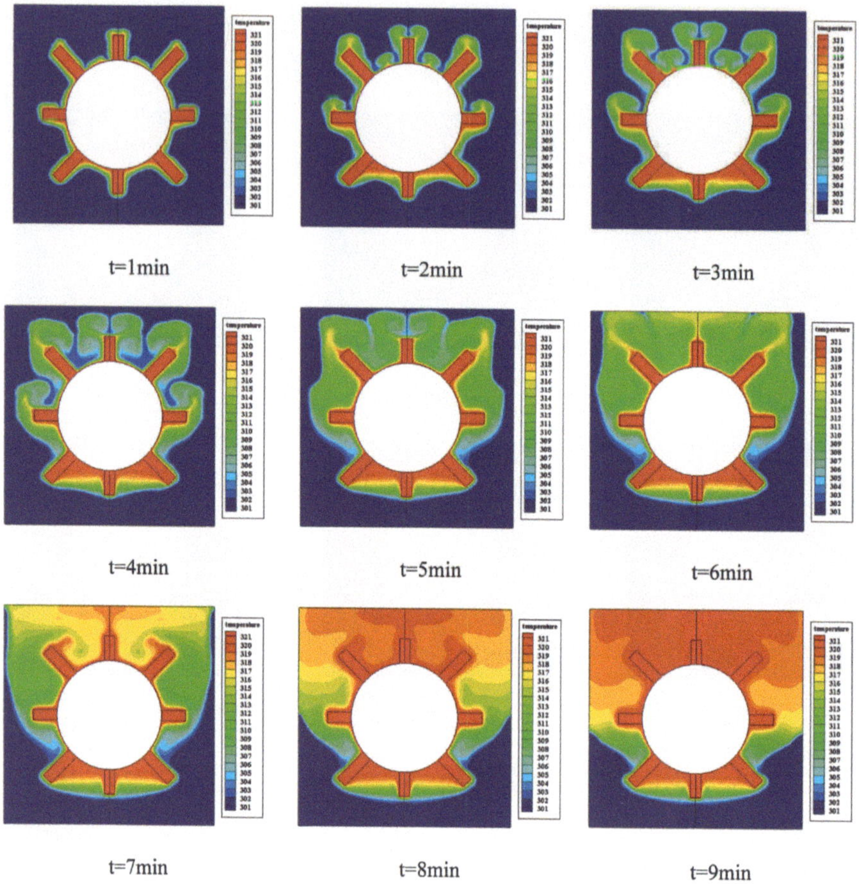

FIGURE 6.4

Variation of Temperature during melting for case 1.

Heat transfer rate of case 3 is greater than other cases in initial time of charging process (before 7 min), but inverse observation is shown when time progress.

6.2 Charging of NEPCM in a two-dimensional thermal storage unit

6.2.1 Definition of the problem

The schematic diagram of the current computational model shape has been illustrated in Figs. 6.14 and 6.15. The working fluid is with CuO/n-octadecane paraffin. Three configurations of fins are simulated. The details of three cases are given in

FIGURE 6.5

Variation of Streamline during melting for case 1.

FIGURE 6.6

Variation of Liquid fraction during melting for case 2.

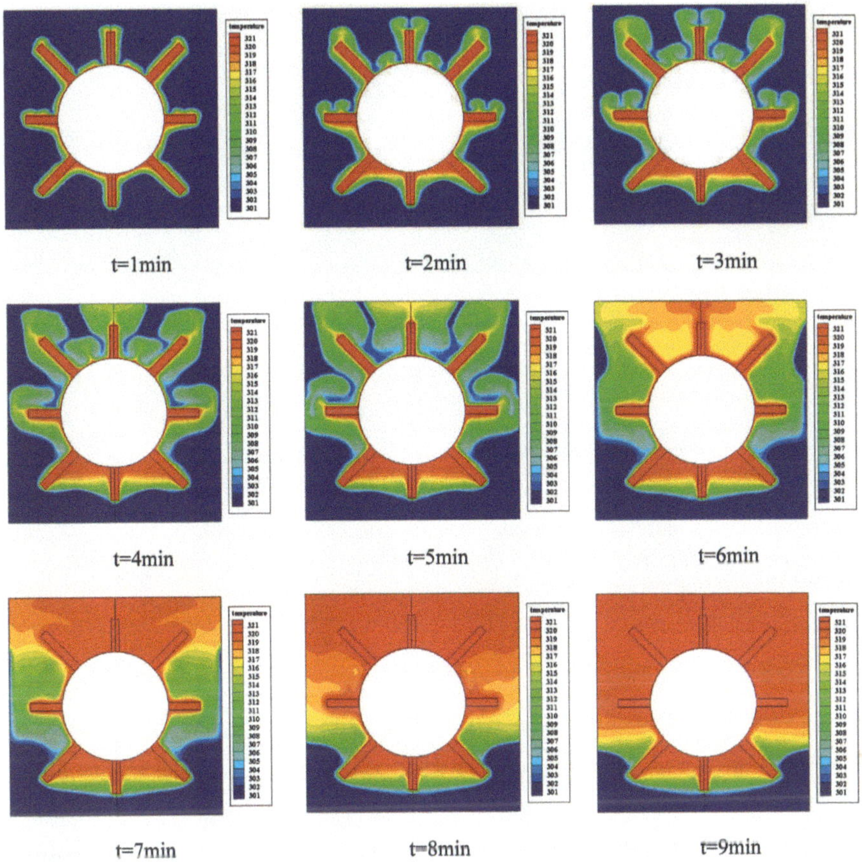

FIGURE 6.7

Variation of Temperature during melting for case 2.

Table 6.2. At the initial time, all fluids exist in solidus temperature [2]. The circular and square cylinders were hot and adiabatic, respectively.

In this chapter, the formula of $S_{gen,total}, S_{gen,th}, S_{gen,f}$ are

$$S_{gen,total} = S_{gen,th} + S_{gen,f}$$

$$= \frac{k_{nf}}{T^2}\left[\left(\frac{\partial T}{\partial x}\right)^2 + \left(\frac{\partial T}{\partial y}\right)^2\right] + \frac{\mu_{nf}}{T}\left\{\begin{array}{l} 2\left[\left(\frac{\partial u_x}{\partial x}\right)^2 + \left(\frac{\partial u_y}{\partial y}\right)^2\right] + \\ \left(\frac{\partial u_x}{\partial y} + \frac{\partial u_y}{\partial x}\right)^2 \end{array}\right\} \tag{6.14}$$

FIGURE 6.8

Variation of Streamline during melting for case 2.

6.2.2 Effects of active parameters

The numerical outputs are obtained for the current charging process of NEPCM with emphasis on the effect of entropy generation and heat transfer behaviors. FVM has been selected to present the number of simulations and is carried out

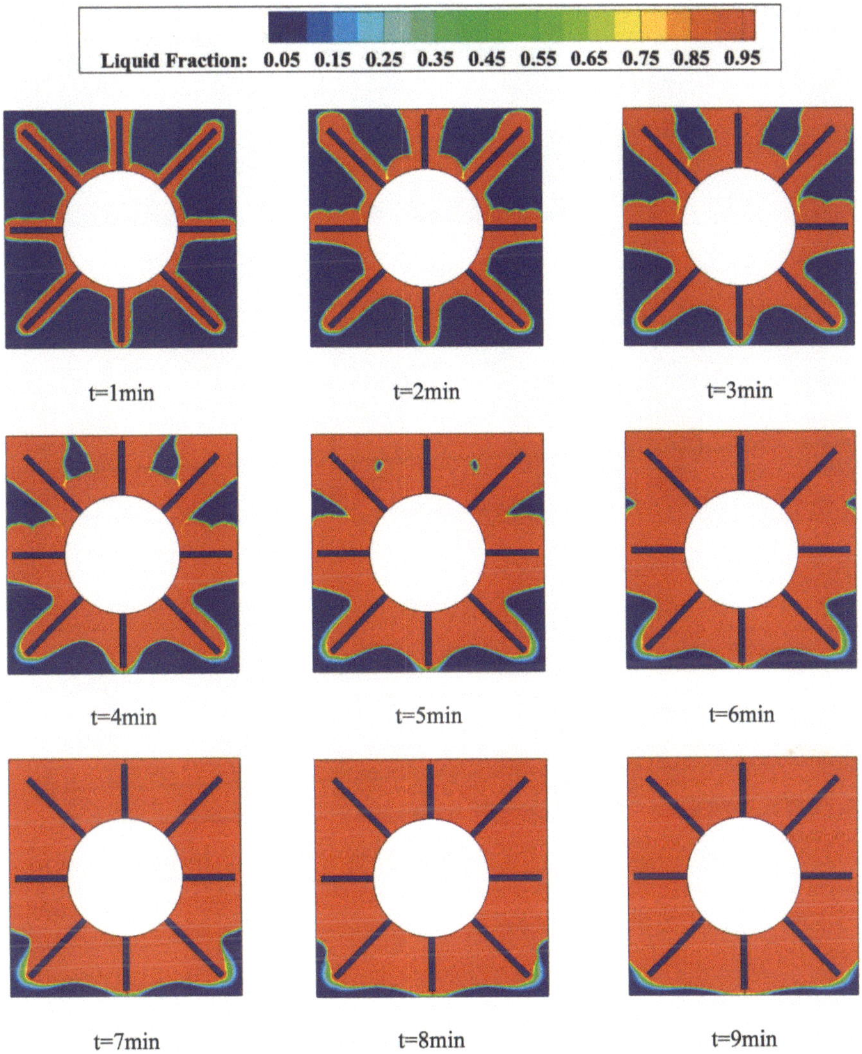

FIGURE 6.9

Variation of Liquid fraction during melting for case 3.

for various fins configuration. The outcomes are explained in detail in Figs. 6.16–6.18 in terms of stream functions, isotherms, frictional and thermal entropy generation. In the initial time, the temperature is kept at solidus temperature. In lower time, conduction can affect the process. So, $S_{gen,f}$ augments in this time.

FIGURE 6.10

Variation of Temperature during melting for case 3.

After that time, convection becomes more effective and do not allow conduction to take part in melting. Temperature augments and $S_{gen,th}$ reduces. Thermal plumes generate when convection becomes stronger. Vortex becomes stronger during this process. Bottom region is the last zone in which melting occurs. Melting front reaches to outer side sooner when length of fins enhances. $S_{gen,f}$ contours have more complex shape in case 3 than others. Greatest melting rate is observed for case 3. This signifies that Buoyancy force is aided more strongly by using fins for case 3. It is noticed that $S_{gen,th}$ is mainly concentrated near the hot fins.

FIGURE 6.11

Variation of Streamline during melting for case 3.

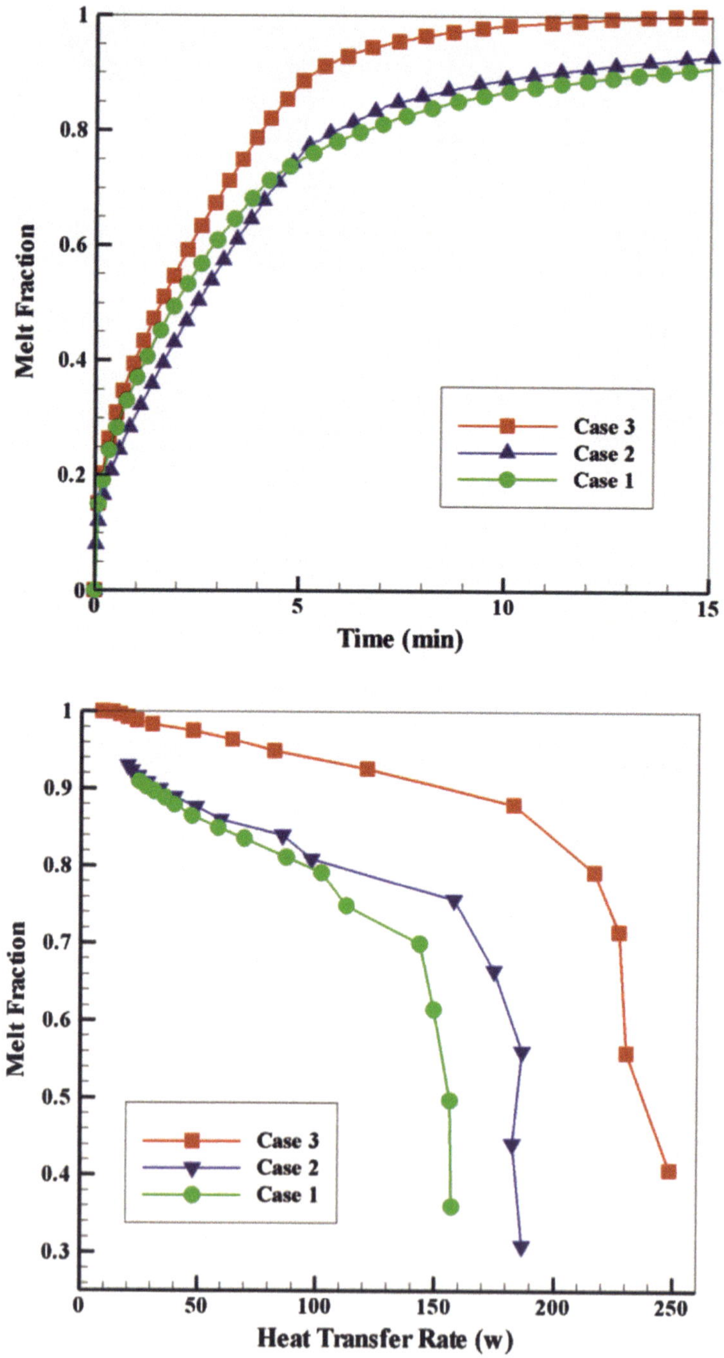

FIGURE 6.12

Average liquid fraction and heat transfer rate for various cases.

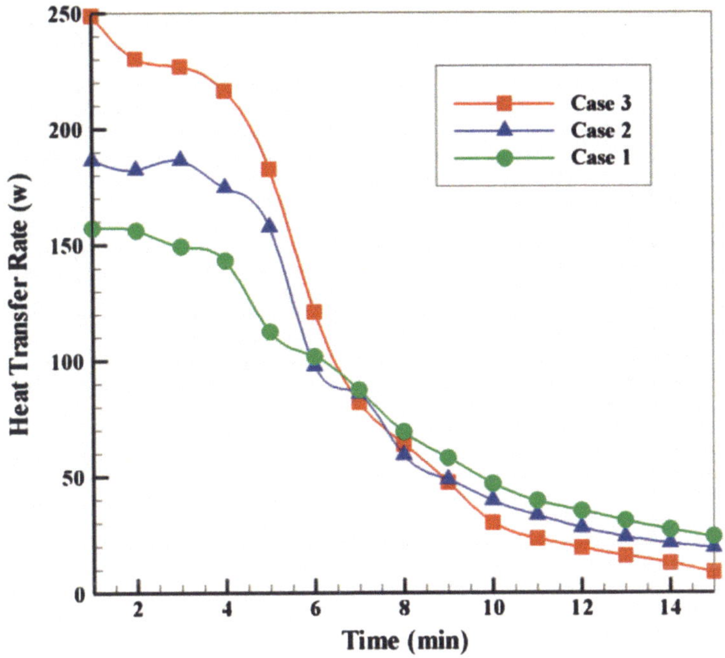

FIGURE 6.12 cont'd.

Fig. 6.19 shows the comparisons of entropy component and melt fraction among various cases. $S_{gen,f}$ is significantly lower compared to another component. The addition of time decreases the $S_{gen,th}$ while melt fraction enhances. Case 3 provides the maximum melt fraction.

6.3 Entropy generation for NEPCM melting process inside a heat storage system

6.3.1 Definition of the problem

The aim of this chapter is to carry out numerical investigation of transient, two-dimensional melting process inside the cavity with V-shaped fins [3]. The cavity is filled with PCM-based copper oxide nanoparticle. This geometry can be fins in Figs. 6.20 and 6.21. In this chapter, we consider $L = 0.04$ m, $D = 0.02$ m, $l = 0.01$m, $t = 0.5$ mm. The concentration of copper oxide is 0.05. Several other authors investigated about application of nanomaterial [4–63].

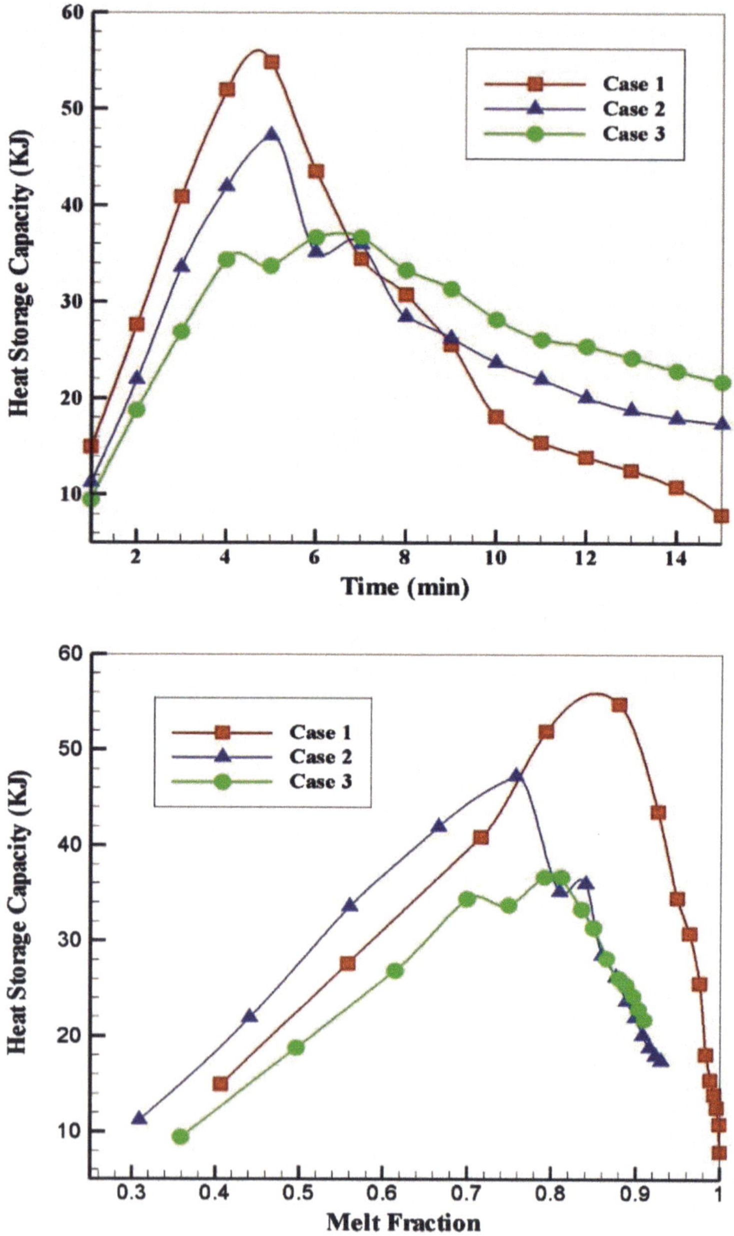

FIGURE 6.13

Heat storage capacity for various cases.

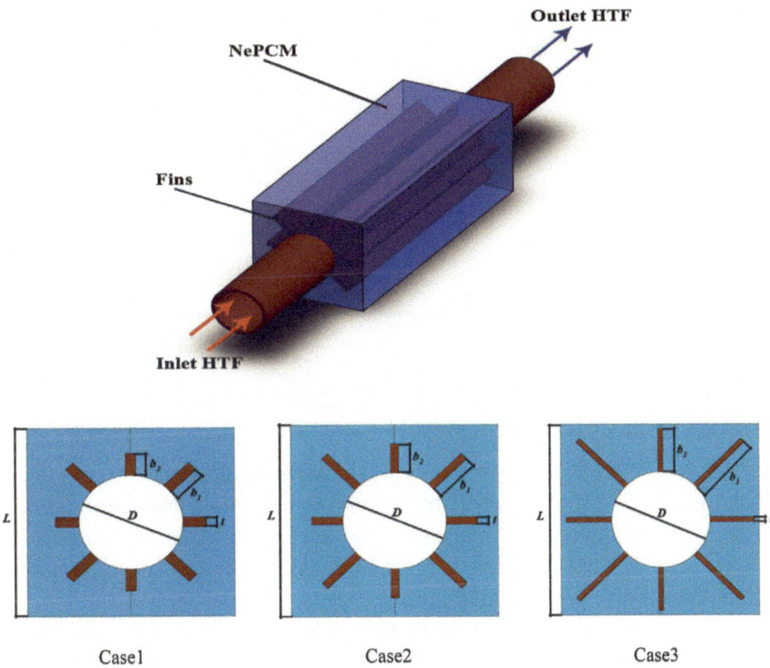

FIGURE 6.14

Finned enclosure for saving energy.

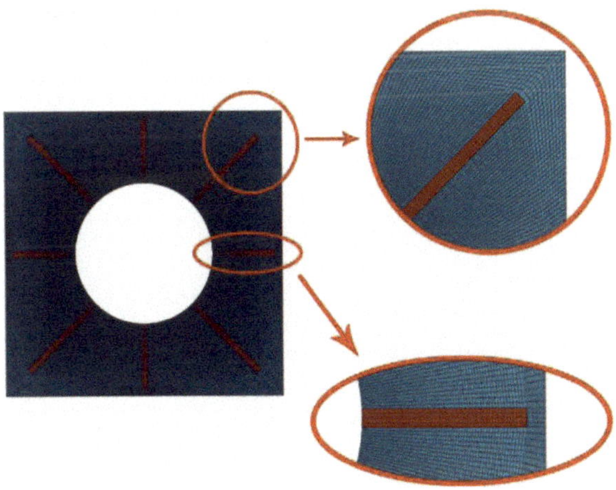

FIGURE 6.15

Example for utilized grid.

Table 6.2 Different cases in this chapter.

	Case 1	Case 2	Case 3
b1	4.5	6	9
b2	6.75	9	13.5
a	2	1.5	1
A_1	9	9	9
A_2	13.5	13.5	13.5
$4(A_1+A_2)$	90	90	90

6.3.2 Effects of active parameters

This section has been devoted to the analysis of melting process in an enclosure with V-shaped fins. Not only melt fraction and temperature but also entropy generation and exergy have been analyzed. The system is filled with n-octadecane paraffin in which CuO has been dispersed. The solution of equations was calculated using finite volume method.

Both first and second law analysis during process was illustrated in Figs. 6.22–6.25. In lower time, the buoyancy is not enough strong to affect the melting. In greater time, convective mechanism affects the melting front and its shape becomes more complex. Vortex becomes stronger as the melting front reaches the outer surface. The middle up of the enclosure converts to liquid phase sooner than the middle down. The region near the down wall is the final space in which the remaining solid converts to liquid. This phenomena is due to absence of rotating eddies in that region. As temperature enhances, thermal entropy generation detracts as depicted in contours. Exergy loss has same treatment, too. Liquid fraction, $S_{gen,f}$, and $S_{gen,th}$ are shown in Fig. 6.26. As time progresses, average temperature enhances and melt fraction increases until it reaches its maximum value. Thermal entropy generation detracts with rise of melt fraction. Frictional component has on maximum point in beginning times.

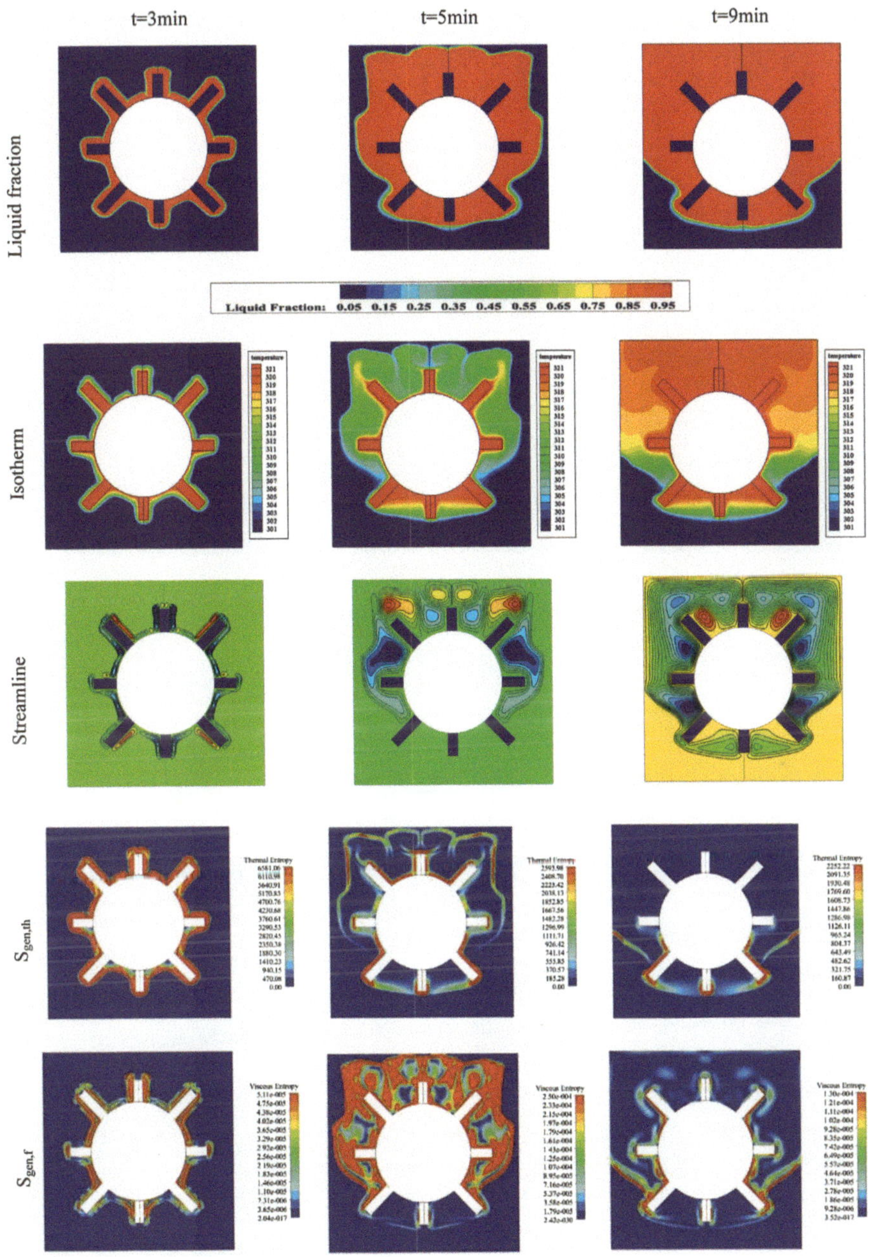

FIGURE 6.16

Contours for various outputs of case 1.

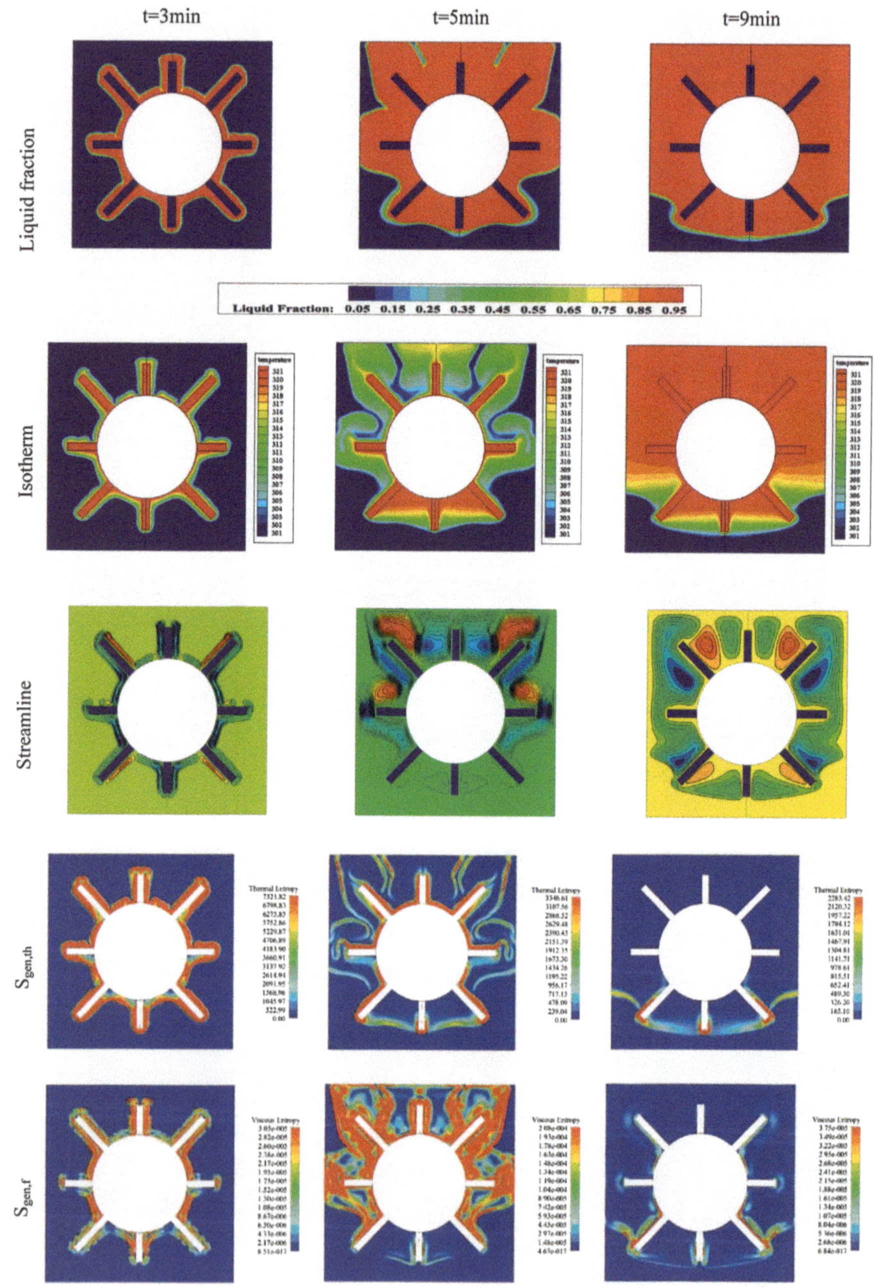

FIGURE 6.17

Contours for various outputs of case 2.

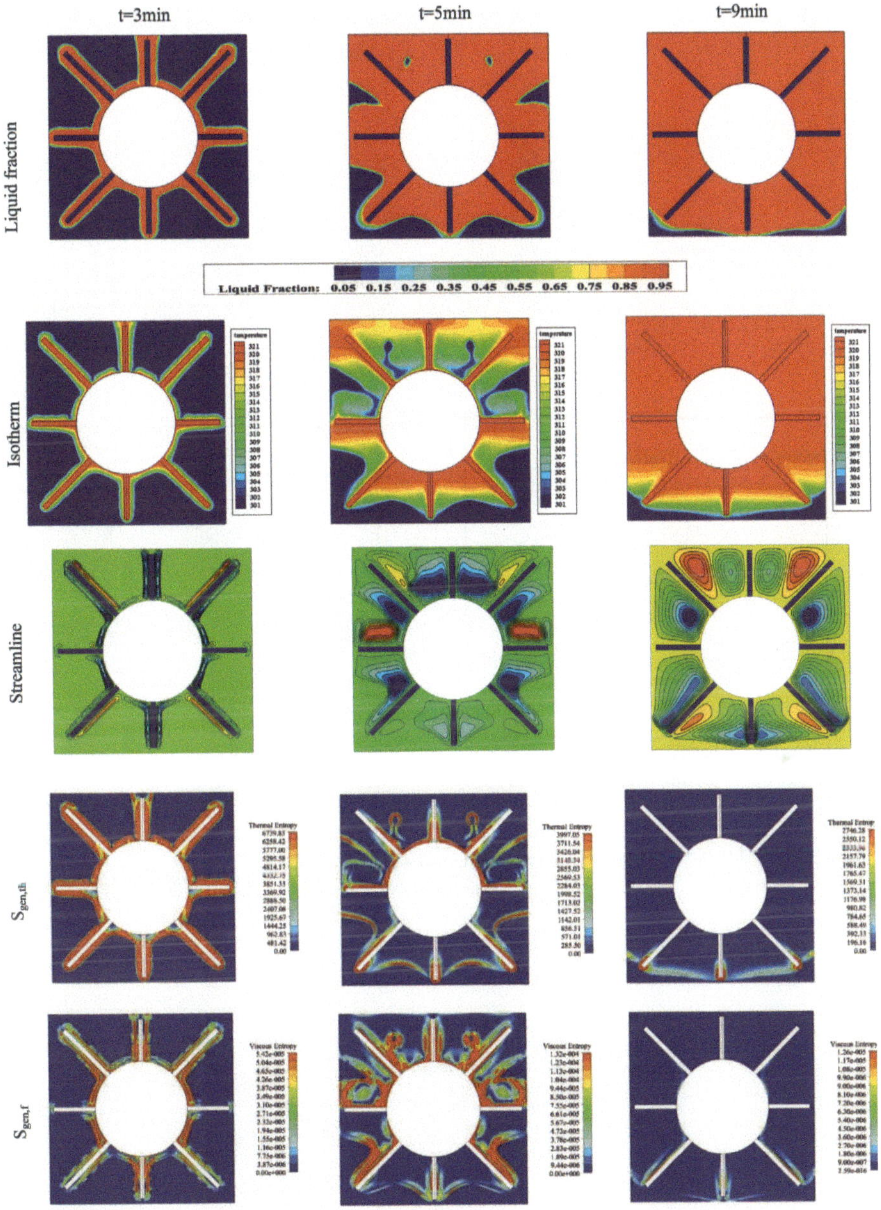

FIGURE 6.18

Contours for various outputs of case 3.

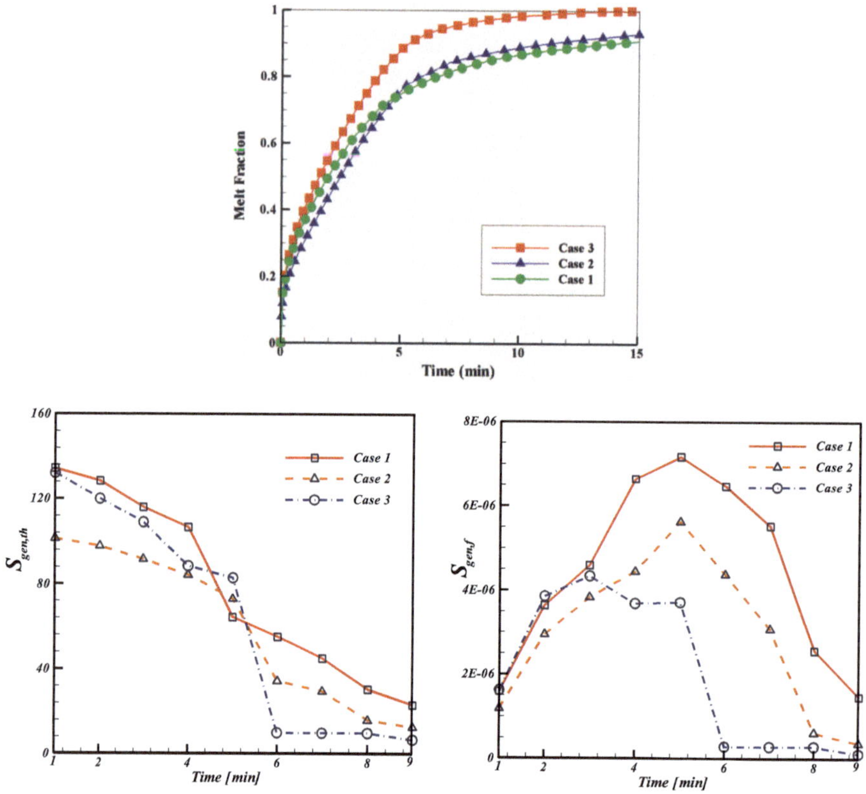

FIGURE 6.19

Melt fraction, entropy generation components for different cases.

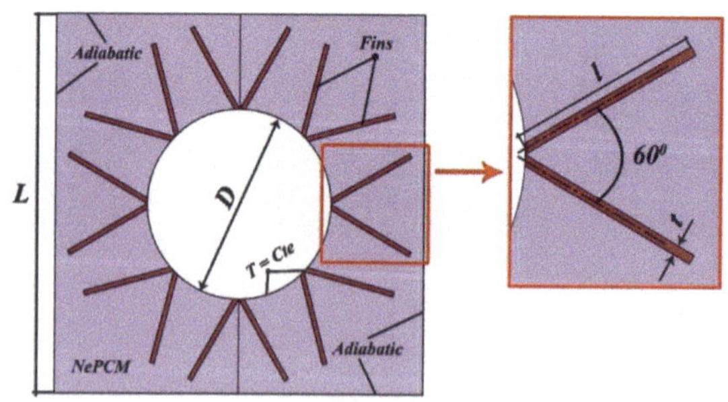

FIGURE 6.20

Enclosure with V-shaped fins.

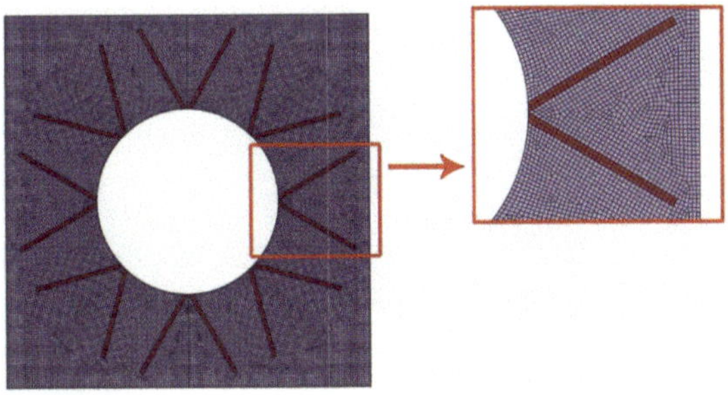

FIGURE 6.21

Example for grid.

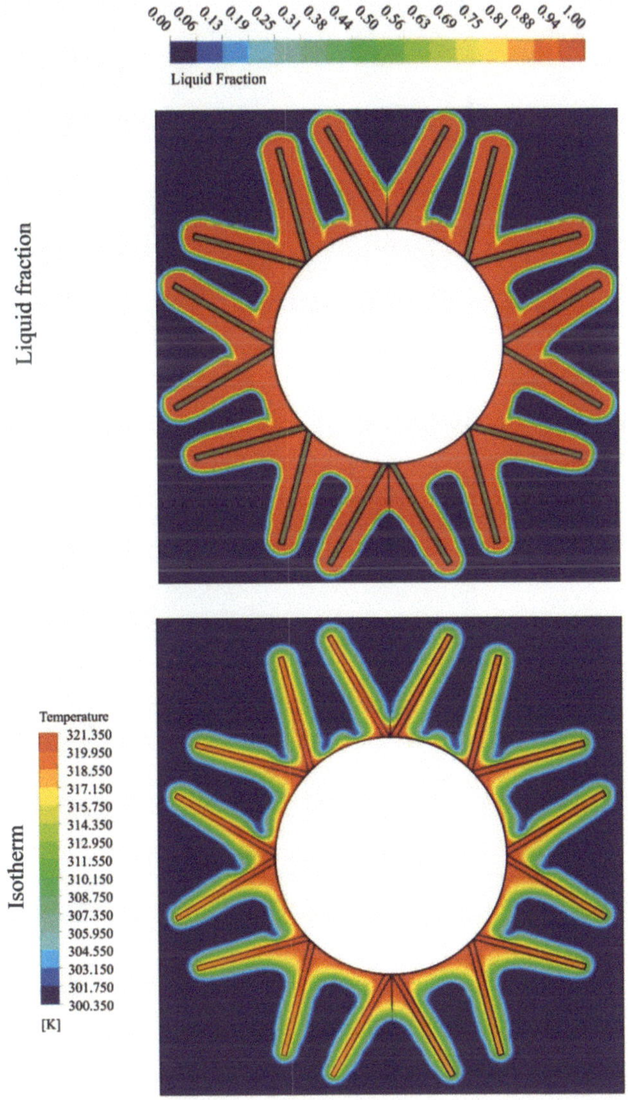

FIGURE 6.22

Contour plots of t = 60 s.

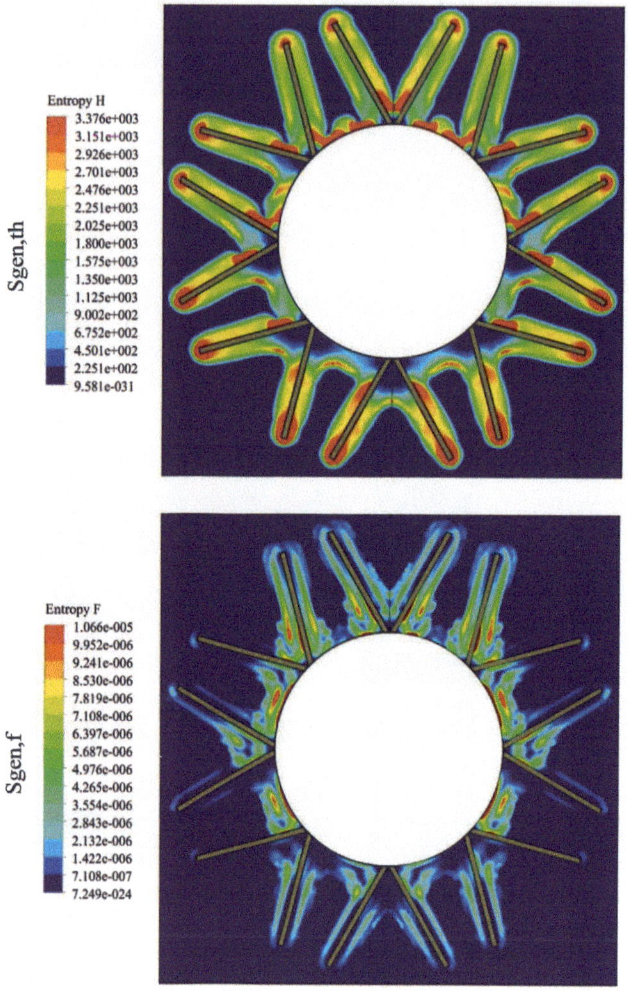

FIGURE 6.22 cont'd.

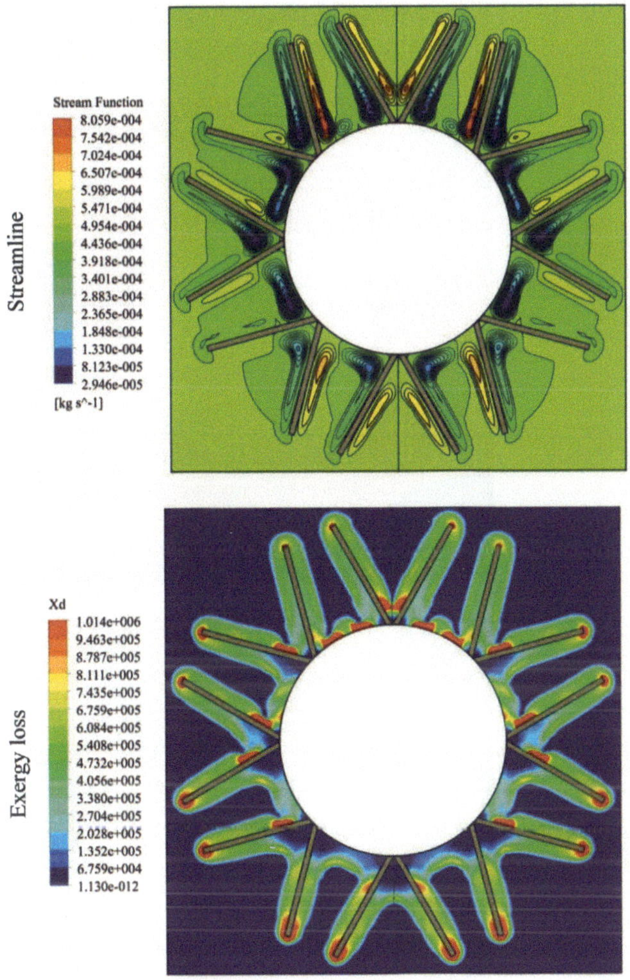

FIGURE 6.22 cont'd.

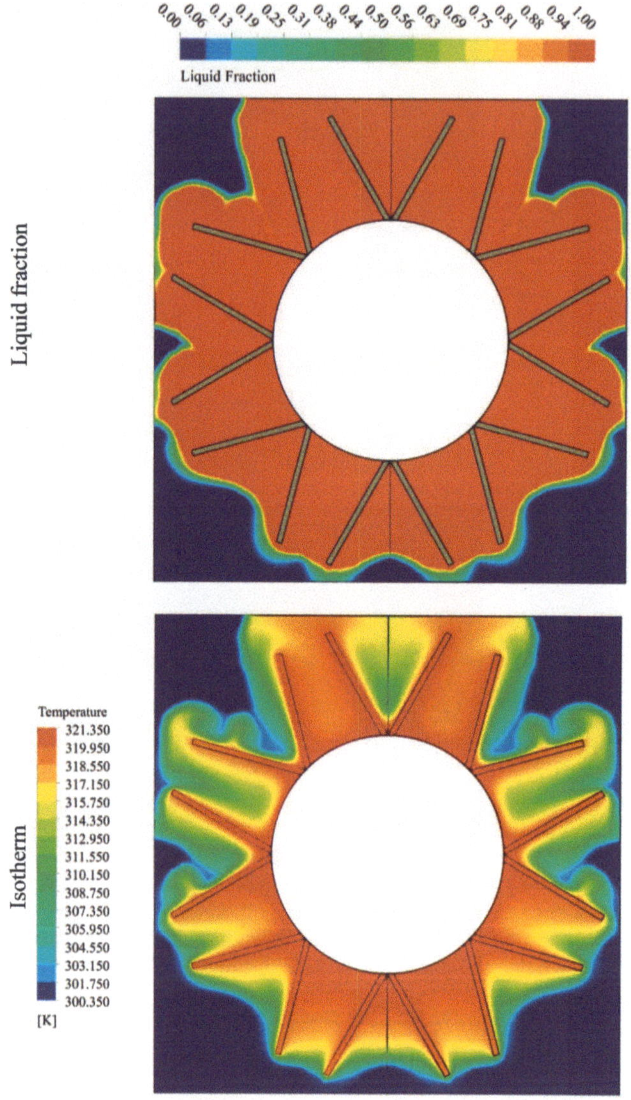

FIGURE 6.23

Contour plots of t = 180 s.

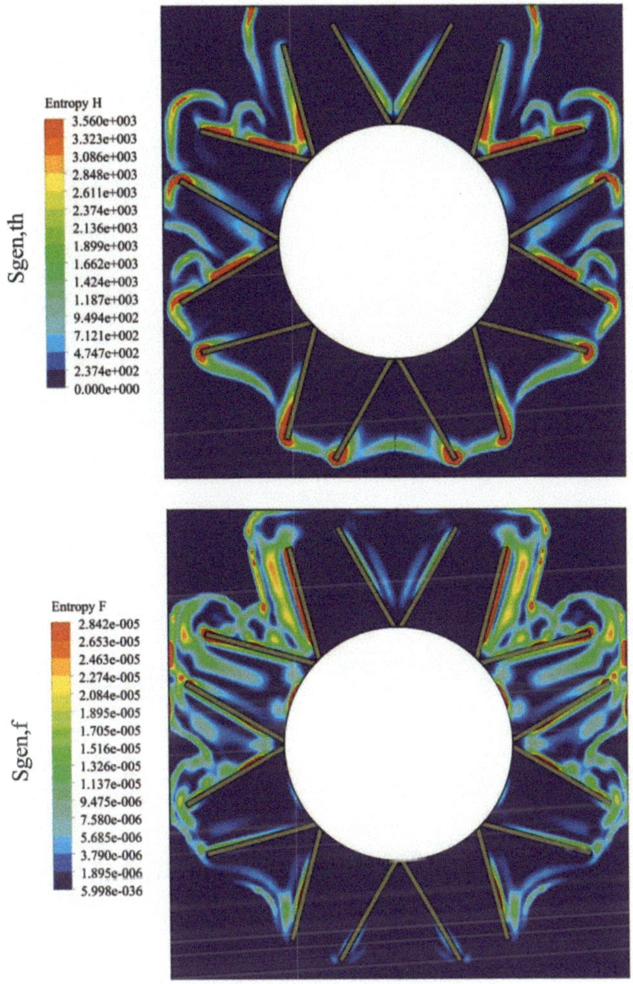

FIGURE 6.23 cont'd.

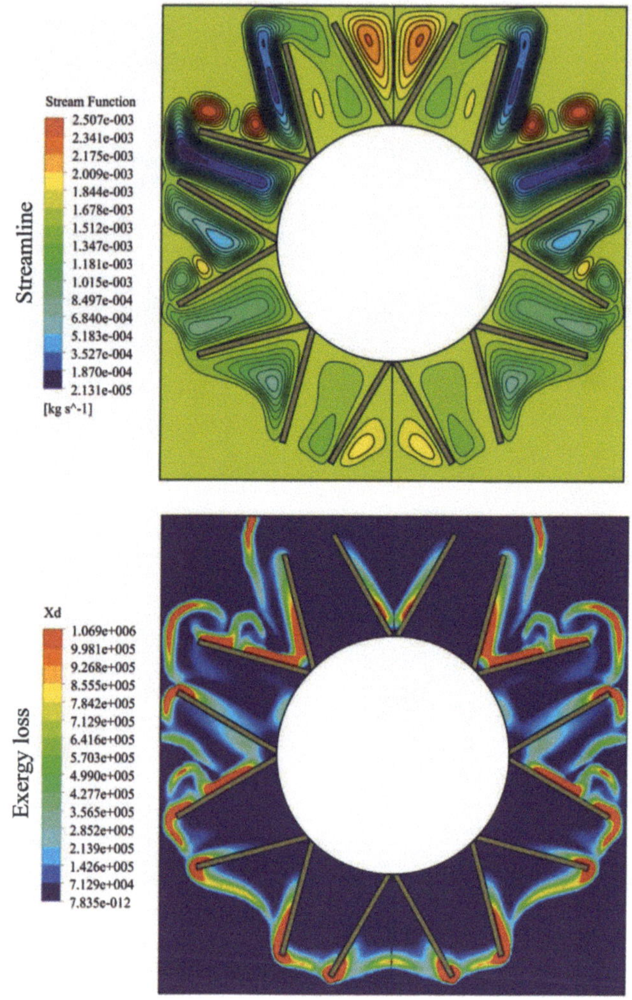

FIGURE 6.23 cont'd.

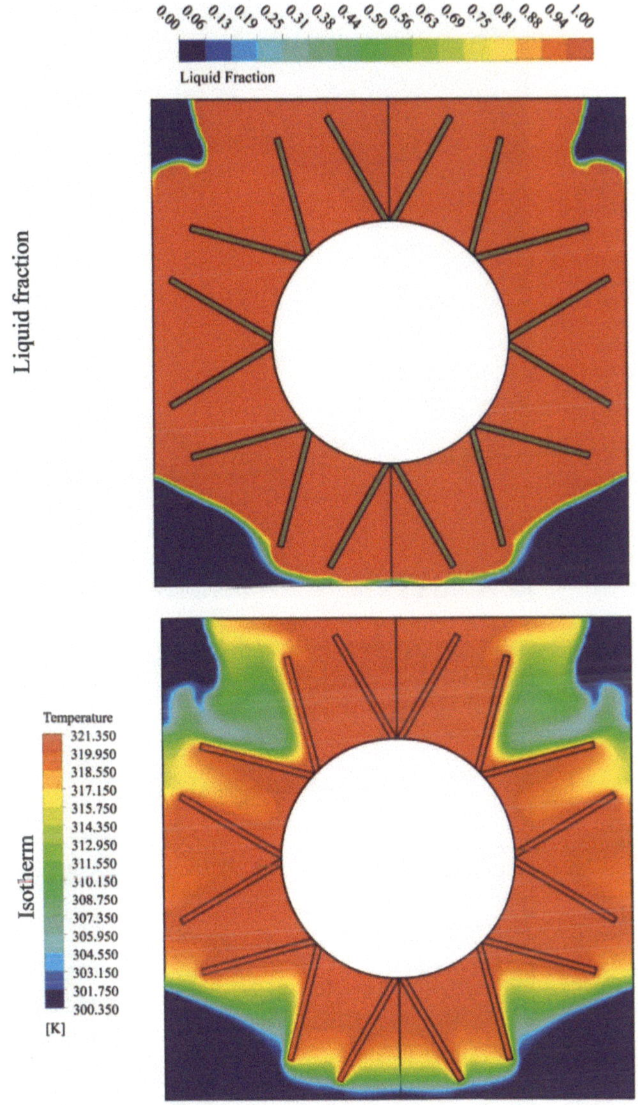

FIGURE 6.24

Contour plots of t = 300 s.

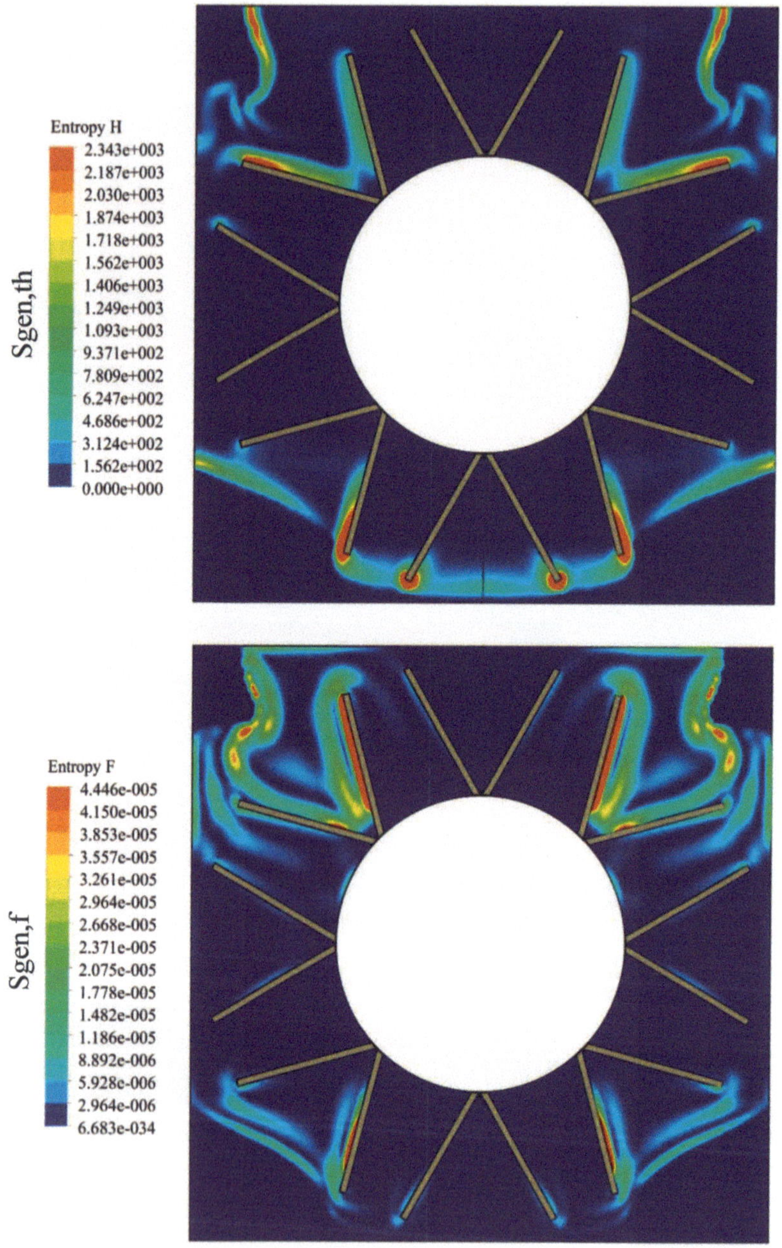

FIGURE 6.24 cont'd.

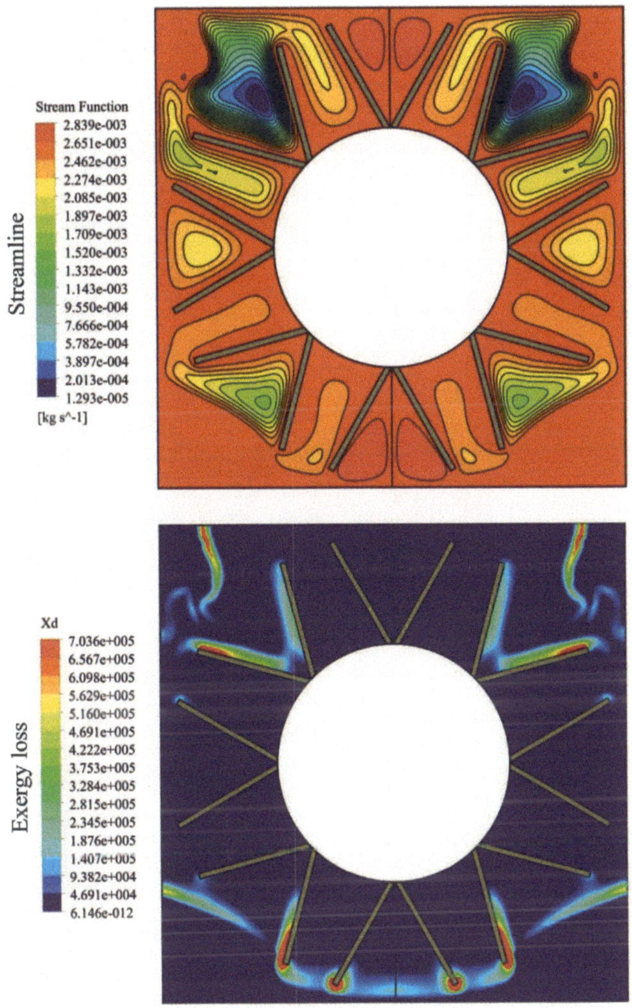

FIGURE 6.24 cont'd.

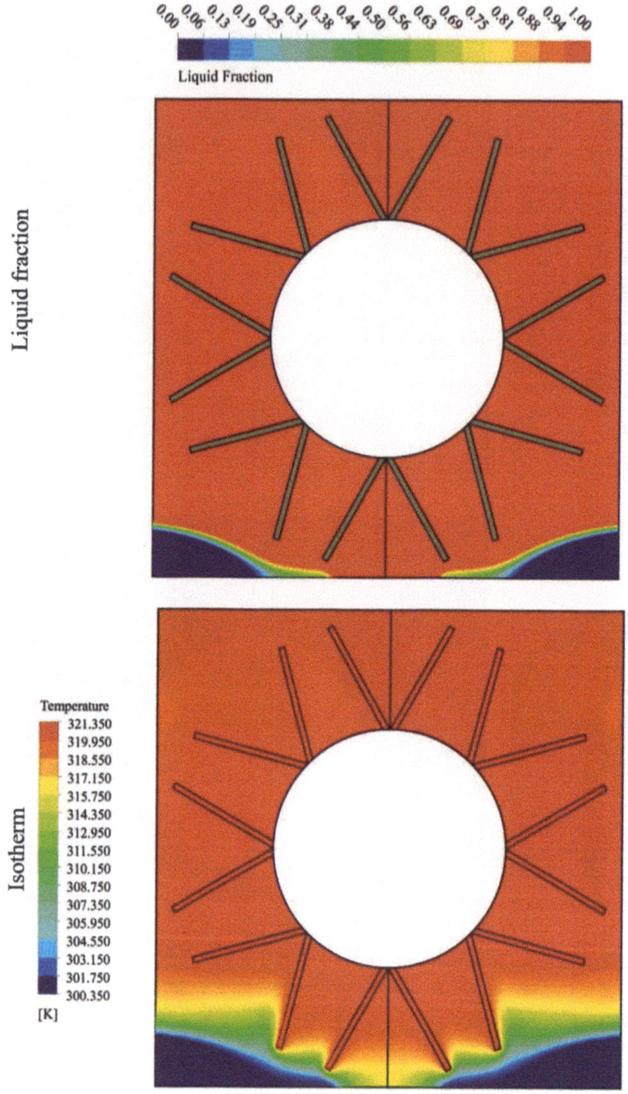

FIGURE 6.25

Contour plots of t = 540 s.

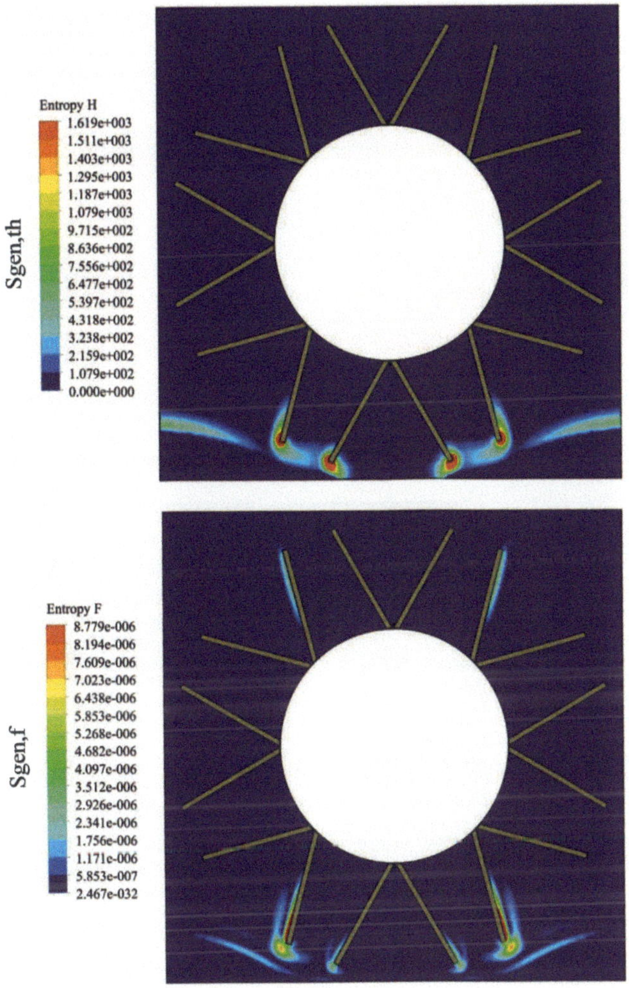

FIGURE 6.25 cont'd.

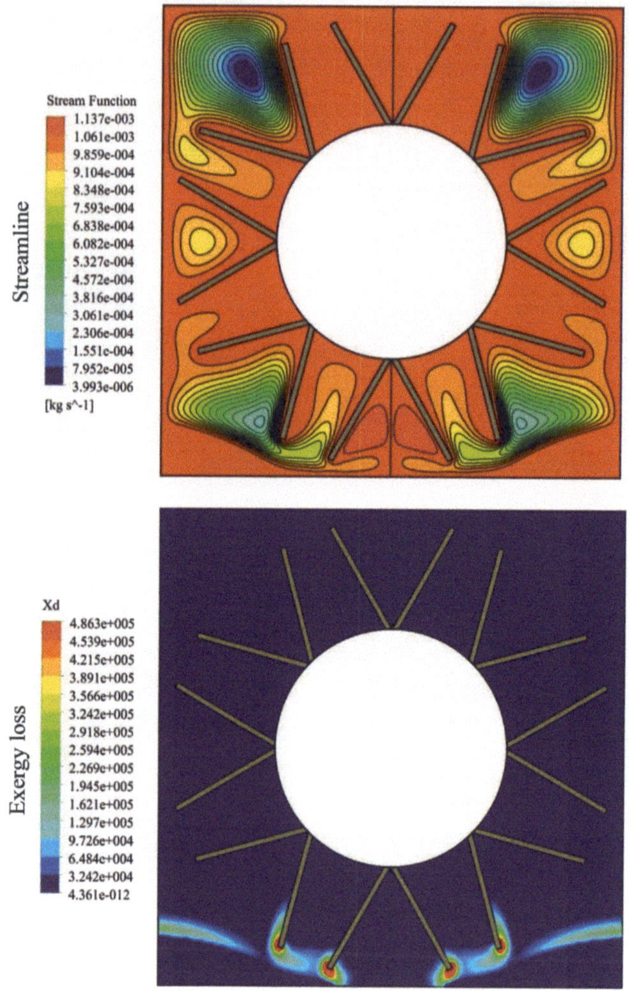

FIGURE 6.25 cont'd.

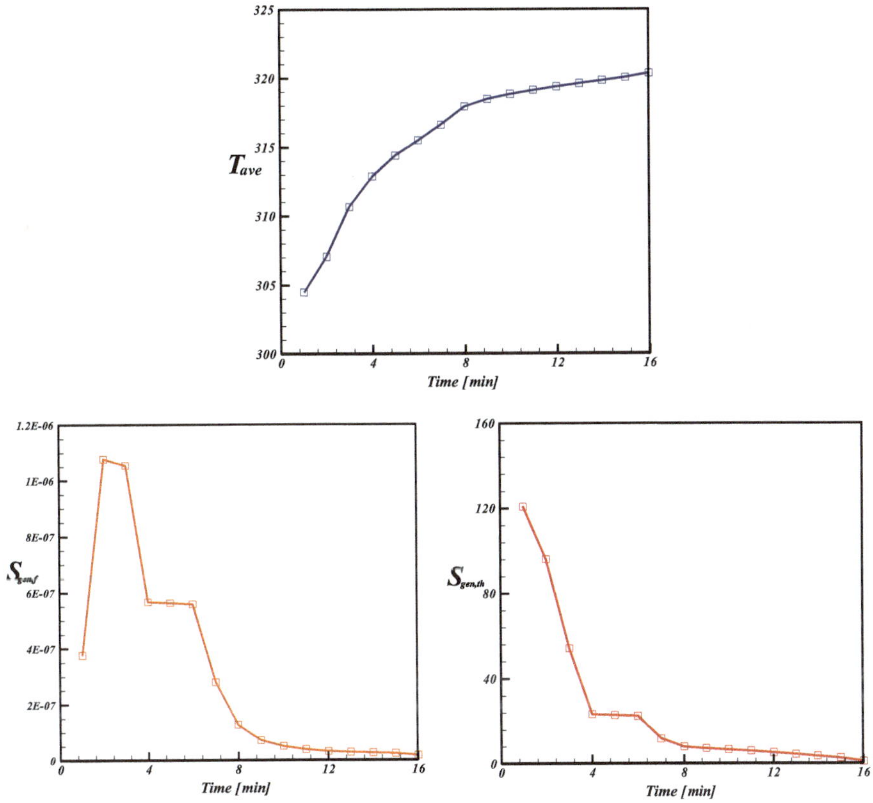

FIGURE 6.26

Variation of temperature and entropy generation during melting process.

References

[1] M. Sheikholeslami, M. Jafaryar, A. Shafee, Z. Li, Simulation of nanoparticles application for expediting melting of PCM inside a finned enclosure, Physica A: Statistical Mechanics and its Applications 523 (2019) 544–556.

[2] M. Sheikholeslami, M. Jafaryar, A. Shafee, Z. Li, Hydrothermal and second law behavior for charging of NEPCM in a two dimensional thermal storage unit, Chinese Journal of Physics 58 (2019) 244–252.

[3] M. Sheikholeslami, M. Jafaryar, A. Shafee, Z. Li, Analyze of entropy generation for NEPCM melting process inside a heat storage system, Microsystem Technologies (2019), https://doi.org/10.1007/s00542-019-04301-w.

[4] M. Sheikholeslami, A. Arabkoohsar, I. Khan, A. Shafee, Z. Li, Impact of Lorentz forces on Fe_3O_4-water ferrofluid entropy and exergy treatment within a permeable semi annulus, Journal of Cleaner Production 221 (2019) 885–898.

[5] M. Sheikholeslami, K. Vajravelu, M.M. Rashidi, Forced convection heat transfer in a semi annulus under the influence of a variable magnetic field, International Journal of Heat and Mass Transfer 92 (2016) 339–348.
[6] M. Sheikholeslami, Magnetic field influence on CuO -H_2O nanofluid convective flow in a permeable cavity considering various shapes for nanoparticles, International Journal of Hydrogen Energy 42 (2017) 19611–19621.
[7] M. Sheikholeslami, Numerical simulation for solidification in a LHTESS by means of Nano-enhanced PCM, Journal of the Taiwan Institute of Chemical Engineers 86 (2018) 25–41.
[8] Y. Qin, H. He, X. Ou, T. Bao, Experimental study on darkening water-rich mud tailings for accelerating desiccation, Journal of Cleaner Production (2019) 118235, https://doi.org/10.1016/j.jclepro.2019.118235.
[9] Y. Qin, J.E. Hiller, D. Meng, Linearity between pavement thermophysical properties and surface temperatures, Journal of Materials in Civil Engineering (2019), https://doi.org/10.1061/(ASCE)MT.1943-5533.0002890.
[10] Y. Qin, J. Luo, Z. Chen, G. Mei, L.-E. Yan, Measuring the albedo of limited-extent targets without the aid of known-albedo masks, Solar Energy 171 (2018) 971–976.
[11] Y. Qin, A review on the development of cool pavements to mitigate urban heat island effect, Renewable and Sustainable Energy Reviews 52 (2015) 445–459.
[12] M. Sheikholeslami, T. Hayat, A. Alsaedi, Numerical study for external magnetic source influence on water based nanofluid convective heat transfer, International Journal of Heat and Mass Transfer 106 (2017) 745–755.
[13] M. Sheikholeslami, T. Hayat, A. Alsaedi, S. Abelman, Numerical analysis of EHD nanofluid force convective heat transfer considering electric field dependent viscosity, International Journal of Heat and Mass Transfer 108 (2017) 2558–2565.
[14] M. Sheikholeslami, M.A. Sheremet, A. Shafee, I. Tlili, Simulation of nanoliquid thermogravitational convection within a porous chamber imposing magnetic and radiation impacts, Physica A: Statistical Mechanics and its Applications (2020), https://doi.org/10.1016/j.physa.2019.124058.
[15] M. Sheikholeslami, Effect of uniform suction on nanofluid flow and heat transfer over a cylinder, Journal of the Brazilian Society of Mechanical Sciences and Engineering 37 (2015) 1623–1633.
[16] M. Sheikholeslami, Lattice Boltzmann Method simulation of MHD non-Darcy nanofluid free convection, Physica B 516 (2017) 55–71.
[17] M. Sheikholeslami, Numerical approach for MHD Al_2O_3-water nanofluid transportation inside a permeable medium using innovative computer method, Computer Methods in Applied Mechanics and Engineering 344 (2019) 306–318.
[18] M. Sheikholeslami, A.N. Keshteli, H. Babazadeh, Nanoparticles favorable effects on performance of thermal storage units, Journal of Molecular Liquids 300 (2020) 112329.
[19] X. Ma, M. Sheikholeslami, M. Jafaryar, A. Shafee, T. Nguyen-Thoi, Z. Li, Solidification inside a clean energy storage unit utilizing phase change material with copper oxide nanoparticles, Journal of Cleaner Production (2019), https://doi.org/10.1016/j.jclepro.2019.118888.
[20] M. Sheikholeslami, Numerical modeling of nano enhanced PCM solidification in an enclosure with metallic fin, Journal of Molecular Liquids 259 (2018) 424–438.
[21] M. Sheikholeslami, New computational approach for exergy and entropy analysis of nanofluid under the impact of Lorentz force through a porous media, Computer Methods in Applied Mechanics and Engineering 344 (2019) 319–333.

[22] M. Sheikholeslami, Finite element method for PCM solidification in existence of CuO nanoparticles, Journal of Molecular Liquids 265 (2018) 347–355.
[23] M. Sheikholeslami, M. Jafaryar, A. Shafee, Z. Li, R.-ul Haq, Heat transfer of nanoparticles employing innovative turbulator considering entropy generation, International Journal of Heat and Mass Transfer 136 (2019) 1233–1240.
[24] M. Sheikholeslami, Solidification of NEPCM under the effect of magnetic field in a porous thermal energy storage enclosure using CuO nanoparticles, Journal of Molecular Liquids 263 (2018) 303–315.
[25] M. Sheikholeslami, Influence of magnetic field on Al_2O_3-H_2O nanofluid forced convection heat transfer in a porous lid driven cavity with hot sphere obstacle by means of LBM, Journal of Molecular Liquids 263 (2018) 472–488.
[26] M. Sheikholeslami, O. Mahian, Enhancement of PCM solidification using inorganic nanoparticles and an external magnetic field with application in energy storage systems, Journal of Cleaner Production 215 (2019) 963–977.
[27] M. Sheikholeslami, B. Rezaeianjouybari, M. Darzi, A. Shafee, Z. Li, T.K. Nguyen, Application of nano-refrigerant for boiling heat transfer enhancement employing an experimental study, International Journal of Heat and Mass Transfer 141 (2019) 974–980.
[28] M. Sheikholeslami, M. Jafaryar, M. Hedayat, A. Shafee, Z. Li, T.K. Nguyen, M. Bakouri, Heat transfer and turbulent simulation of nanomaterial due to compound turbulator including irreversibility analysis, International Journal of Heat and Mass Transfer 137 (2019) 1290–1300.
[29] M. Sheikholeslami, Application of Darcy law for nanofluid flow in a porous cavity under the impact of Lorentz forces, Journal of Molecular Liquids 266 (2018) 495–503.
[30] M. Sheikholeslami, S.A. Shehzad, Z. Li, A. Shafee, Numerical modeling for alumina nanofluid magnetohydrodynamic convective heat transfer in a permeable medium using Darcy law, International Journal of Heat and Mass Transfer 127 (2018) 614–622.
[31] M. Sheikholeslami, Z. Li, A. Shafee, Lorentz forces effect on NEPCM heat transfer during solidification in a porous energy storage system, International Journal of Heat and Mass Transfer 127 (2018) 665–674.
[32] M. Sheikholeslami, M. Darzi, Z. Li, Experimental investigation for entropy generation and exergy loss of nano-refrigerant condensation process, International Journal of Heat and Mass Transfer 125 (2018) 1087–1095.
[33] M. Sheikholeslami, S.A. Shehzad, Z. Li, Water based nanofluid free convection heat transfer in a three dimensional porous cavity with hot sphere obstacle in existence of Lorenz forces, International Journal of Heat and Mass Transfer 125 (2018) 375–386.
[34] M. Sheikholeslami, M. Jafaryar, Z. Li, Nanofluid turbulent convective flow in a circular duct with helical turbulators considering CuO nanoparticles, International Journal of Heat and Mass Transfer 124 (2018) 980–989.
[35] M. Sheikholeslami, A. Ghasemi, Solidification heat transfer of nanofluid in existence of thermal radiation by means of FEM, International Journal of Heat and Mass Transfer 123 (2018) 418–431.
[36] M. Sheikholeslami, S.A. Shehzad, CVFEM simulation for nanofluid migration in a porous medium using Darcy model, International Journal of Heat and Mass Transfer 122 (2018) 1264–1271.
[37] M. Sheikholeslami, M. Darzi, M.K. Sadoughi, Heat transfer improvement and pressure drop during condensation of refrigerant-based nanofluid; an experimental procedure, International Journal of Heat and Mass Transfer 122 (2018) 643–650.

[38] M. Sheikholeslami, S.A. Shehzad, Simulation of water based nanofluid convective flow inside a porous enclosure via Non-equilibrium model, International Journal of Heat and Mass Transfer 120 (2018) 1200–1212.
[39] M. Sheikholeslami, M. Seyednezhad, Simulation of nanofluid flow and natural convection in a porous media under the influence of electric field using CVFEM, International Journal of Heat and Mass Transfer 120 (2018) 772–781.
[40] M. Sheikholeslami, H.B. Rokni, Numerical modeling of nanofluid natural convection in a semi annulus in existence of Lorentz force, Computer Methods in Applied Mechanics and Engineering 317 (2017) 419–430.
[41] M. Sheikholeslami, M.M. Bhatti, Forced convection of nanofluid in presence of constant magnetic field considering shape effects of nanoparticles, International Journal of Heat and Mass Transfer 111 (2017) 1039–1049.
[42] M. Sheikholeslami, M. Sadoughi, Mesoscopic method for MHD nanofluid flow inside a porous cavity considering various shapes of nanoparticles, International Journal of Heat and Mass Transfer 113 (2017) 106–114.
[43] M. Sheikholeslami, R. Ellahi, Three dimensional mesoscopic simulation of magnetic field effect on natural convection of nanofluid, International Journal of Heat and Mass Transfer 89 (2015) 799–808.
[44] M. Sheikholeslami, S.A. Shehzad, Thermal radiation of ferrofluid in existence of Lorentz forces considering variable viscosity, International Journal of Heat and Mass Transfer 109 (2017) 82–92.
[45] M. Sheikholeslami, M.M. Bhatti, Active method for nanofluid heat transfer enhancement by means of EHD, International Journal of Heat and Mass Transfer 109 (2017) 115–122.
[46] M. Sheikholeslami, S.A. Shehzad, Magnetohydrodynamic nanofluid convection in a porous enclosure considering heat flux boundary condition, International Journal of Heat and Mass Transfer 106 (2017) 1261–1269.
[47] M. Sheikholeslami, H.B. Rokni, Nanofluid two phase model analysis in existence of induced magnetic field, International Journal of Heat and Mass Transfer 107 (2017) 288–299.
[48] M. Sheikholeslami, M.K. Sadoughi, Simulation of CuO- water nanofluid heat transfer enhancement in presence of melting surface, International Journal of Heat and Mass Transfer 116 (2018) 909–919.
[49] M. Sheikholeslami, H.B. Rokni, Numerical simulation for impact of Coulomb force on nanofluid heat transfer in a porous enclosure in presence of thermal radiation, International Journal of Heat and Mass Transfer 118 (2018) 823–831.
[50] M. Sheikholeslami, S.A. Shehzad, Numerical analysis of $Fe_3O_4-H_2O$ nanofluid flow in permeable media under the effect of external magnetic source, International Journal of Heat and Mass Transfer 118 (2018) 182–192.
[51] M. Sheikholeslami, S.A. Shehzad, CVFEM for influence of external magnetic source on $Fe_3O_4 - H_2O$ nanofluid behavior in a permeable cavity considering shape effect, International Journal of Heat and Mass Transfer 115 (2017) 180–191.
[52] M. Sheikholeslami, M. Seyednezhad, Nanofluid heat transfer in a permeable enclosure in presence of variable magnetic field by means of CVFEM, International Journal of Heat and Mass Transfer 114 (2017) 1169–1180.
[53] M. Sheikholeslami, H.B. Rokni, Simulation of nanofluid heat transfer in presence of magnetic field: a review, International Journal of Heat and Mass Transfer 115 (2017) 1203–1233.

[54] M. Sheikholeslami, T. Hayat, A. Alsaedi, On simulation of nanofluid radiation and natural convection in an enclosure with elliptical cylinders, International Journal of Heat and Mass Transfer 115 (2017) 981–991.

[55] M. Sheikholeslami, B. Houman, Rokni, Melting heat transfer influence on nanofluid flow inside a cavity in existence of magnetic field, International Journal of Heat and Mass Transfer 114 (2017) 517–526.

[56] M. Sheikholeslami, Numerical investigation for CuO-H_2O nanofluid flow in a porous channel with magnetic field using mesoscopic method, Journal of Molecular Liquids 249 (2018) 739–746.

[57] M. Sheikholeslami, Numerical investigation of nanofluid free convection under the influence of electric field in a porous enclosure, Journal of Molecular Liquids 249 (2018) 1212–1221.

[58] M. Sheikholeslami, CuO-water nanofluid flow due to magnetic field inside a porous media considering Brownian motion, Journal of Molecular Liquids 249 (2018) 921–929.

[59] M. Sheikholeslami, A. Zeeshan, Analysis of flow and heat transfer in water based nanofluid due to magnetic field in a porous enclosure with constant heat flux using CVFEM, Computer Methods in Applied Mechanics and Engineering 320 (2017) 68–81.

[60] M. Sheikholeslami, M. Jafaryar, S. Saleem, Z. Li, S. Ahmad, Y. Jiang, Nanofluid heat transfer augmentation and exergy loss inside a pipe equipped with innovative turbulators, International Journal of Heat and Mass Transfer 126 (2018) 156–163.

[61] M. Sheikholeslami, R.-ul Haq, A. Shafee, Z. Li, Y.G. Elaraki, I. Tlili, Heat transfer simulation of heat storage unit with nanoparticles and fins through a heat exchanger, International Journal of Heat and Mass Transfer 135 (2019) 470–478.

[62] M. Sheikholeslami, R.-ul Haq, A. Shafee, Z. Li, Heat transfer behavior of nanoparticle enhanced PCM solidification through an enclosure with V shaped fins, International Journal of Heat and Mass Transfer 130 (2019) 1322–1342.

[63] M. Sheikholeslami, A. Ghasemi, Z. Li, A. Shafee, S. Saleem, Influence of CuO nanoparticles on heat transfer behavior of PCM in solidification process considering radiative source term, International Journal of Heat and Mass Transfer 126 (2018) 1252–1264.

CHAPTER

Air heat exchanger storage unit

7

7.1 Heat transfer simulation during charging of nanoparticle-enhanced PCM within a channel

7.1.1 Definition of the problem

A geometrical picture of the heat storage which was taken into account in current research has been demonstrated in Fig. 7.1. Current study may be practiced in heating and cooling of houses. NEPCM and air are used in outer and inner sides of duct [1]. The hot laminar air flow (Re = 2000) which can be the result of solar energy acts as heating source for melting process. So, the temperature of air reduces in the outlet of duct. The top wall is adiabatic. The entire above layer carries a mixture of n-octadecane paraffin and CuO nanoparticles (see Table 7.1). Four selecting points are considered in outer layer to show the behavior of NEPCM during charging process. To analyze the melting process in the upper layer, the following equations can be expressed.

$$\nabla \cdot \vec{V} = 0 \tag{7.1}$$

$$\frac{\partial u}{\partial t} + \frac{\nabla P}{\rho_{nf}} + \nabla u \cdot \vec{V} = \mu_{nf}\frac{1}{\rho_{nf}}\nabla^2 u + u\,\frac{(1-\lambda)^2}{\lambda^3+\varepsilon}C \tag{7.2}$$

$$\frac{\partial v}{\partial t} + \vec{V} \cdot \nabla v = \frac{1}{\rho_{nf}}\left(-\nabla P + \mu_{nf}\nabla^2 v\right) + C\,v\,\frac{(1-\lambda)^2}{\lambda^3+\varepsilon} \tag{7.3}$$

$$\frac{\partial T}{\partial t} + \vec{V} \cdot \nabla T + \frac{\partial(\rho L \lambda)_{nf}}{\partial t} = \frac{k_{nf}}{(\rho C_p)_{nf}}\nabla^2 T \tag{7.4}$$

We employed single-phase model for estimating properties of NEPCM:

$$\mu_{nf} = \frac{\mu_f}{(1-\phi)^{2.5}} \tag{7.5}$$

$$(C_p)_{nf} = \left[(\rho C_p)_f(1-\phi) + (\rho C_p)_s\phi\right]/\rho_{nf} \tag{7.6}$$

$$\rho_{nf} = (1-\phi)\rho_f + \rho_s\phi \tag{7.7}$$

Nanofluid in Heat Exchangers for Mechanical Systems. https://doi.org/10.1016/B978-0-12-821923-2.00007-5

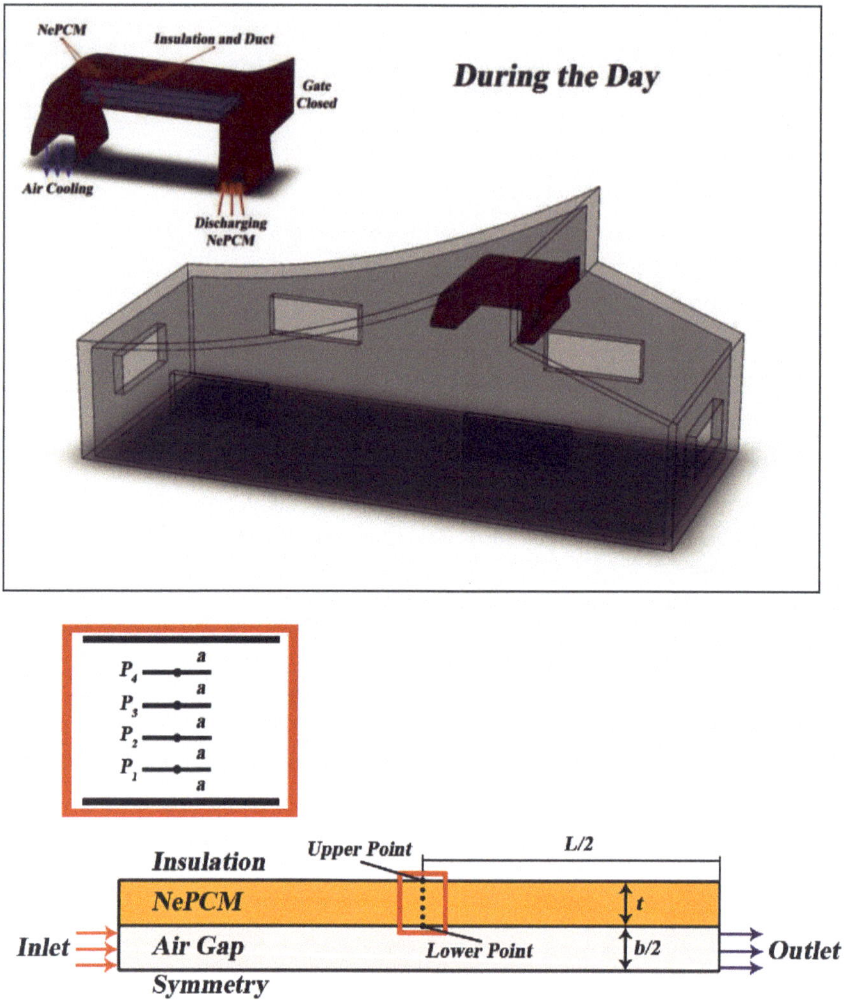

L	1 m
b	0.05 m
t	0.025 m
a	5 mm

FIGURE 7.1

Heat storage unit with air heat exchanger.

Table 7.1 Characteristics of paraffin and CuO as nanoparticles.

Property	N-octadecane		CuO nanoparticle
	Solid	Liquid	
ρ [kg/m^3]	865	770	6500
C_p [j/kg K]	1934	2196	540
k [w/mK]	0.358	0.148	18
μ × 10^3 [Pa.s]	–	3.85	–
β × 10^5 [K^{-1}]	–	91	29
L [j/kg]	–	243,500	–
Fusion point [°C]	28	28	–

$$k_{nf} = k_f \frac{2k_f - 2\phi KK + k_p}{2k_f + \phi KK + k_p}, KK = k_f - k \tag{7.8}$$

$$(\rho L)_{nf} = (1-\phi)(\rho L)_f \tag{7.9}$$

Enthalpy can be calculated according to the following equations, take note that h_{ref} is enthalpy at $T_{ref} = 273K$.

$$H_e = h + \gamma L \tag{7.10}$$

$$h = h_{ref} + \int_{T_{ret}}^{T} (C_p)_{nf} dT \tag{7.11}$$

$$\gamma = \begin{cases} 0 & T < T_s \\ \dfrac{T - T_s}{T_l - T_s} & T_s < T < T_l \\ 1 & T < T_l \end{cases} \tag{7.12}$$

7.1.2 Effects of active parameters

Section 7.3 of this chapter exhibits numerical simulation for melting due to hot air flow in a duct. Air flows inside the inner layer, and outer layer is filled with NEPCM. This process has been take place during the day when the sun exists in sky. Fine volume method has been used for current paper by employing ANSYS Fluent 18.1. In this software, we selected SIMPLE, PRESTO, second-order upwind approach for pressure–velocity coupling, pressure correction, and advection terms, respectively. As exhibited in Fig. 7.2, current code was authenticated by making a comparison with the previous paper. We proved that our outputs are in great agreement with previous experimental data.

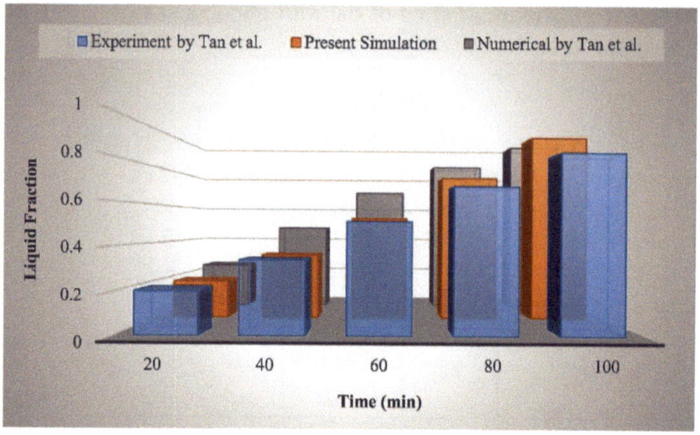

FIGURE 7.2

Verification of current code [1].

Temperatures of NEPCM and air can be influenced by dispersing nanoparticle. These effects are shown in Figs. 7.3–7.5. The temperature close to the upper wall is lower than inner sides. This is due to flow of hot air inside the inner duct. Higher volume fraction of nanofluid causes greater conduction, and air reaches to final temperature in lower time. Moreover, NEPCM reaches to melting point sooner than pure PCM. As depicted in Fig. 7.5, boundary layer thickness reduces with increase in ϕ. As time passes, temperature of NEPCM enhances to reach the melting point. Nusselt number in air side is reported in Fig. 7.6. By increasing concentration of NEPCM, outlet temperature of air enhances. It is not a deniable fact that dispersing CuO causes to more heat transfer and a greater volume of the upper layer affected is by heat conduction. Therefore, to reach full melting, there is no need to further heat transfer with air side. Consequently, rate of heat transfer in air side has reverse behavior as time passes. Liquid fraction of NEPCM and heat storage capacity with respect to ϕ was demonstrated in Figs. 7.7 and 7.8. Melting in presence of nanoparticles finishes sooner than pure paraffin. Heat storage profile has one maximum point, and this extremum shift to greater time by adding nanoparticles.

7.2 Nanoparticle application for heat transfer and irreversibility analysis in an air conditioning unit

7.2.1 Definition of the problem

Copper oxide enhanced paraffin melting process within an air heat exchanger has been investigated as shown in Fig. 7.9. The duct was extended to 1000 mm from the entrance, and thickness of each layer of NEPCM is 25 mm. The air gap as inner duct has 50 mm width. Air becomes warmer due to sunlight and enters the inner duct to melt the NEPCM. The regime of air flow is laminar, and current problem can be

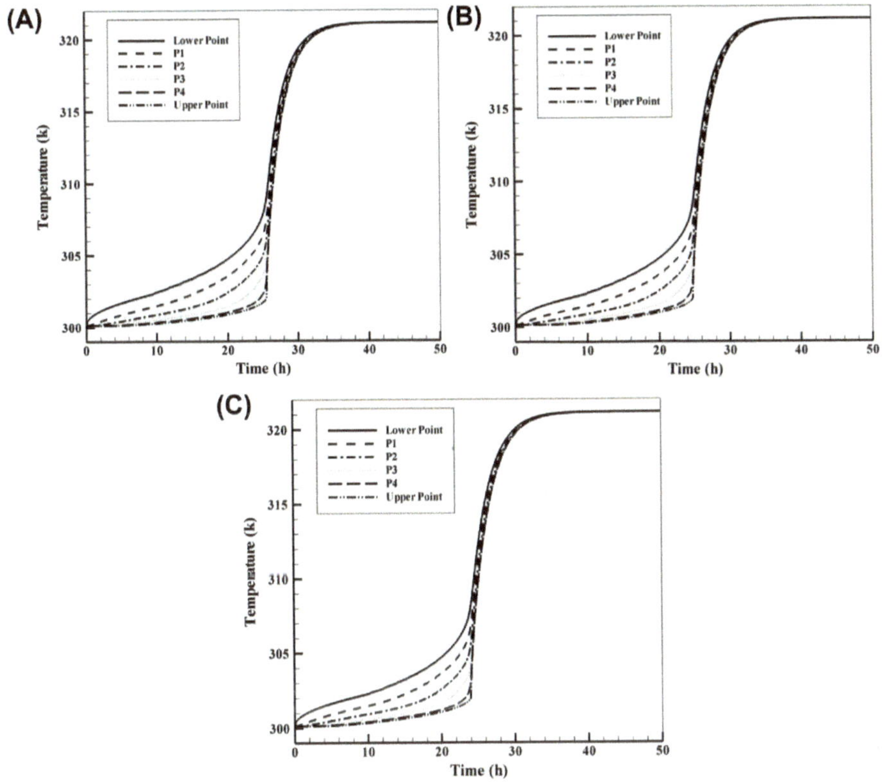

FIGURE 7.3

Temperature profile during melting process for various concentration of nanofluid. (A) $\phi = 0$; (B) $\phi = 0.025$; (C) $\phi = 0.05$.

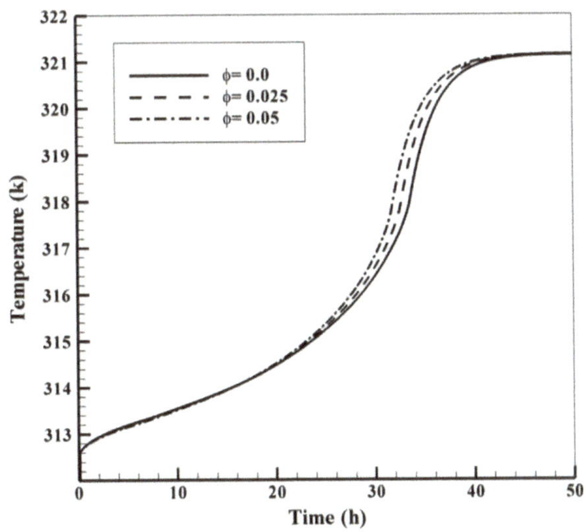

FIGURE 7.4

Air outlet temperature variation during melting process.

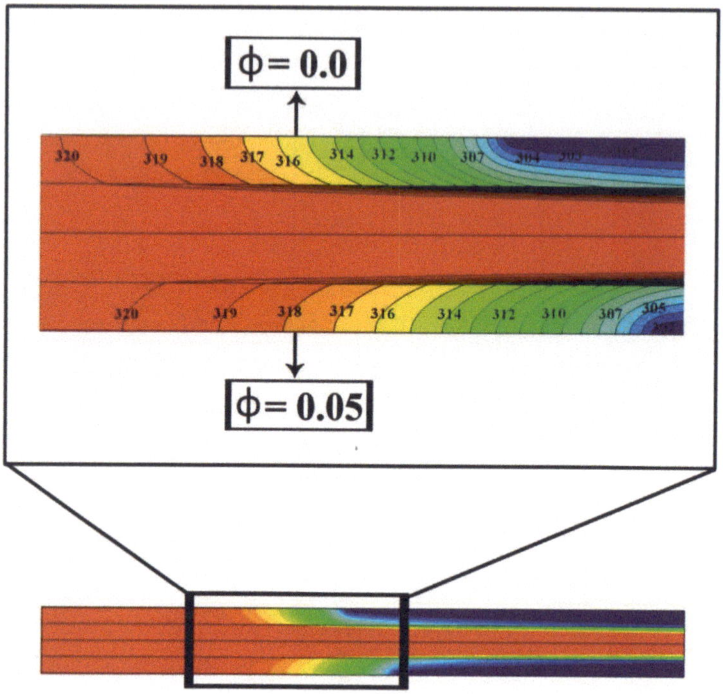

FIGURE 7.5

Isotherms at t = 20 h for various volume fractions of nanofluid.

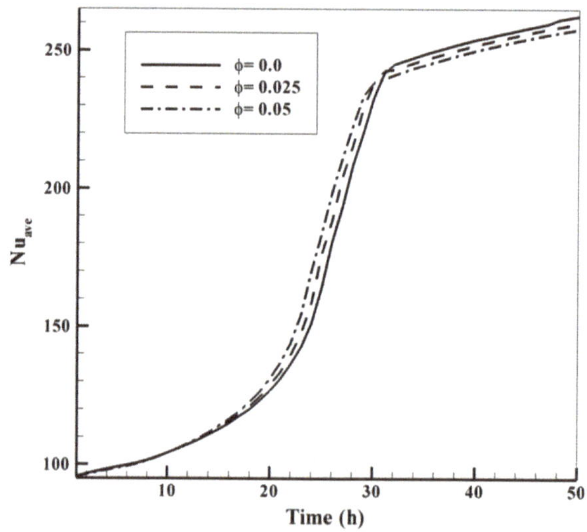

FIGURE 7.6

Nu_{ave} in air side during melting process.

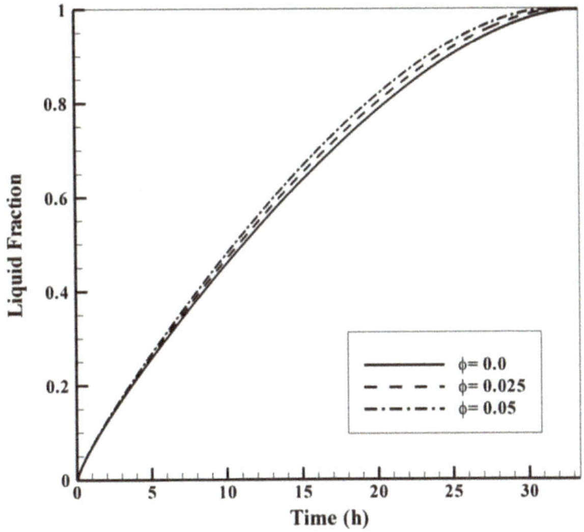

FIGURE 7.7

Melt fraction profile during charging process.

practiced for cooling of building. Similar to previous publication [2], copper oxide and n-octadecane were chosen as nanomaterial and PCM. To model such phase change duration, enthalpy porosity model [2] with the help of ANSYS Fluent 18.1 has been applied. To achieve our selected approaches in detail, Ref. [3] should be read.

7.2.2 Effects of active parameters

In the current numerical simulation, melting of copper oxide enhanced paraffin inside an air conditioning unit was considered. To achieve the superior performance, nanoparticles were involved because new carrier fluid has greater storage capacity. Low deviation proves nice agreement.

Contours of thermal irreversibility, temperature, and Γ for pure n-octadecane were depicted in Figs. 7.10–7.12. To show the impact of dispersing copper oxide in to paraffin, outcomes were illustrated in Figs. 7.13–7.15. To remove the mismatch of demand and supply of energy, heat storage was improved via various ways and in the current article, nanoparticles has been selected. With mixing the pure PCM with nanopowders, greater stored energy can be achieved. During 20 h of charging, no sensible influence on contours was observed with adding nanoparticles. While, after 30 h of melting, the influence of ϕ becomes more important and rate of melting enhances with rise of ϕ. Melting process starts from the section in which inlet hot air enters and then melting front progresses along the duct to reach the outlet section. Contours of entropy has more variation in zones where melt front exists and no observable changes occurs for remain domain. Temperature difference

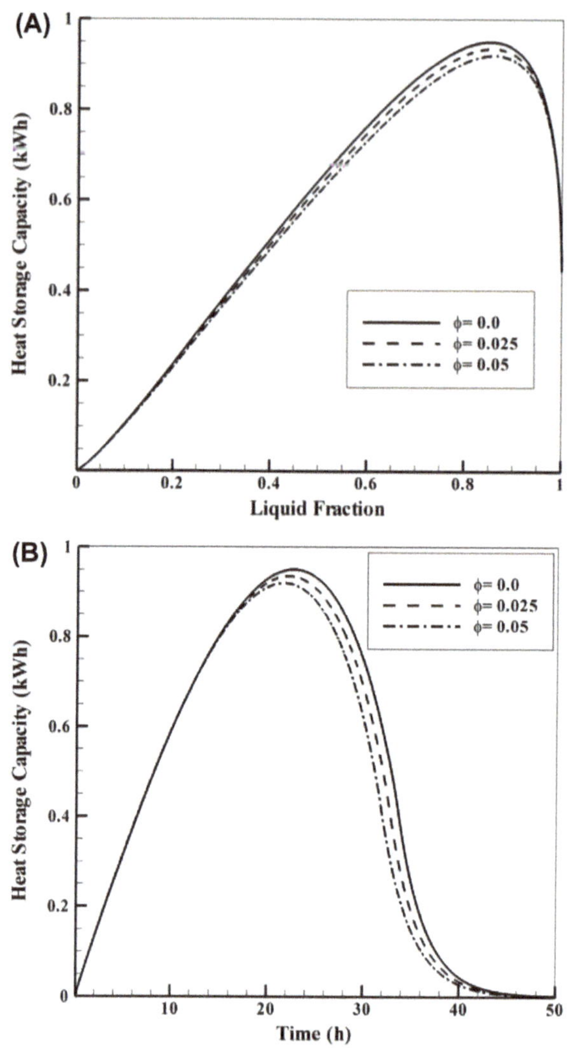

FIGURE 7.8

Heat storage capacity during charging. (A) changing with liquid fraction; (B) changing with time.

between NEPCM layer and air become zero in region far away from melt front. The charging process goes on until the whole NEPCM melted. The time needed to finish the NEPCM charging is a significant factor in measuring the amount of heat absorbed. Charging time declines with dispersing nanomaterial, and similar behavior was reported for entropy generation. As distance from the inner wall augments, the temperature of NEPCM reduces. By dispersing more nanoparticles, the required external flux for melting reduces, so, air temperature at outlet augments.

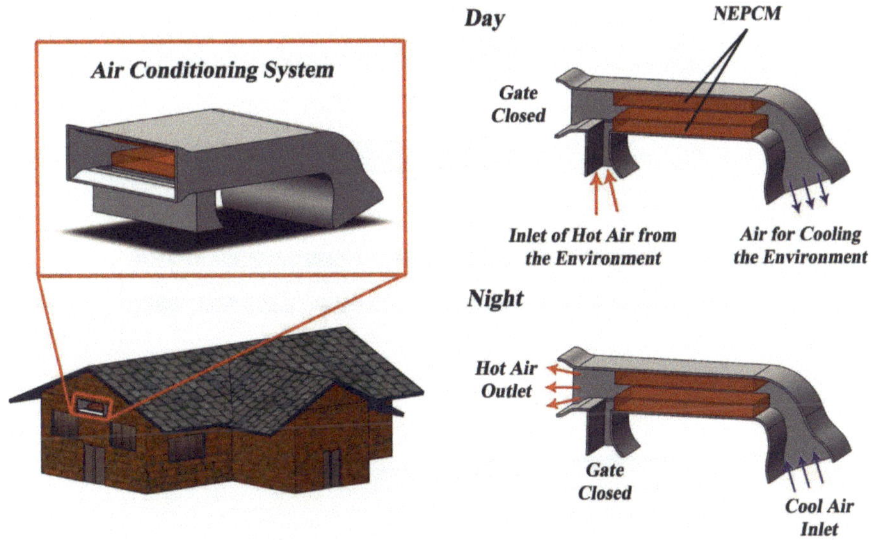

FIGURE 7.9

System for building ventilation.

Changing ϕ and time of duration can affect the average values of scrutinized factors as depicted in Fig. 7.16. As time progresses, impact of dispersing cooper oxide on S_{gen} and T_{ave} enhances. Lower S_{gen} and higher T_{ave} are the outcomes of augmenting ϕ and at initial time, no significant impact on T_{ave} was reported. As a consequence of greater conduction, liquid fraction reaches to its maximum value in lower time with incorporating copper oxide nanoparticles.

7.3 Solidification inside a clean energy storage unit utilizing phase change material with copper oxide nanoparticles

7.3.1 Definition of the problem

The schematic configuration of current system was illustrated in Fig. 7.17. Usage of renewable and alternative energy is the effective way to confront endangering issues such as reduction of fossil fuels and in this chapter, application of solar heating unit in discharge mode has been analyzed [4]. Two layers are filled with paraffin which is mixed with copper oxide nanoparticle (see Table 7.1 to find the properties), and air flows through the middle layer. At night, the discharging process occurs and cold air entering from surrounding becomes warmer due to release of heat from PCM layers. Six cases have been analyzed numerically. Taking into consideration that gravity effects appear when liquid region exists, but in solidification process, it can be neglected and this force has no practical impact on discharging. The geometric

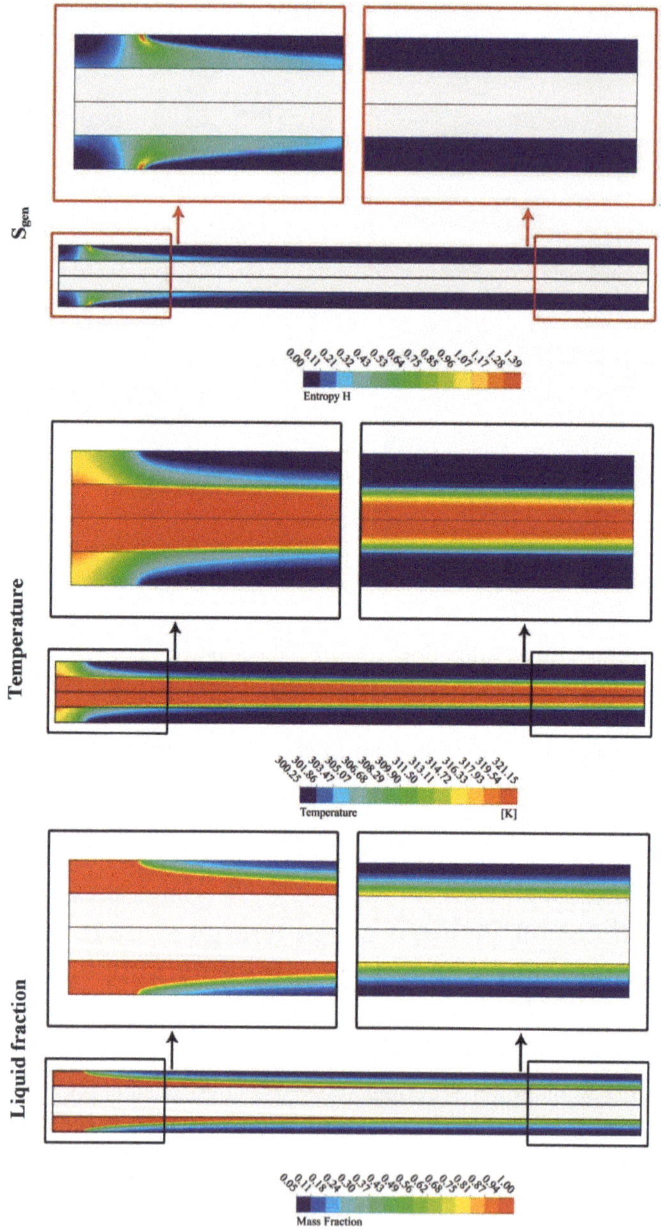

FIGURE 7.10

Variation of Γ, S_{gen}, T at $\phi = 0.0$ when t = 10h.

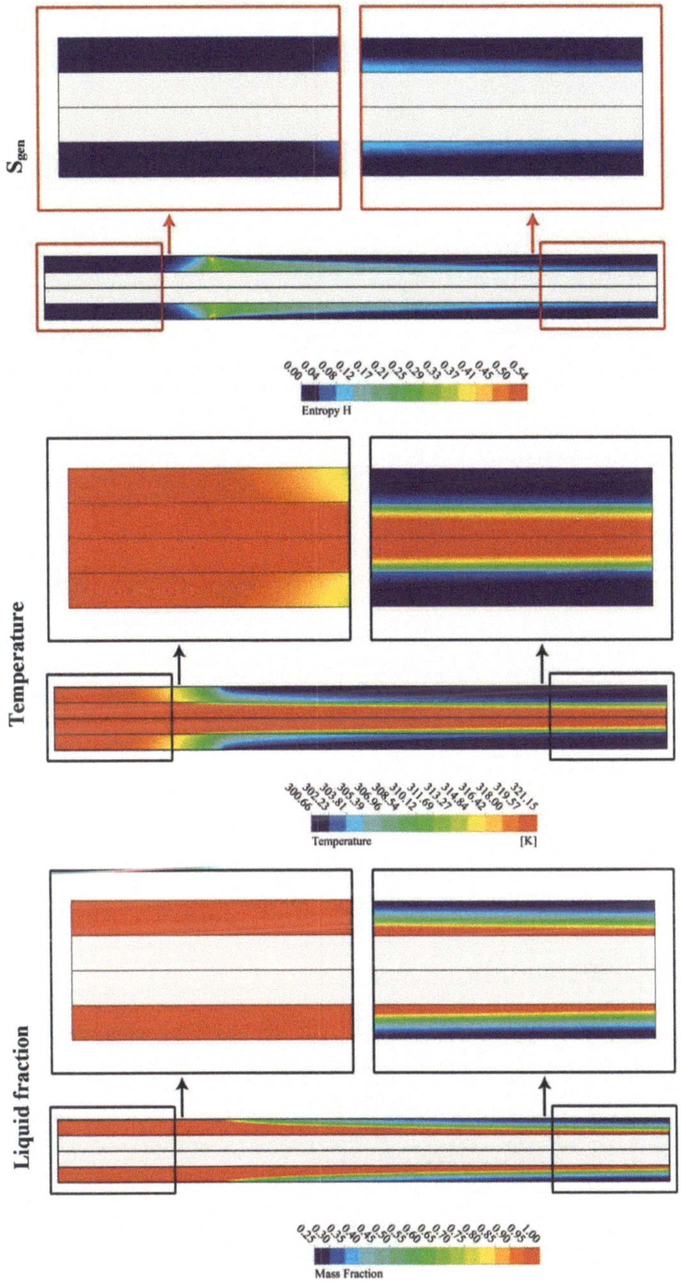

FIGURE 7.11

Variation of Γ, S_{gen}, T at $\phi = 0.0$ when t = 20h.

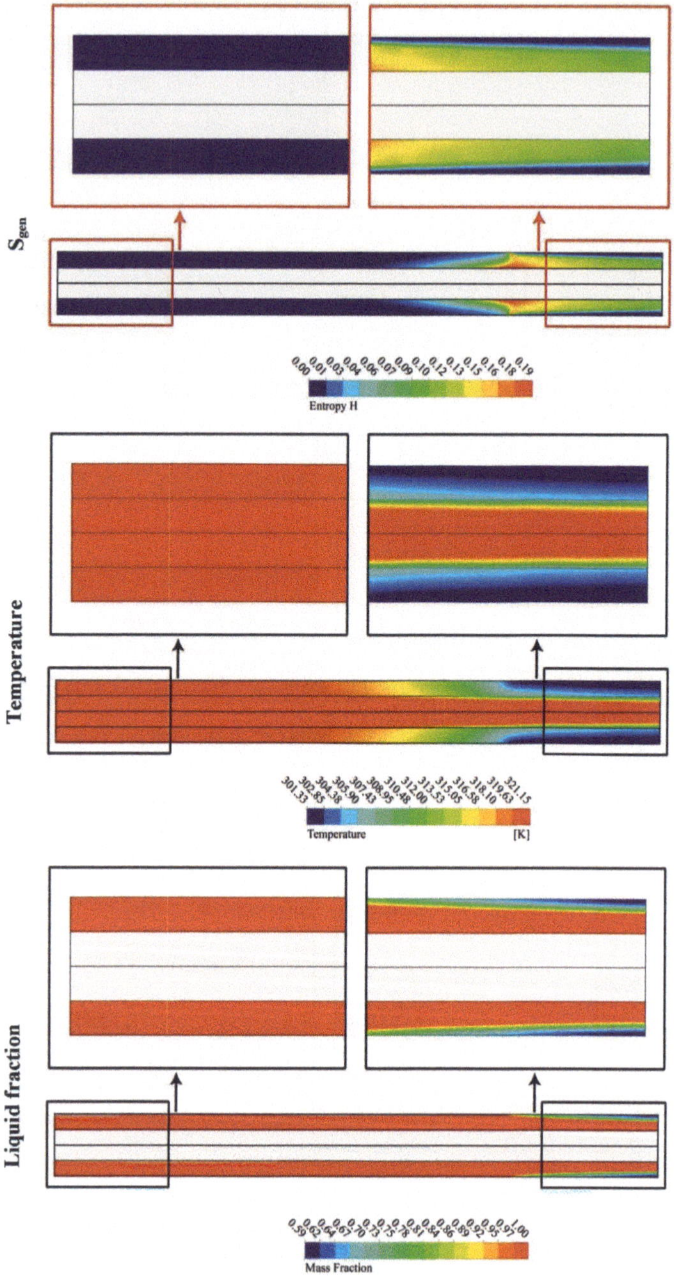

FIGURE 7.12

Variation of Γ, S_{gen}, T at $\phi = 0.0$ when t = 30h.

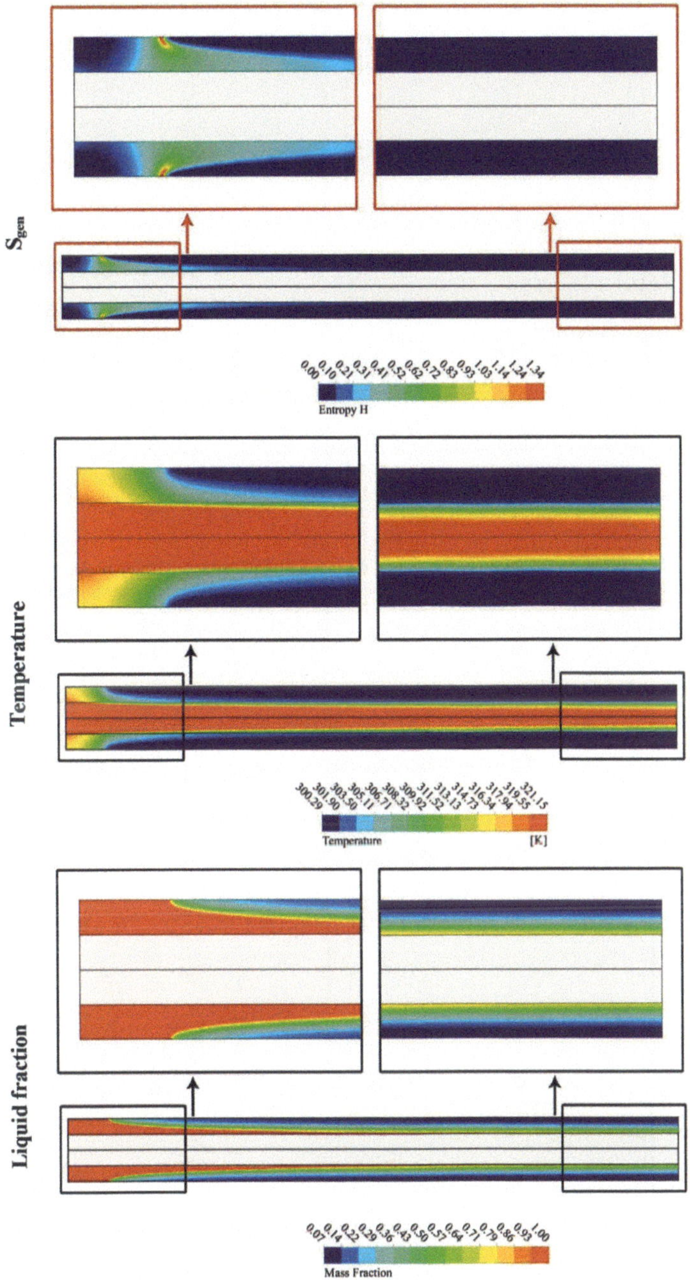

FIGURE 7.13

Variation of Γ, S_{gen}, T at $\phi = 0.05$ when t = 10h.

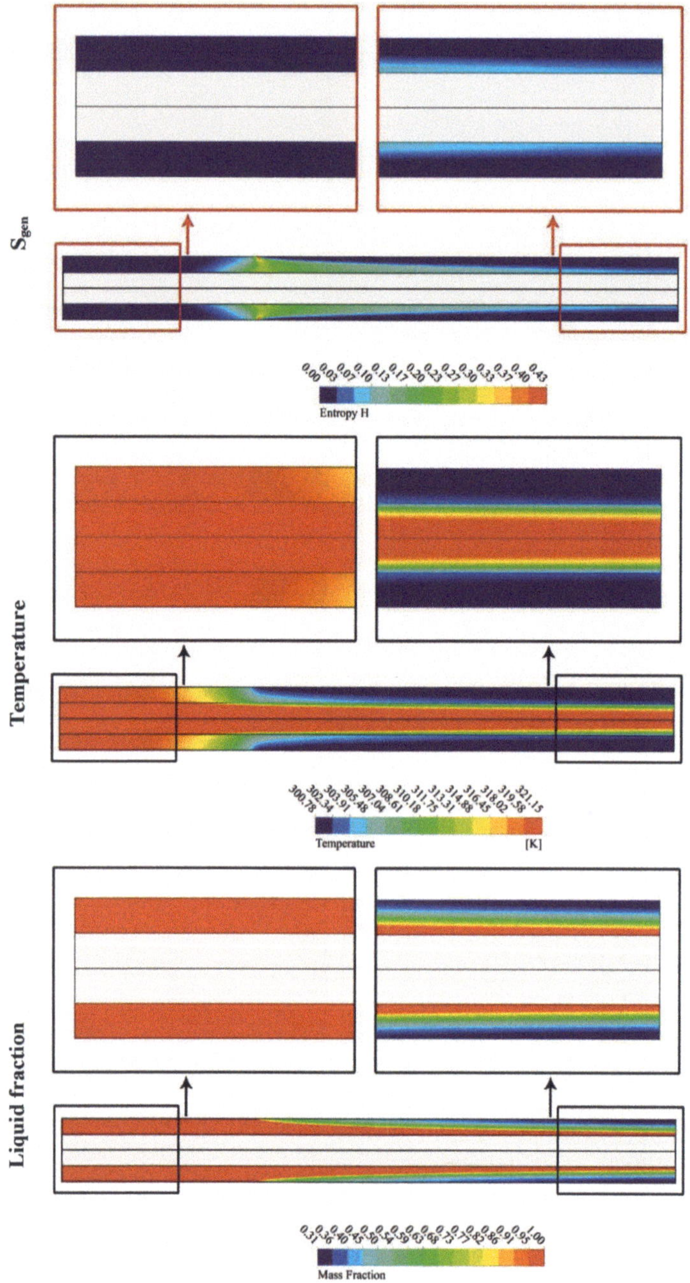

FIGURE 7.14

Variation of Γ, S_{gen}, T at $\phi = 0.05$ when t = 20h.

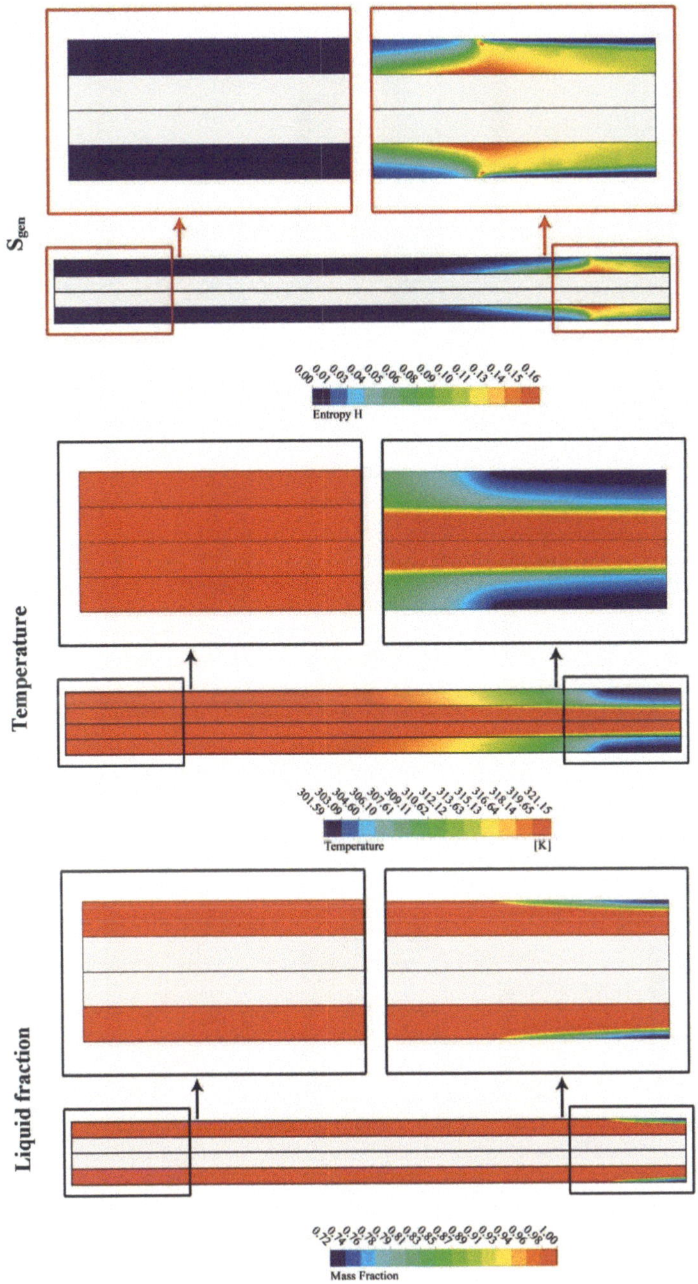

FIGURE 7.15

Variation of Γ, S_{gen}, T at $\phi = 0.05$, when t = 30h.

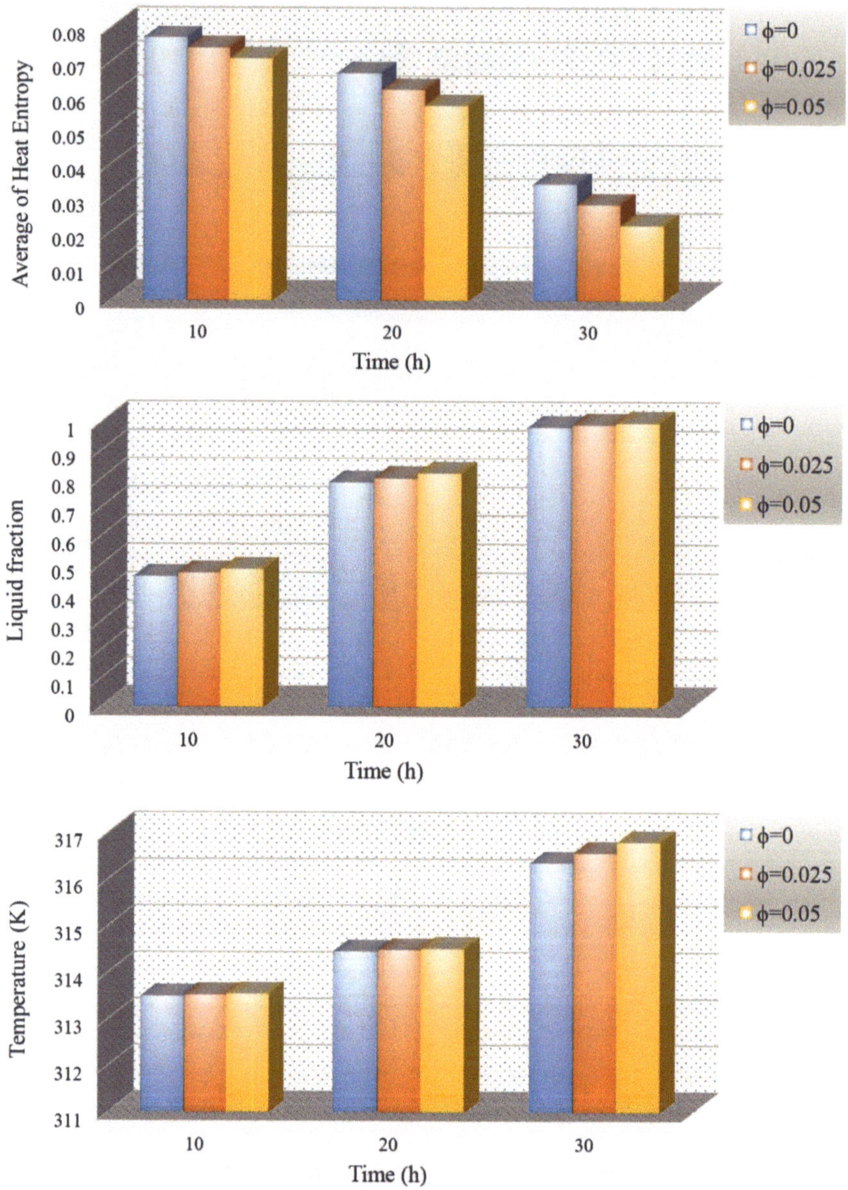

FIGURE 7.16

Average values of important factors in various times with different ϕ.

parameters and formula for creating wavy surface have been mentioned in the first figure. Samples of mesh for different values of a are illustrated. Various articles have been published about passive techniques of heat transfer augmentation [5–61].

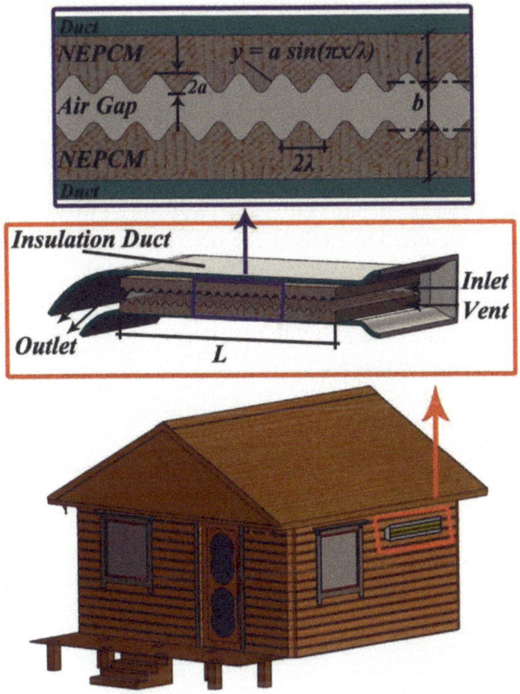

L	1 m
λ	22.222 mm
a	0, 5, 10 mm
l	25 mm
h	50 mm

	a	ϕ
Case 1	flat	0
Case 2	flat	0.05
Case 3	0.005	0
Case 4	0.005	0.05
Case 5	0.01	0
Case 6	0.01	0.05

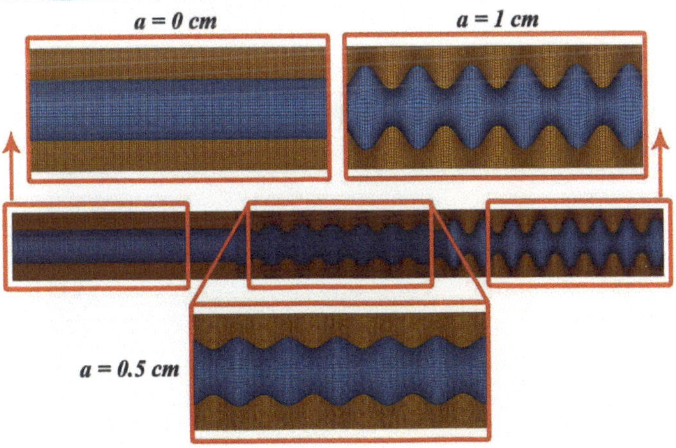

FIGURE 7.17

Present unit and related sample.

7.3.2 Effects of active parameters

To decrease energy consumption in ventilation system of building, performance of such a system should be increased. Utilizing PCM in solar heat storage unit can employ in a building to control the room temperature and decreasing energy required for space heating and supports the employ of renewable energy. In this chapter, usage of wavy duct and nanosized particles has been considered as a promising way for enhancing discharging of paraffin. Two-dimensional numerical modeling has been developed to analyze the performance of the unit. Such processes make the air warmer due to conversion of liquid to solid.

The time needed for discharging the PCM layers should be calculated to clarify the performance of the unit. To show the variation of solidification time, Eq. (7.13) and Fig. 7.18 have been presented.

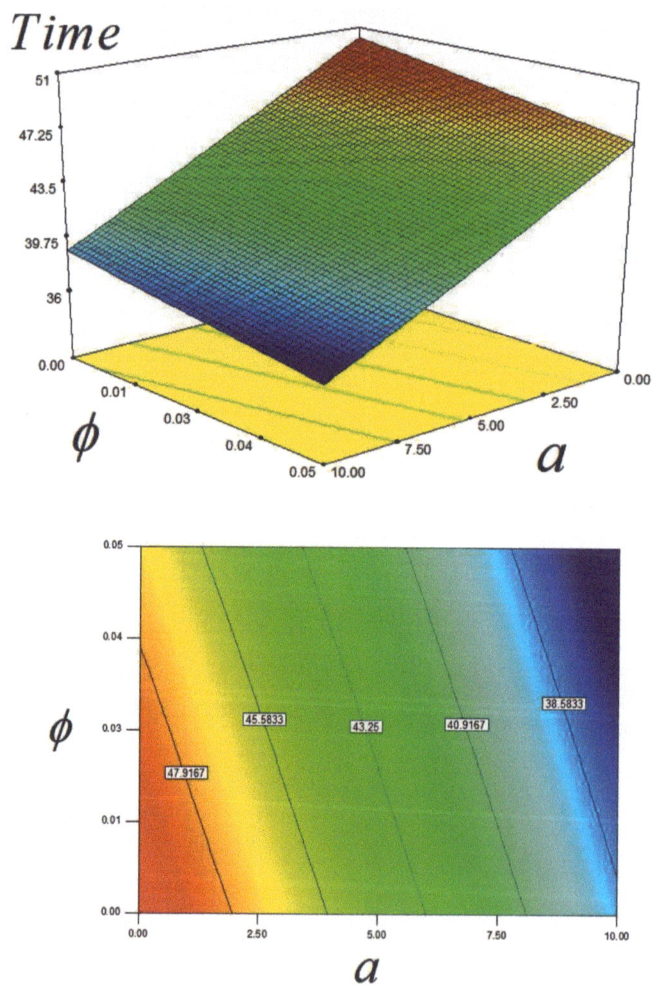

FIGURE 7.18

Impact of a and ϕ on discharging time.

$$Time = 42.95 - 5.55a - 1.45\phi + 0.15a\phi + 0.15a^2 \tag{7.13}$$

In modern industries, utilizing paraffin is common owing to its capable absorbing heat, but its conductivity should be improved to have the better performance. Changing the geometry and type of PCM plays an effective role in solidification time. Both factors have positive effect on reduction of solidification time, and this function declines with rise of a and ϕ. This is attributed to domination of conduction mechanism. Mixing nanomaterial becomes more effective about 0.4% when highest a is utilized. NEPCM discharging finishes 1.29 times faster with utilizing wavy walls. The steeper slope can be seen for a, so a is more effective than ϕ, and it can be concluded from this graph. The most interesting output is improvement of discharging performance about 28% with use of wavy duct and nanoparticles.

Figs. 7.19–7.21 depict the distributions of S_{gen} and Lf while solidification. To decline the mismatch between the heat demand and supply, a combination of

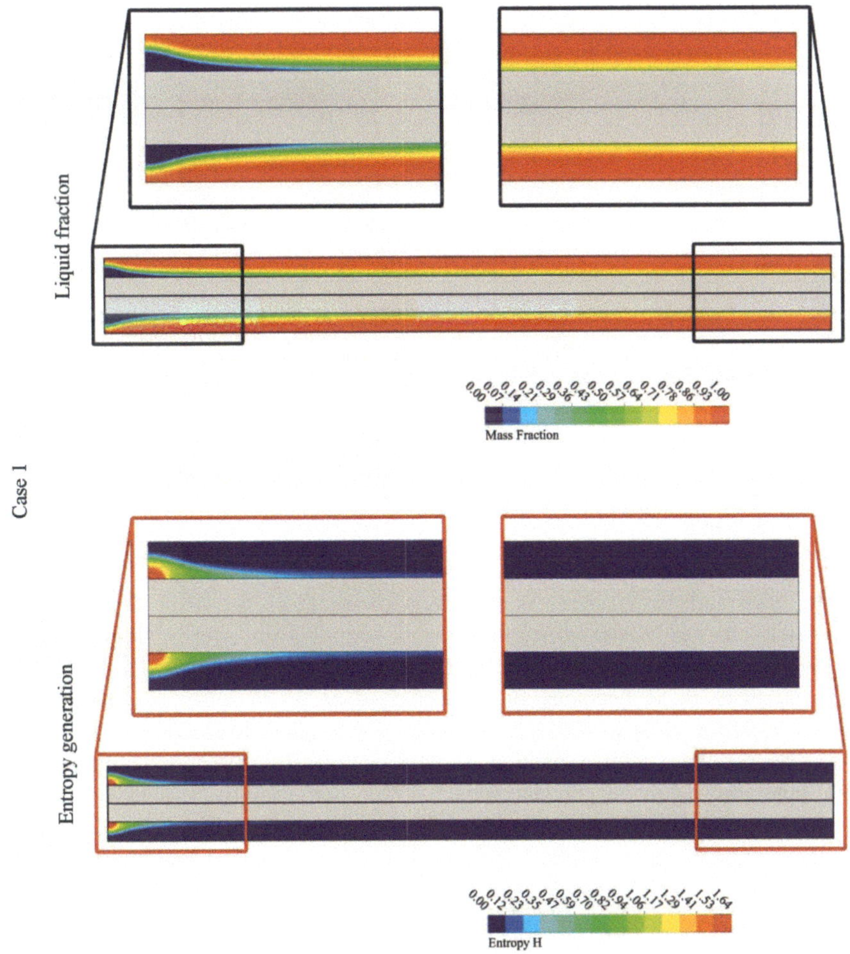

FIGURE 7.19

Presentation of S_{gen} and Lf at 5 h.

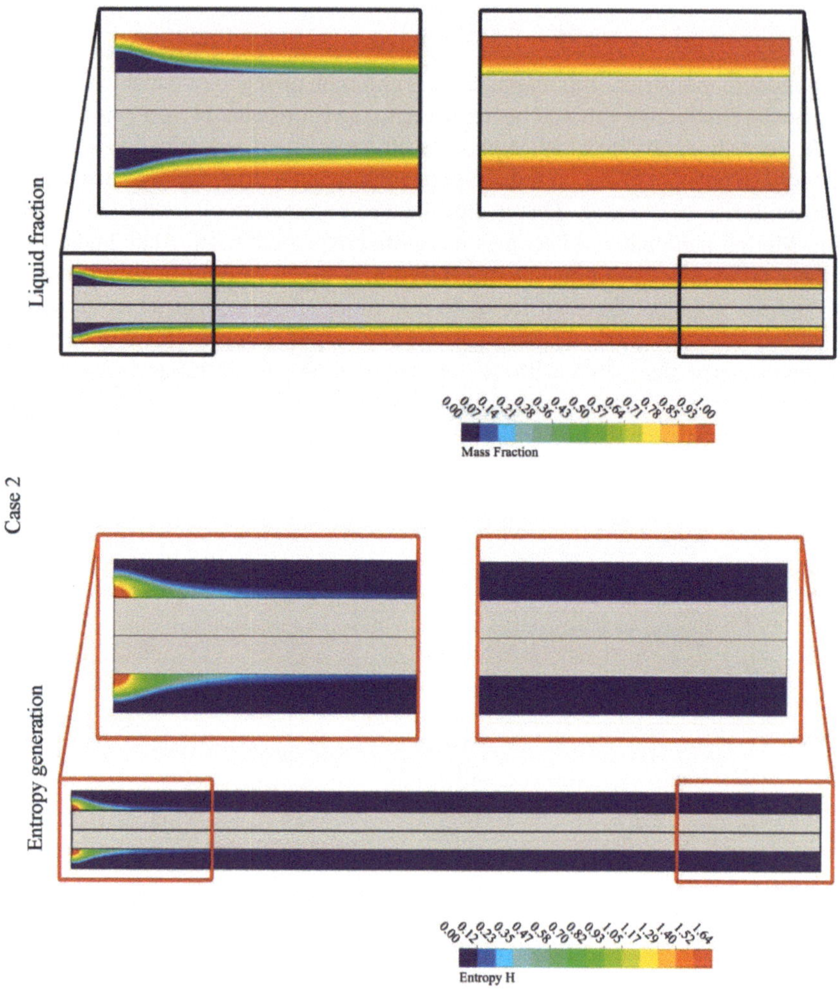

FIGURE 7.19 cont'd.

nanoparticles and wavy surface is necessary for the TES unit. After the paraffin is completely charged, the solidification process begins. Thin layer of liquid paraffin has been shaped, while the discharging and length of this layer reduce as amplitude augments. The PCM becomes fully discharged at 36.2h when $a = 10mm$ and $\phi = 0.05$ and for cases with $a = 5mm$ and $\phi = 0$, it takes 50.3h. When$a = 10mm$, dispersing nanofluid with a fraction of 0.05 declines the discharging time about 6.9%, and this reduction percentage is 6.56% when straight channel is utilized. When a changes from 0 to 10 mm, solidification time reduces about 22.66% in the presence of nanomaterial, while it declines about 22.98% when $\phi = 0$. As time augments, the region which has the greatest entropy generation is shifted to outlet side,

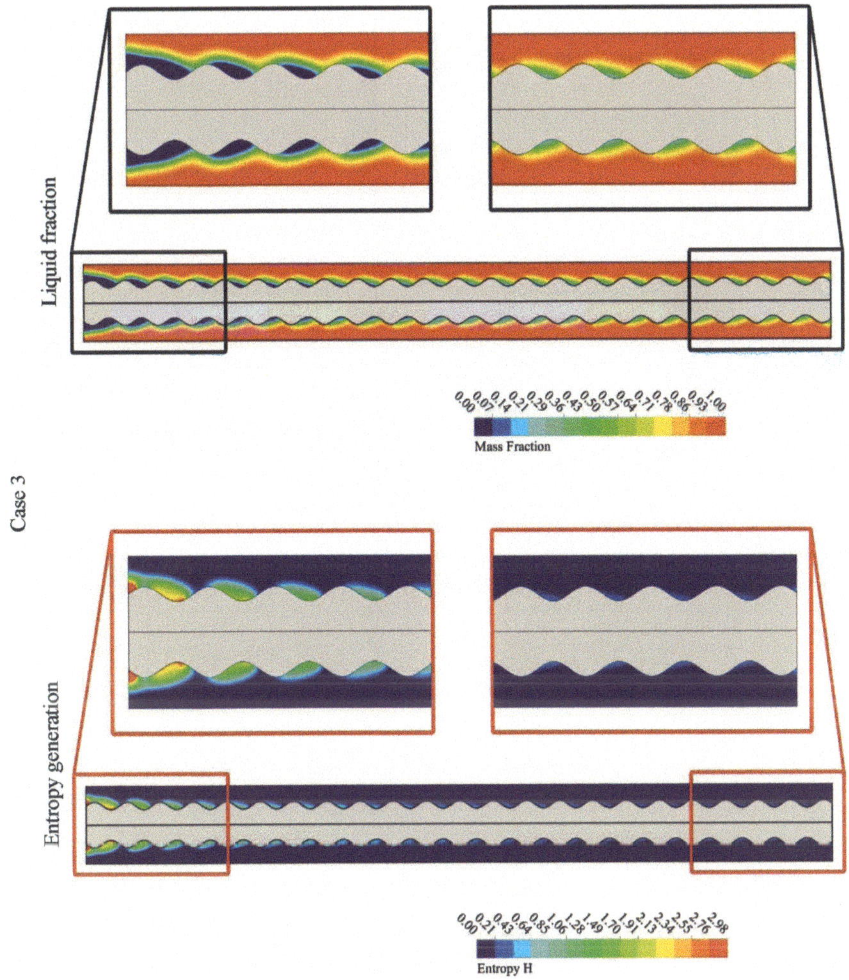

FIGURE 7.19 cont'd.

and this zone is near the region in which the remaining liquid is converting to solid. The minimum volume of liquid PCM exists for $a = 10$ mm in all time stages, and this is owing to greater conduction mode with rise of a. After 15h of discharging, the greatest S_{gen} is obtained due to highest temperature gradient. Solidification starts in a region near the walls of inner duct in which cold air moves, then solid front progresses and reaches the region near the outlet of air. As time progresses, paraffin becomes colder and converts to solid phase, and this solid layer does not allow heat flux to reach the outer region.

To illustrate the variation of S_{gen} and Lf, Fig. 7.22 has been depicted. According to heat flux distribution, in initial time, cases with nanoparticles have greater heat

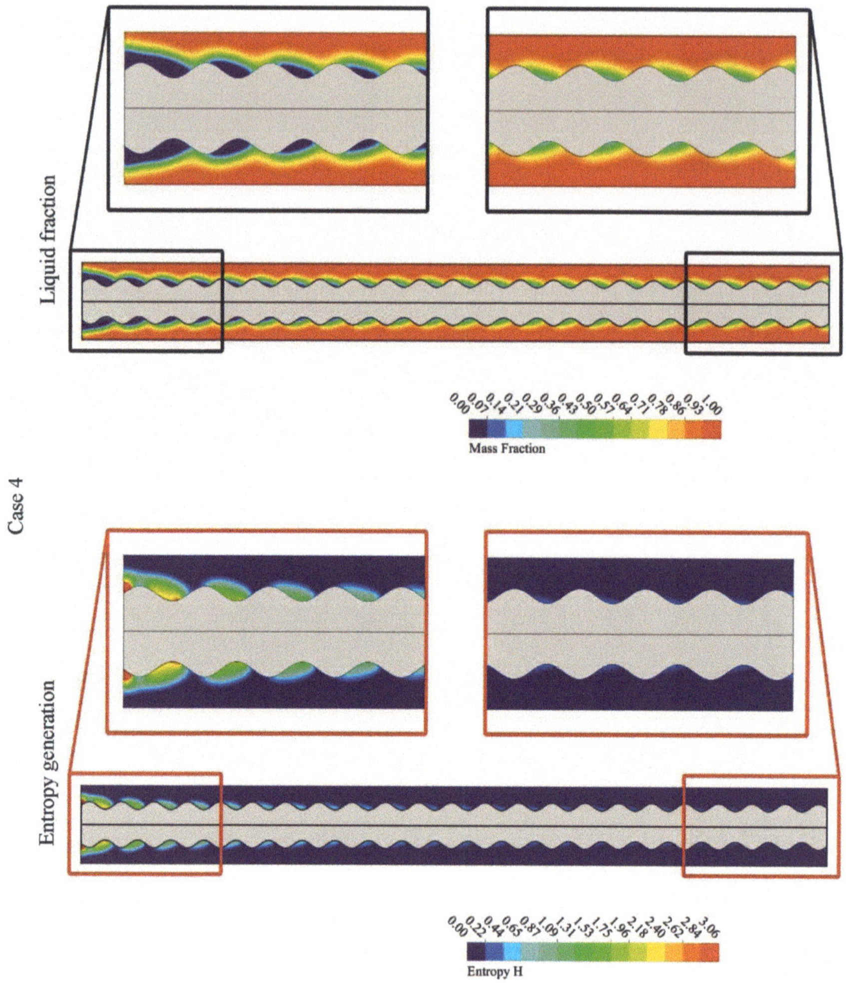

FIGURE 7.19 cont'd.

flux, while with progress of time, reverse trend has been reported. This observation occurs because freezing layer reduces the rate of heat transfer, so, as solidification progress heat flux declines. In addition, using wavy channel can reduce the heat flux because the solidification accelerates with rise of amplitude of the sinusoidal wall. Both dispersing nanosized material and imposing wavy duct help the solidification, and liquid fraction declines with augment of a and ϕ. Such impacts are more visible in greater time. Dispersing nanosized powders to paraffin declines

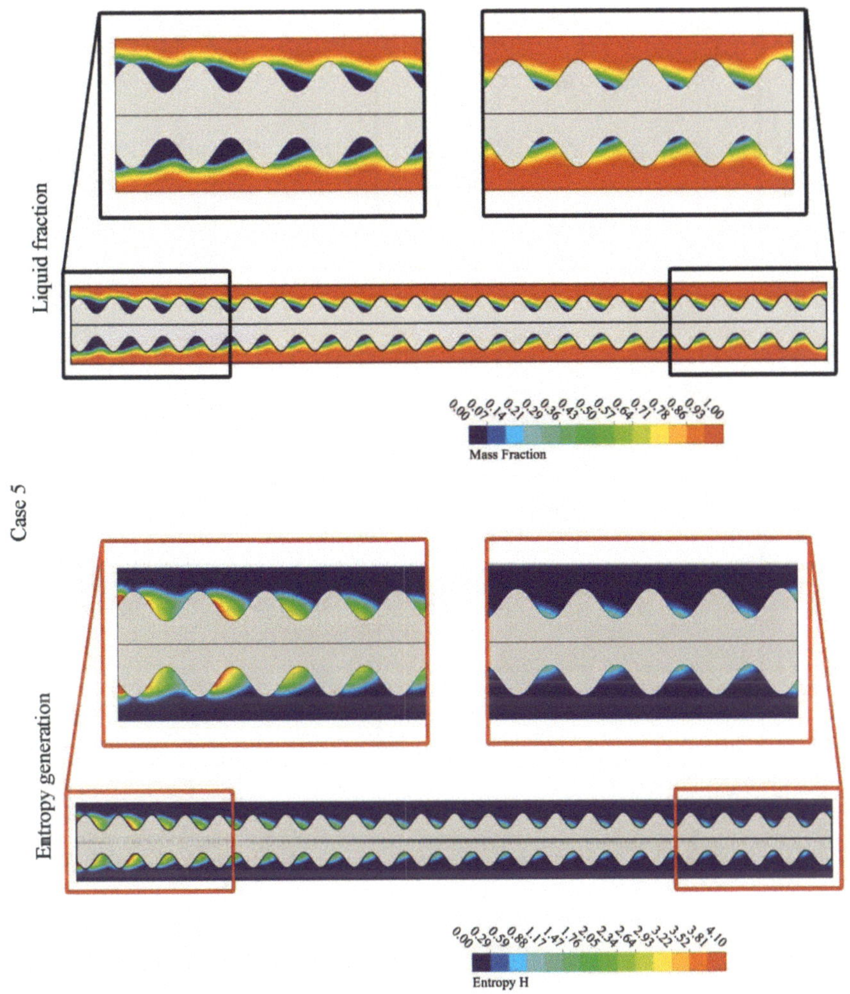

FIGURE 7.19 cont'd.

the entropy generation, and this trend is more pronounced at t = 15 h. In this regard, mixing nanoparticles with PCM is helpful not only in view of expedition in solidification but also reduction in entropy generation. Influence of ϕ on S_{gen} in final time is greater than that in the initial time. Utilizing wavy wall makes S_{gen} to increase, and maximum increment observed for t = 15 h. During discharging, at first entropy generation augments, then it declines, and finally it reaches to lower values in comparison to initial time.

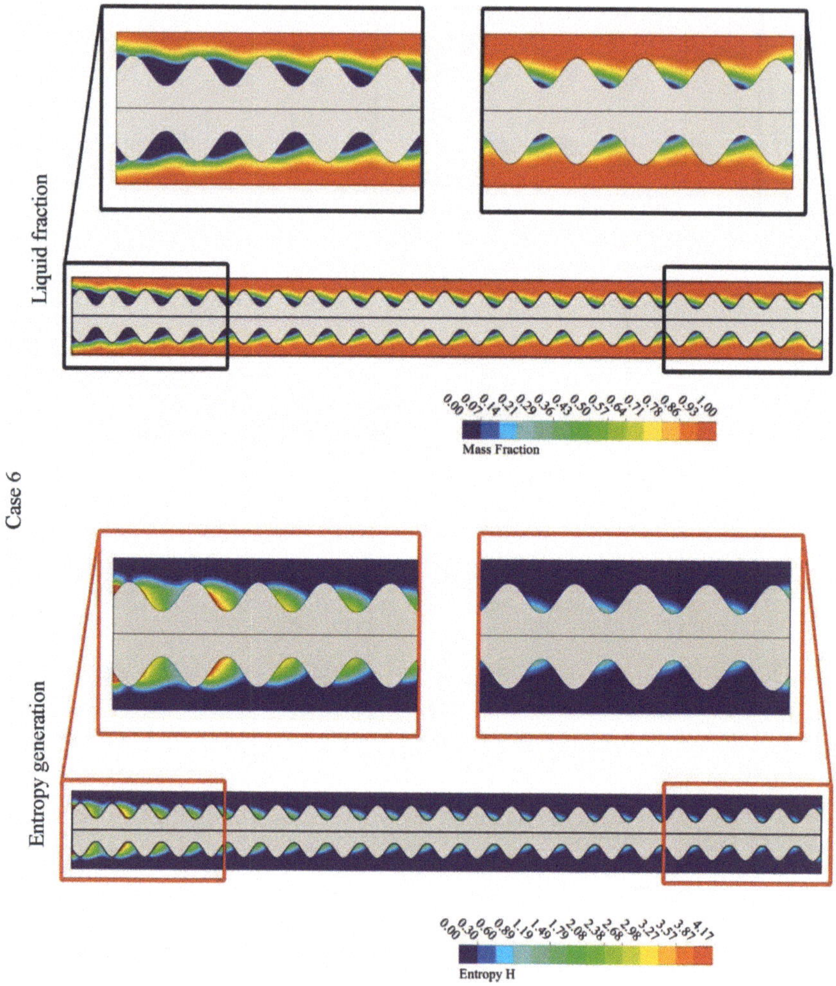

FIGURE 7.19 cont'd.

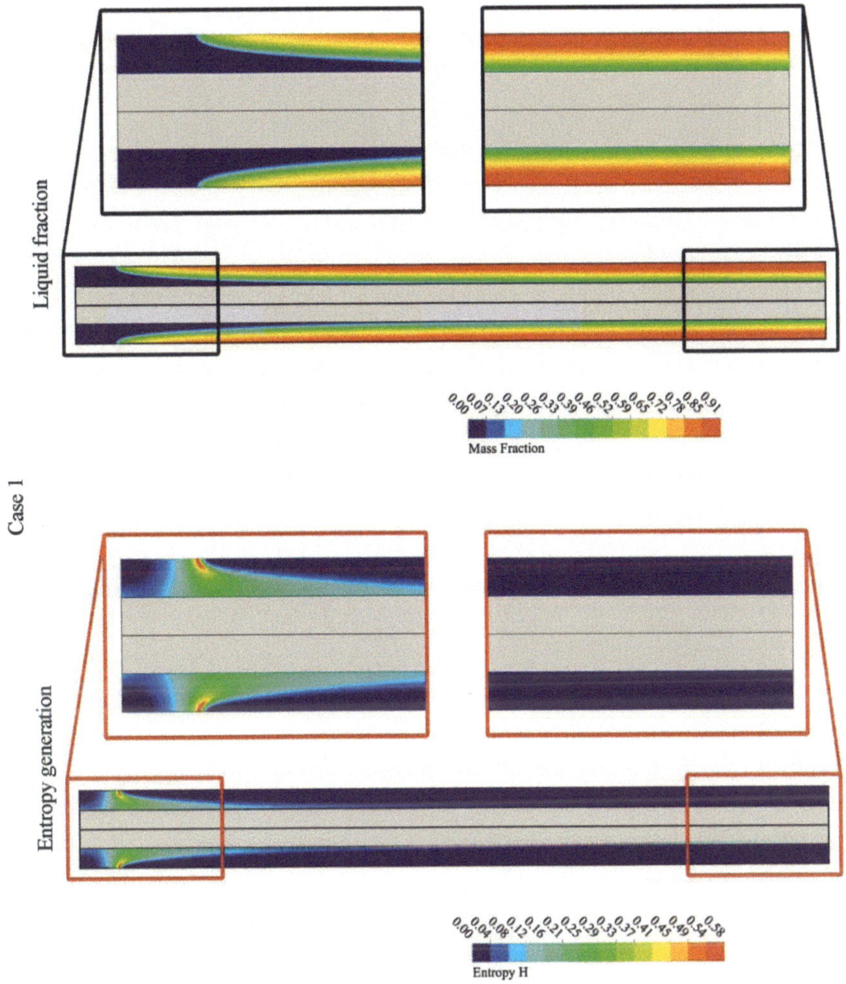

FIGURE 7.20

Presentation of S_{gen} and Lf at 15 h.

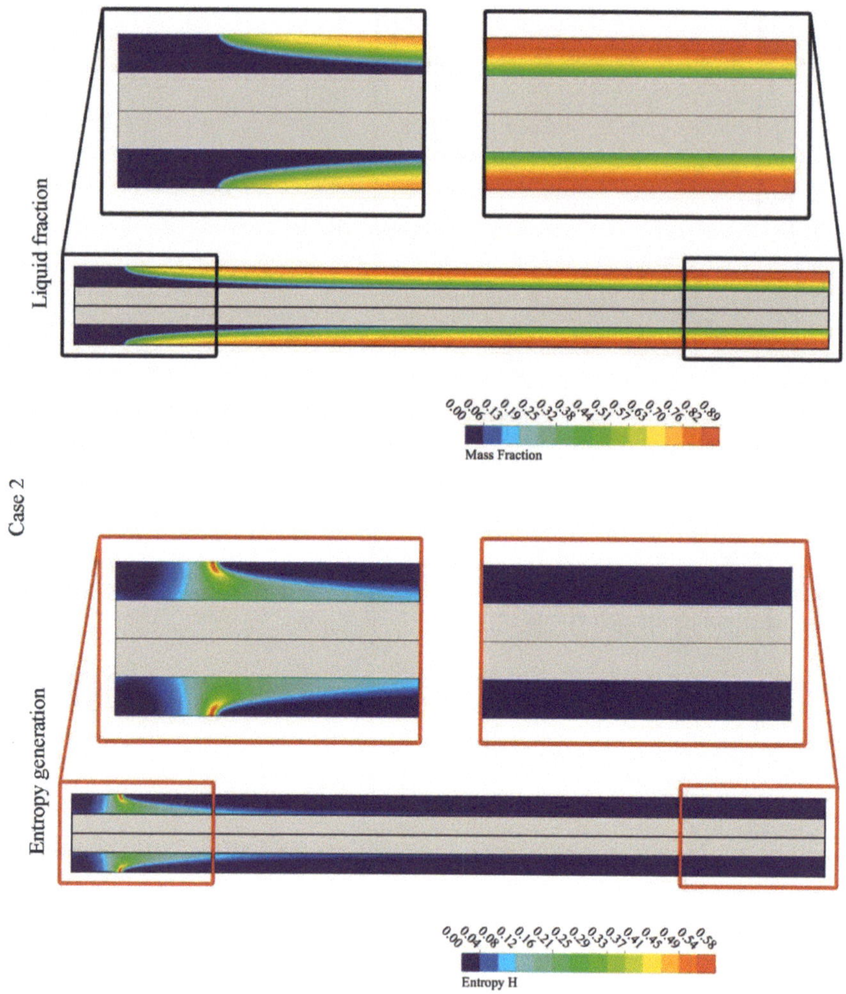

FIGURE 7.20 cont'd.

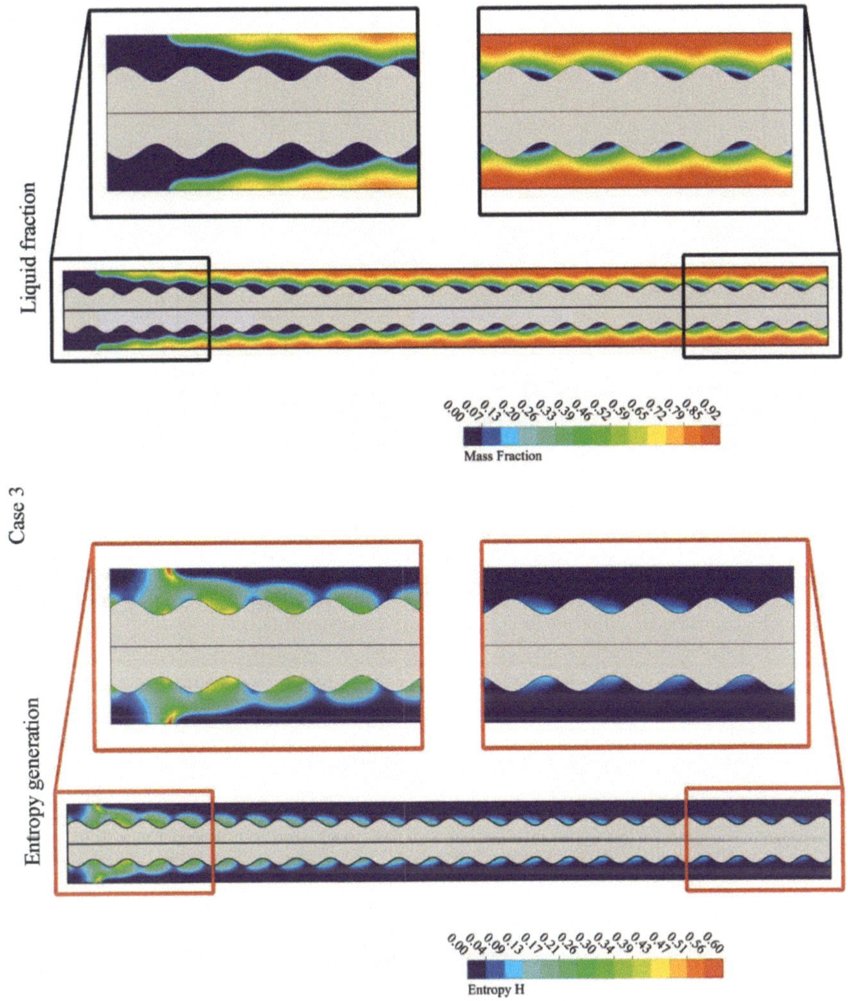

FIGURE 7.20 cont'd.

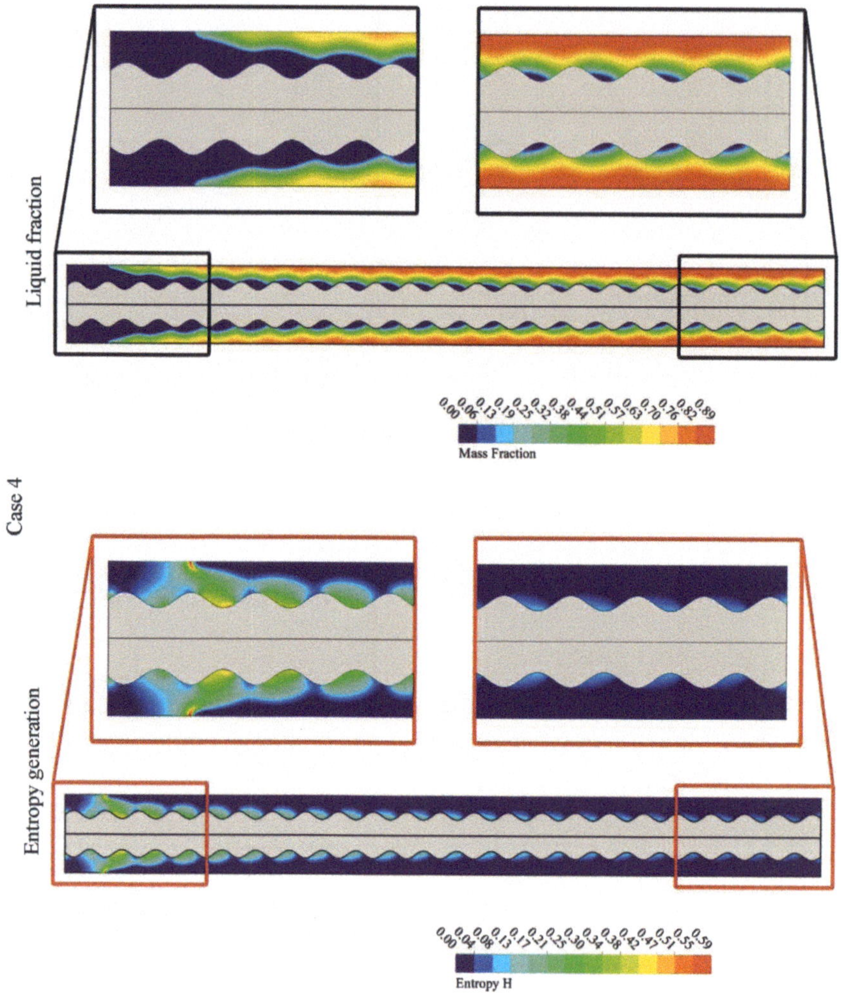

FIGURE 7.20 cont'd.

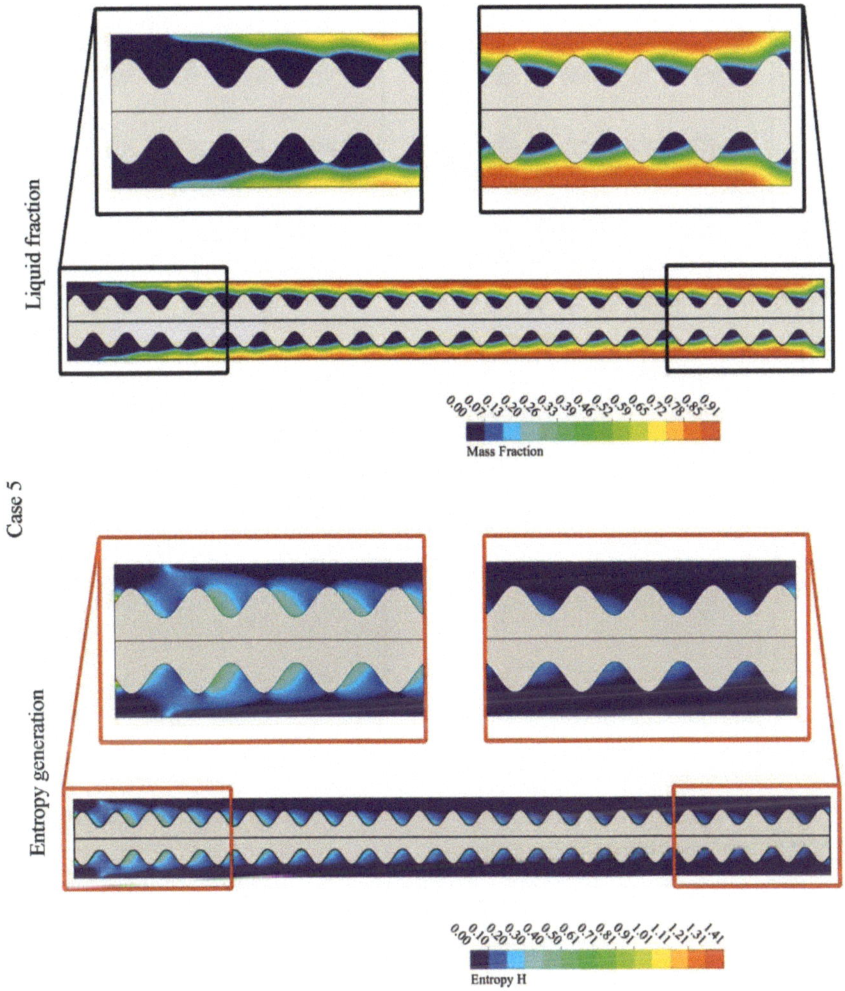

FIGURE 7.20 cont'd.

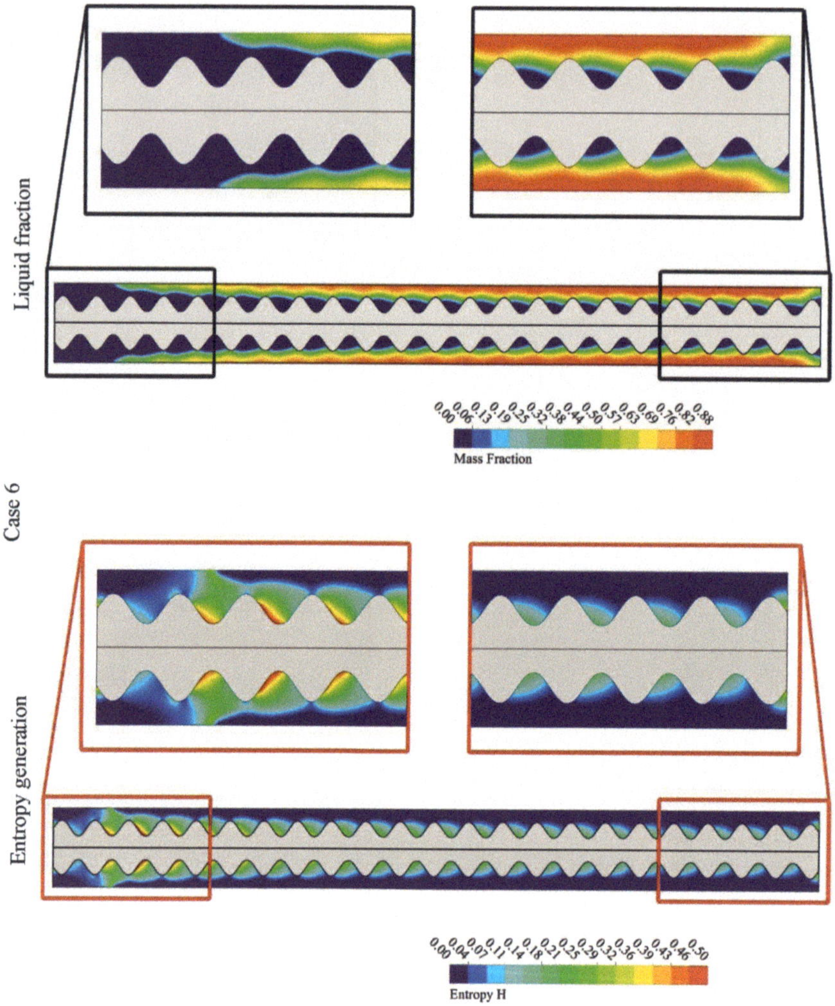

FIGURE 7.20 cont'd.

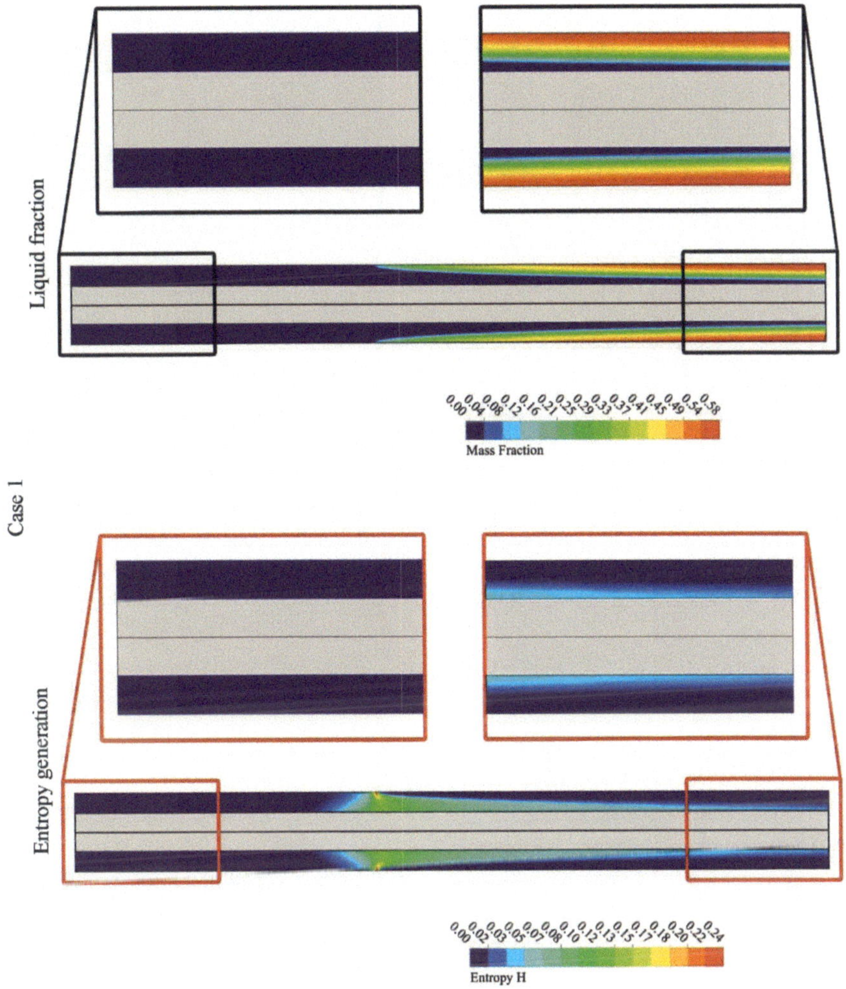

FIGURE 7.21

Presentation of S_{gen} and Lf at 35 h.

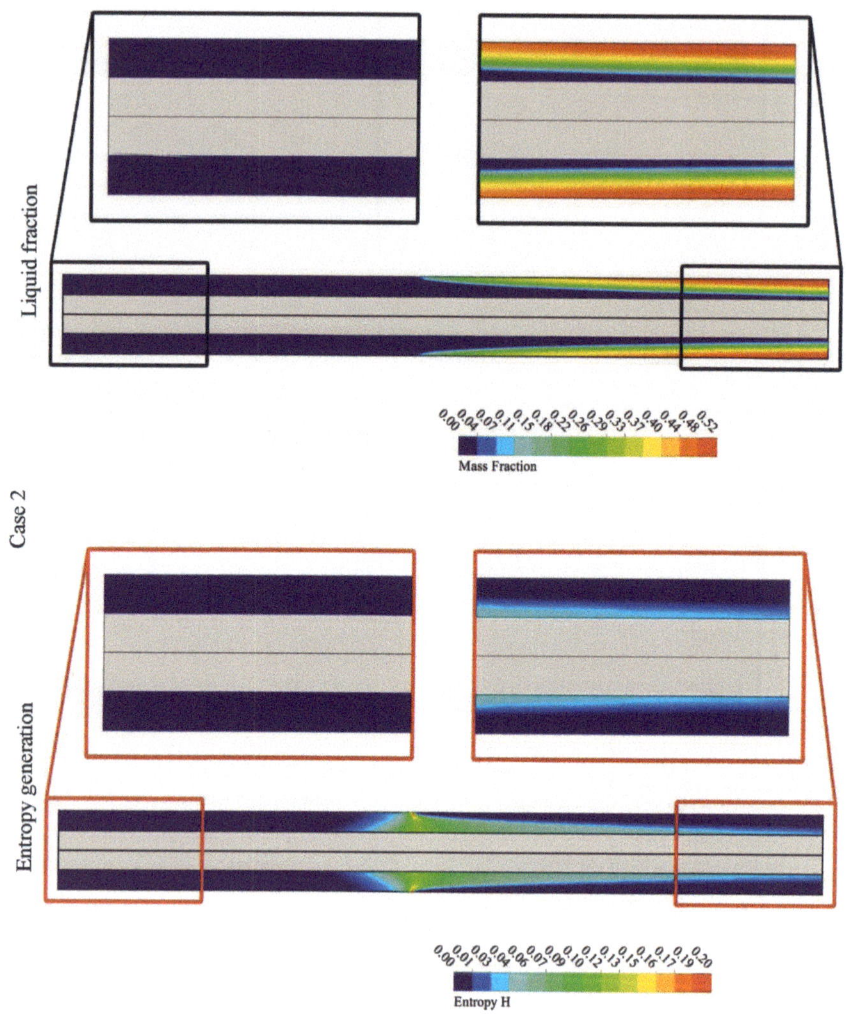

FIGURE 7.21 cont'd.

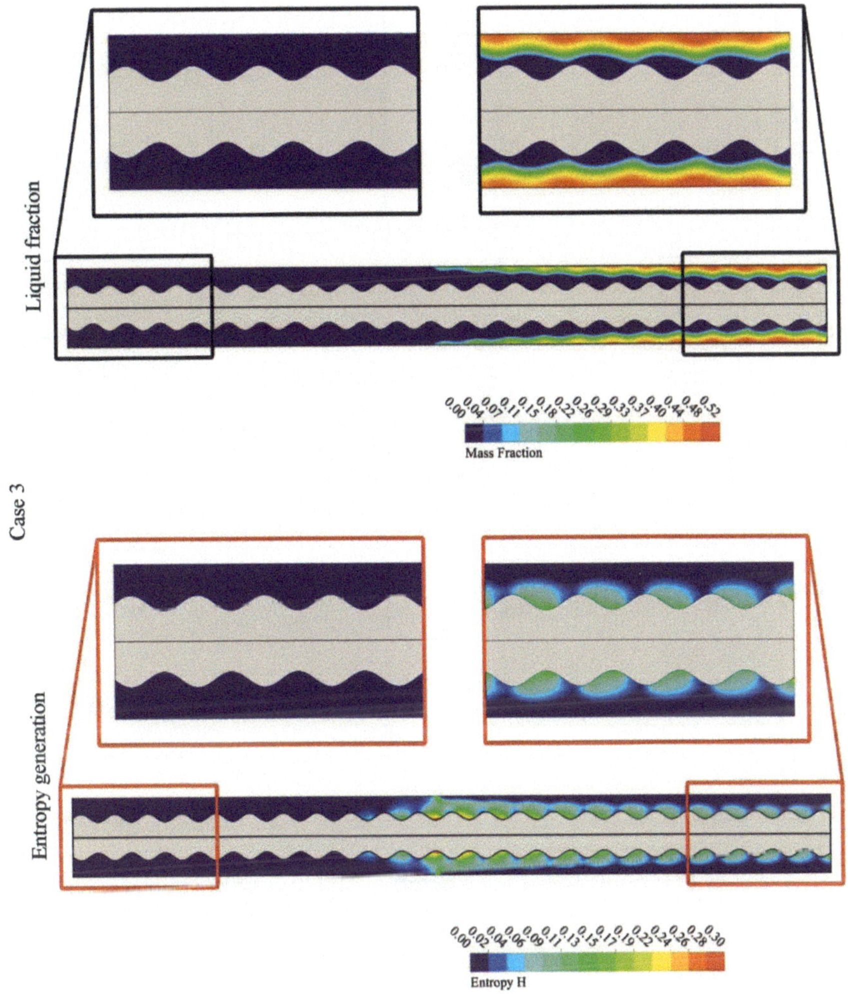

FIGURE 7.21 cont'd.

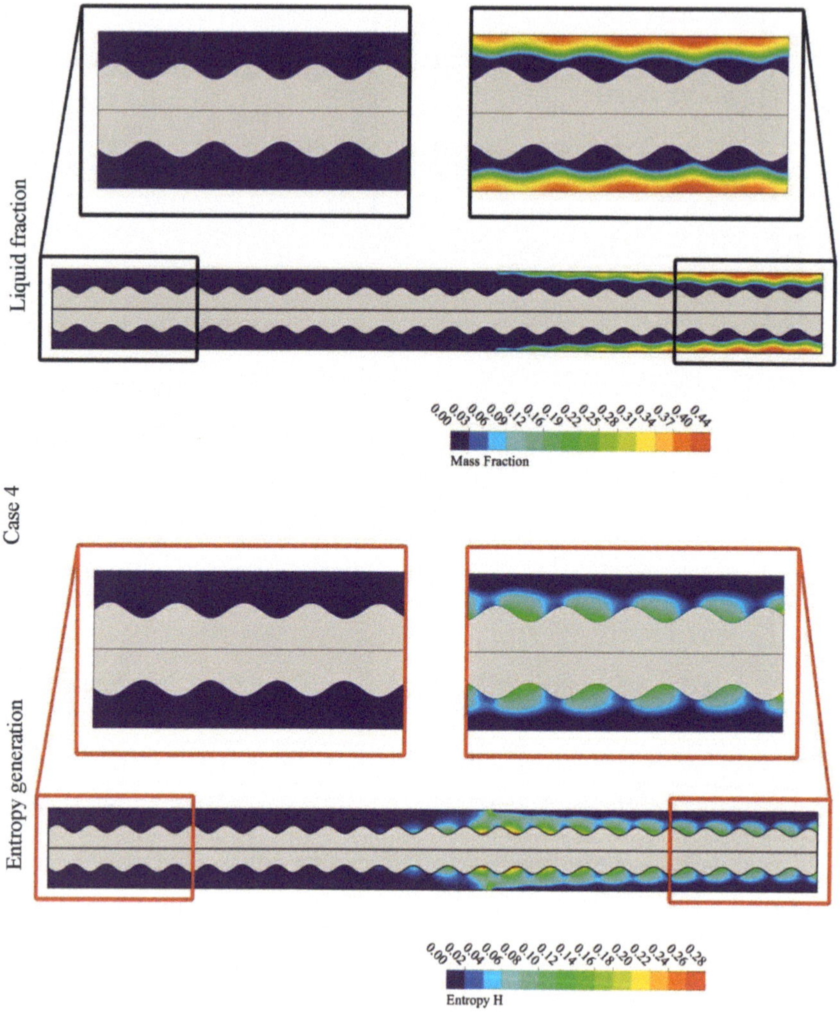

FIGURE 7.21 cont'd.

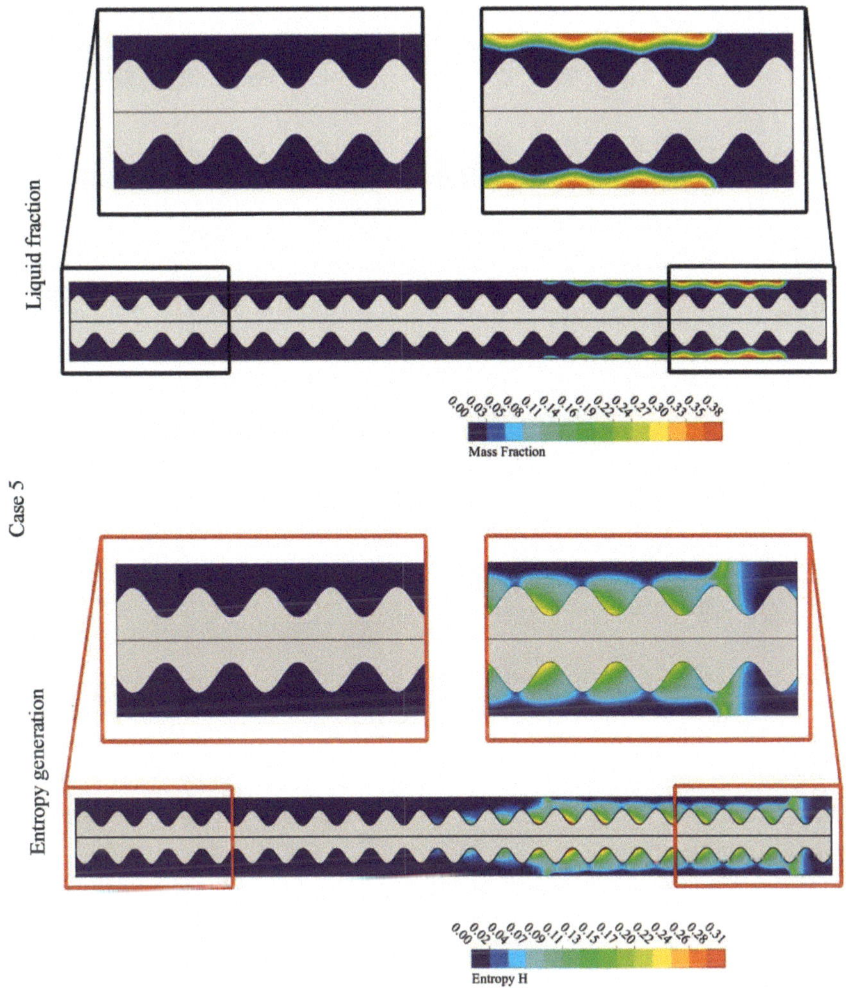

FIGURE 7.21 cont'd.

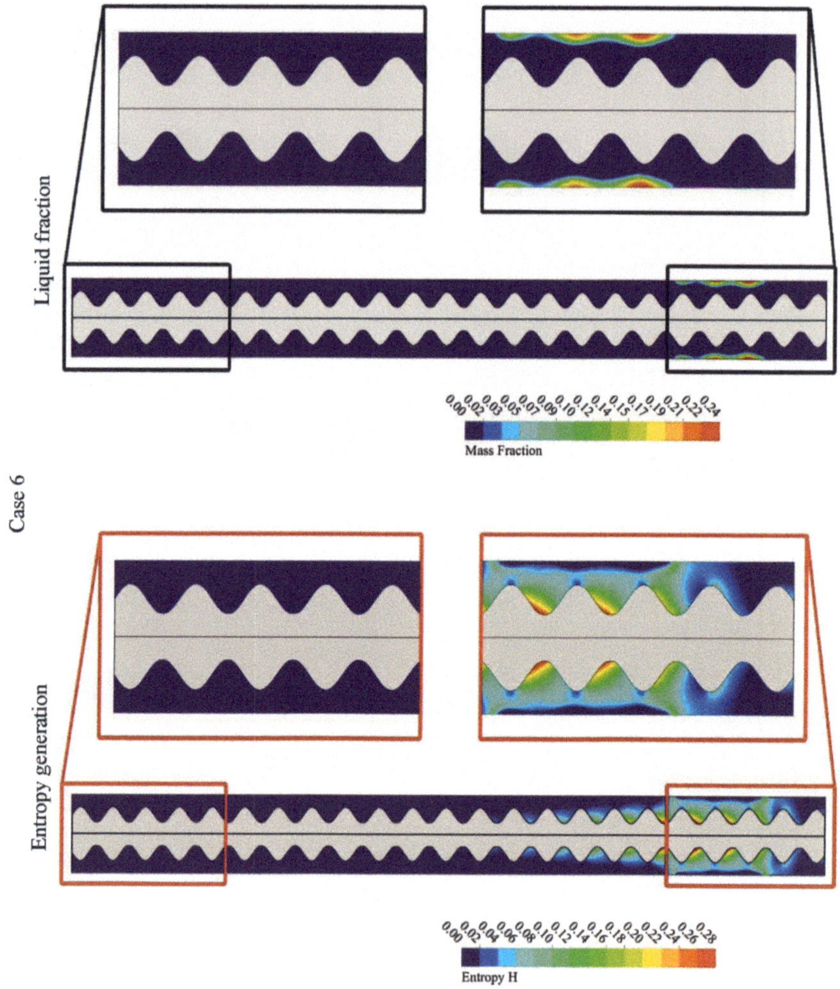

FIGURE 7.21 cont'd.

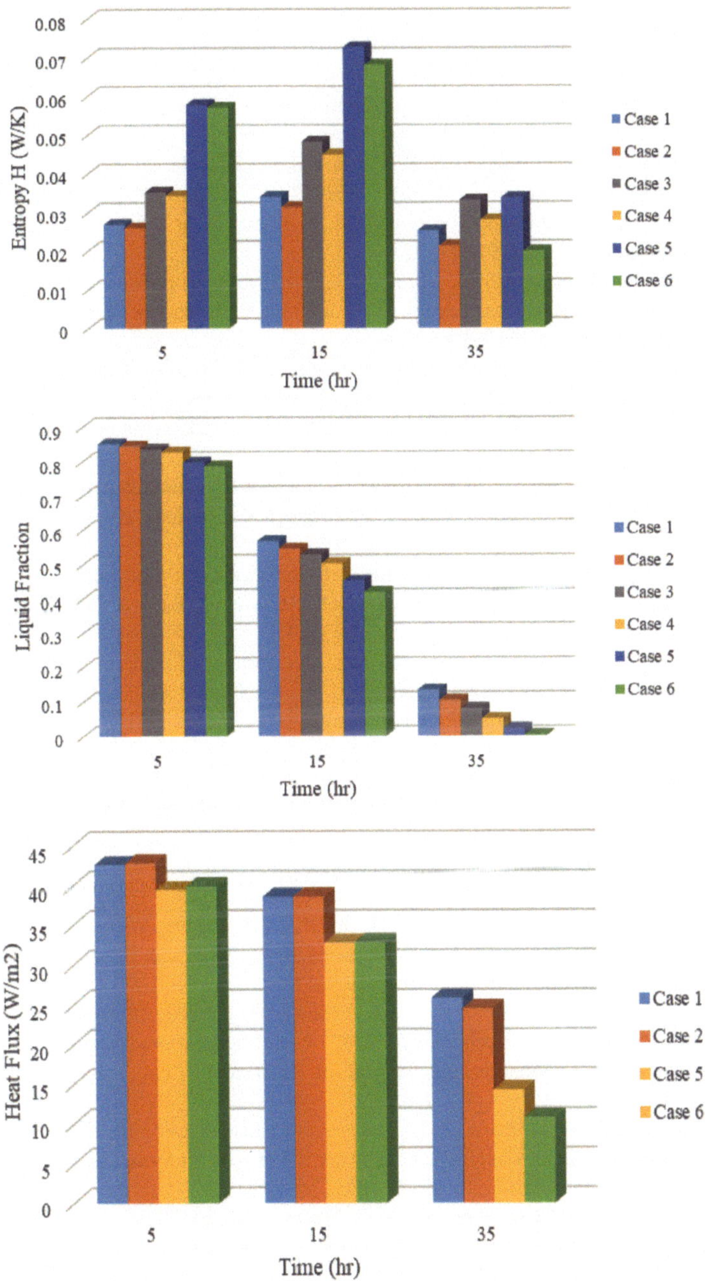

FIGURE 7.22

Variation of S_{gen}, Lf, and heat flux during solidification.

References

[1] M. Sheikholeslami, A. Zareei, M. Jafaryar, A. Shafee, Z. Li, S. Amor, I. Tlili, Heat transfer simulation during charging of nanoparticle enhanced PCM within a channel, Physica A: Statistical Mechanics and its Applications 525 (2019) 557–565.

[2] Q. Xiong, M. Vaseghi, J.A. Ali, S. Mustafa Hamad, M. Jafaryar, M. Sheikholeslami, A. Shafee, T.A. Alkanhal, T. Nguyen-Thoi, I. Tlili, Z. Li, Nanoparticle application for heat transfer and irreversibility analysis in an air conditioning unit, Journal of Molecular Liquids 292 (2019) 111372.

[3] M. Sheikholeslami, R. Ellahi, Three dimensional mesoscopic simulation of magnetic field effect on natural convection of nanofluid, International Journal of Heat and Mass Transfer 89 (2015) 799–808.

[4] X. Ma, M. Sheikholeslami, M. Jafaryar, A. Shafee, T. Nguyen-Thoi, Z. Li, Solidification inside a clean energy storage unit utilizing phase change material with copper oxide nanoparticles, Journal of Cleaner Production (2019), https://doi.org/10.1016/j.jclepro.2019.118888.

[5] Y. Qin, H. He, X. Ou, T. Bao, Experimental study on darkening water-rich mud tailings for accelerating desiccation, Journal of Cleaner Production (2019) 118235, https://doi.org/10.1016/j.jclepro.2019.118235.

[6] Y. Qin, J.E. Hiller, D. Meng, Linearity between pavement thermophysical properties and surface temperatures, Journal of Materials in Civil Engineering (2019), https://doi.org/10.1061/(ASCE)MT.1943-5533.0002890.

[7] Y. Qin, J. Luo, Z. Chen, G. Mei, L.-E. Yan, Measuring the albedo of limited-extent targets without the aid of known-albedo masks, Solar Energy 171 (2018) 971–976.

[8] Y. Qin, A review on the development of cool pavements to mitigate urban heat island effect, Renewable and Sustainable Energy Reviews 52 (2015) 445–459.

[9] Y. Qin, Y. He, J.E. Hiller, G. Mei, A new water-retaining paver block for reducing runoff and cooling pavement, Journal of Cleaner Production 199 (2018) 948–956.

[10] M. Sheikholeslami, Z. Li, A. Shafee, Lorentz forces effect on NEPCM heat transfer during solidification in a porous energy storage system, International Journal of Heat and Mass Transfer 127 (2018) 665–674.

[11] M. Sheikholeslami, M. Jafaryar, S. Saleem, Z. Li, A. Shafee, Y. Jiang, Nanofluid heat transfer augmentation and exergy loss inside a pipe equipped with innovative turbulators, International Journal of Heat and Mass Transfer 126 (2018) 156–163.

[12] M. Sheikholeslami, A. Ghasemi, Z. Li, A. Shafee, S. Saleem, Influence of CuO nanoparticles on heat transfer behavior of PCM in solidification process considering radiative source term, International Journal of Heat and Mass Transfer 126 (2018) 1252–1264.

[13] M. Sheikholeslami, M. Darzi, Z. Li, Experimental investigation for entropy generation and exergy loss of nano-refrigerant condensation process, International Journal of Heat and Mass Transfer 125 (2018) 1087–1095.

[14] M. Sheikholeslami, S.A. Shehzad, Z. Li, Water based nanofluid free convection heat transfer in a three dimensional porous cavity with hot sphere obstacle in existence of Lorenz forces, International Journal of Heat and Mass Transfer 125 (2018) 375–386.

[15] M. Sheikholeslami, M. Jafaryar, Z. Li, Nanofluid turbulent convective flow in a circular duct with helical turbulators considering CuO nanoparticles, International Journal of Heat and Mass Transfer 124 (2018) 980–989.

[16] M. Sheikholeslami, A. Ghasemi, Solidification heat transfer of nanofluid in existence of thermal radiation by means of FEM, International Journal of Heat and Mass Transfer 123 (2018) 418–431.
[17] M. Sheikholeslami, S.A. Shehzad, CVFEM simulation for nanofluid migration in a porous medium using Darcy model, International Journal of Heat and Mass Transfer 122 (2018) 1264–1271.
[18] M. Sheikholeslami, M. Darzi, M.K. Sadoughi, Heat transfer improvement and pressure drop during condensation of refrigerant-based nanofluid; an experimental procedure, International Journal of Heat and Mass Transfer 122 (2018) 643–650.
[19] M. Sheikholeslami, S.A. Shehzad, Simulation of water based nanofluid convective flow inside a porous enclosure via Non-equilibrium model, International Journal of Heat and Mass Transfer 120 (2018) 1200–1212.
[20] M. Sheikholeslami, M. Seyednezhad, Simulation of nanofluid flow and natural convection in a porous media under the influence of electric field using CVFEM, International Journal of Heat and Mass Transfer 120 (2018) 772–781.
[21] M. Sheikholeslami, H.B. Rokni, Numerical simulation for impact of Coulomb force on nanofluid heat transfer in a porous enclosure in presence of thermal radiation, International Journal of Heat and Mass Transfer 118 (2018) 823–831.
[22] M. Sheikholeslami, S.A. Shehzad, Numerical analysis of $Fe_3O_4-H_2O$ nanofluid flow in permeable media under the effect of external magnetic source, International Journal of Heat and Mass Transfer 118 (2018) 182–192.
[23] M. Sheikholeslami, M.K. Sadoughi, Simulation of CuO- water nanofluid heat transfer enhancement in presence of melting surface, International Journal of Heat and Mass Transfer 116 (2018) 909–919.
[24] M. Sheikholeslami, H.B. Rokni, Simulation of nanofluid heat transfer in presence of magnetic field: a review, International Journal of Heat and Mass Transfer 115 (2017) 1203–1233.
[25] M. Sheikholeslami, T. Hayat, A. Alsaedi, On simulation of nanofluid radiation and natural convection in an enclosure with elliptical cylinders, International Journal of Heat and Mass Transfer 115 (2017) 981–991.
[26] M. Sheikholeslami, S.A. Shehzad, CVFEM for influence of external magnetic source on $Fe_3O_4 - H_2O$ nanofluid behavior in a permeable cavity considering shape effect, International Journal of Heat and Mass Transfer 115 (2017) 180–191.
[27] M. Sheikholeslami, M. Seyednezhad, Nanofluid heat transfer in a permeable enclosure in presence of variable magnetic field by means of CVFEM, International Journal of Heat and Mass Transfer 114 (2017) 1169–1180.
[28] M. Sheikholeslami, H.B. Rokni, Melting heat transfer influence on nanofluid flow inside a cavity in existence of magnetic field, International Journal of Heat and Mass Transfer 114 (2017) 517–526.
[29] M. Sheikholeslami, M. Sadoughi, Mesoscopic method for MHD nanofluid flow inside a porous cavity considering various shapes of nanoparticles, International Journal of Heat and Mass Transfer 113 (2017) 106–114.
[30] M. Sheikholeslami, M.M. Bhatti, Forced convection of nanofluid in presence of constant magnetic field considering shape effects of nanoparticles, International Journal of Heat and Mass Transfer 111 (2017) 1039–1049.
[31] M. Sheikholeslami, M.M. Bhatti, Active method for nanofluid heat transfer enhancement by means of EHD, International Journal of Heat and Mass Transfer 109 (2017) 115–122.

[32] M. Sheikholeslami, S.A. Shehzad, Thermal radiation of ferrofluid in existence of Lorentz forces considering variable viscosity, International Journal of Heat and Mass Transfer 109 (2017) 82–92.
[33] M. Sheikholeslami, H.B. Rokni, Nanofluid two phase model analysis in existence of induced magnetic field, International Journal of Heat and Mass Transfer 107 (2017) 288–299.
[34] M. Sheikholeslami, S.A. Shehzad, Magnetohydrodynamic nanofluid convection in a porous enclosure considering heat flux boundary condition, International Journal of Heat and Mass Transfer 106 (2017) 1261–1269.
[35] M. Sheikholeslami, M.A. Sheremet, A. Shafee, I. Tlili, Simulation of nanoliquid thermogravitational convection within a porous chamber imposing magnetic and radiation impacts, Physica A: Statistical Mechanics and its Applications (2020), https://doi.org/10.1016/j.physa.2019.124058.
[36] M. Sheikholeslami, A. Nematpour Keshteli, Houman Babazadeh, Nanoparticles favorable effects on performance of thermal storage units, Journal of Molecular Liquids 300 (2020) 112329.
[37] A. Shafee, M. Sheikholeslami, M. Jafaryar, H. Babazadeh, Utilizing copper oxide nanoparticles for expedition of solidification within a storage system, Journal of Molecular Liquids 302 (2020) 112371, https://doi.org/10.1016/j.molliq.2019.112371.
[38] M. Sheikholeslami, Effect of uniform suction on nanofluid flow and heat transfer over a cylinder, Journal of the Brazilian Society of Mechanical Sciences and Engineering 37 (2015) 1623–1633.
[39] M. Sheikholeslami, Lattice Boltzmann Method simulation of MHD non-Darcy nanofluid free convection, Physica B 516 (2017) 55–71.
[40] M. Sheikholeslami, Magnetic field influence on CuO -H_2O nanofluid convective flow in a permeable cavity considering various shapes for nanoparticles, International Journal of Hydrogen Energy 42 (2017) 19611–19621.
[41] M. Sheikholeslami, Numerical simulation for solidification in a LHTESS by means of Nano-enhanced PCM, Journal of the Taiwan Institute of Chemical Engineers 86 (2018) 25–41.
[42] M. Sheikholeslami, Numerical modeling of Nano enhanced PCM solidification in an enclosure with metallic fin, Journal of Molecular Liquids 259 (2018) 424–438.
[43] M. Sheikholeslami, Solidification of NEPCM under the effect of magnetic field in a porous thermal energy storage enclosure using CuO nanoparticles, Journal of Molecular Liquids 263 (2018) 303–315.
[44] M. Sheikholeslami, Influence of magnetic field on Al_2O_3-H_2O nanofluid forced convection heat transfer in a porous lid driven cavity with hot sphere obstacle by means of LBM, Journal of Molecular Liquids 263 (2018) 472–488.
[45] M. Sheikholeslami, New computational approach for exergy and entropy analysis of nanofluid under the impact of Lorentz force through a porous media, Computer Methods in Applied Mechanics and Engineering 344 (2019) 319–333.
[46] M. Sheikholeslami, Numerical approach for MHD Al_2O_3-water nanofluid transportation inside a permeable medium using innovative computer method, Computer Methods in Applied Mechanics and Engineering 344 (2019) 306–318.
[47] M. Sheikholeslami, O. Mahian, Enhancement of PCM solidification using inorganic nanoparticles and an external magnetic field with application in energy storage systems, Journal of Cleaner Production 215 (2019) 963–977.

[48] M. Sheikholeslami, B. Rezaeianjouybari, M. Darzi, A. Shafee, Z. Li, T.K. Nguyen, Application of nano-refrigerant for boiling heat transfer enhancement employing an experimental study, International Journal of Heat and Mass Transfer 141 (2019) 974–980.

[49] M. Sheikholeslami, R.-UL. Haq, A. Shafee, Z. Li, Heat transfer behavior of Nanoparticle enhanced PCM solidification through an enclosure with V shaped fins, International Journal of Heat and Mass Transfer 130 (2019) 1322–1342.

[50] M. Sheikholeslami, S.A. Shehzad, Z. Li, A. Shafee, Numerical modeling for alumina nanofluid magnetohydrodynamic convective heat transfer in a permeable medium using Darcy law, International Journal of Heat and Mass Transfer 127 (2018) 614–622.

[51] M. Sheikholeslami, Numerical investigation of nanofluid free convection under the influence of electric field in a porous enclosure, Journal of Molecular Liquids 249 (2018) 1212–1221.

[52] M. Sheikholeslami, CuO-water nanofluid flow due to magnetic field inside a porous media considering Brownian motion, Journal of Molecular Liquids 249 (2018) 921–929.

[53] M. Sheikholeslami, Numerical investigation for CuO-H_2O nanofluid flow in a porous channel with magnetic field using mesoscopic method, Journal of Molecular Liquids 249 (2018) 739–746.

[54] M. Sheikholeslami, H.B. Rokni, Numerical modeling of nanofluid natural convection in a semi annulus in existence of Lorentz force, Computer Methods in Applied Mechanics and Engineering 317 (2017) 419–430.

[55] M. Sheikholeslami, A. Zeeshan, Analysis of flow and heat transfer in water based nanofluid due to magnetic field in a porous enclosure with constant heat flux using CVFEM, Computer Methods in Applied Mechanics and Engineering 320 (2017) 68–81.

[56] M. Sheikholeslami, M. Jafaryar, M. Hedayat, A. Shafee, Z. Li, T.K. Nguyen, M. Bakouri, Heat transfer and turbulent simulation of nanomaterial due to compound turbulator including irreversibility analysis, International Journal of Heat and Mass Transfer 137 (2019) 1290–1300.

[57] M. Sheikholeslami, M. Barzegar Gerdroodbary, R. Moradi, A. Shafee, Z. Li, Application of Neural Network for estimation of heat transfer treatment of Al_2O_3- H_2O nanofluid through a channel, Computer Methods in Applied Mechanics and Engineering 344 (2019) 1–12.

[58] M. Sheikholeslami, Application of Darcy law for nanofluid flow in a porous cavity under the impact of Lorentz forces, Journal of Molecular Liquids 266 (2018) 495–503.

[59] M. Sheikholeslami, Finite element method for PCM solidification in existence of CuO nanoparticles, Journal of Molecular Liquids 265 (2018) 347–355.

[60] M. Sheikholeslami, M. Jafaryar, A. Shafee, Z. Li, R-ul Haq, Heat transfer of nanoparticles employing innovative turbulator considering entropy generation, International Journal of Heat and Mass Transfer 136 (2019) 1233–1240.

[61] M. Sheikholeslami, R.-ul Haq, A. Shafee, Z. Li, Y.G. Elaraki, I. Tlili, Heat transfer simulation of heat storage unit with nanoparticles and fins through a heat exchanger, International Journal of Heat and Mass Transfer 135 (2019) 470–478.

CHAPTER

Exhaust heat recovery heat exchanger of gasoline engine

8

8.1 Definition of the problem

As shown in Fig. 8.1, passive technique is utilized to improve nanofluid heat transfer. In this system, a heat exchanger is used for recovery of internal combustion engine. Hot gas moves through the inner pipe and H_2O−CuO nanofluid are considered as working fluids of shell side. Twisted tape has been used in outer side to improve heat transfer. Tables 8.1 and 8.2 depict the characteristics of H_2O and nanoparticles. In this chapter, we suppose that d = 20 mm, D_i = 48 mm, D = 120 mm, D_o = 140 mm, L = 700 mm, L_o = 900 mm, l = 140 mm. Temperature and flow rate of inlet gas are derived from Ref. [1]. Various active and passive methods are developed in recent years [2−63]. Forced convection turbulent nanofluid flow in a heat exchanger can be formulated as

$$\frac{\partial (u_i)}{\partial x_i} = 0 \tag{8.1}$$

$$\frac{\partial}{\partial x_j}(\rho_{nf} u_i u_j) = -\frac{\partial p}{\partial x_i} + \frac{\partial}{\partial x_j}\left(\left(\frac{\partial u_i}{\partial x_j} + \frac{\partial u_j}{\partial x_i}\right)\mu_{nf}\right) + \frac{\partial}{\partial x_j}\left(-\rho_{nf}\overline{u_j' u_i'}\right) \tag{8.2}$$

$$\frac{\partial}{\partial x_i}(\rho_{nf} T u_i) = \frac{\partial}{\partial x_i}\left((\Gamma + \Gamma_t)\frac{\partial T}{\partial x_i}\right), \quad \Gamma(=\mu_{nf} / \mathrm{Pr}_{nf}), \quad {}_t(=\mu_t / Pr_t) \tag{8.3}$$

$\rho_{nf}\overline{u_j' u_i'}$ and μ_t are

$$-\rho_{nf}\overline{u_i' u_j'} = \left(\frac{\partial u_i}{\partial x_j} + \frac{\partial u_j}{\partial x_i}\right)\mu_t - \frac{2}{3}\rho_{nf} k \delta_{ij} - \frac{2}{3}\mu_t \frac{\partial u_k}{\partial x_k}\delta_{ij} \tag{8.4}$$

$$\mu_t = \frac{1}{\varepsilon} k^2 C_\mu \rho_{nf} \tag{8.5}$$

Nanofluid in Heat Exchangers for Mechanical Systems. https://doi.org/10.1016/B978-0-12-821923-2.00008-7

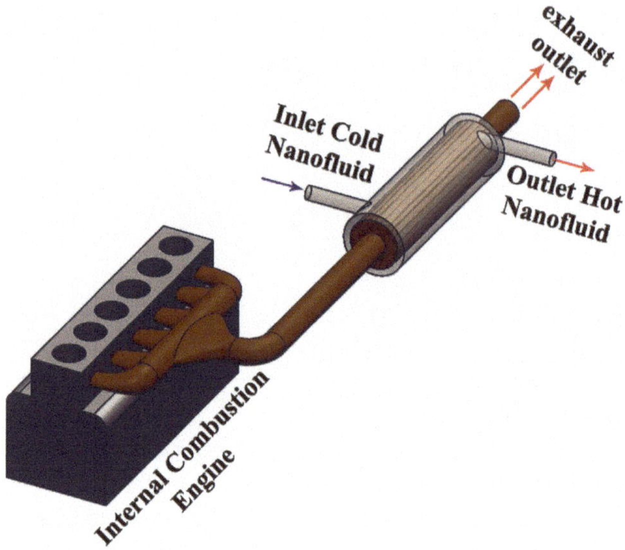

FIGURE 8.1

Current heat exchanger and sample mesh.

Table 8.1 The coefficient values of CuO-Water nanofluid.

Coefficient values	CuO-Water
a_1	−26.5933108
a_2	−0.403818333
a_3	−33.3516805
a_4	−1.915825591
a_5	6.421858E-02
a_6	48.40336955
a_7	−9.787756683
a_8	190.245610009
a_9	10.9285386565
a_{10}	−0.72009983664

Table 8.2 Thermophysical properties of water and nanoparticles.

	$\rho(kg/m^3)$	$C_p(j/kgk)$	$k(W/m.k)$	$\mu(Pa.s)$
Pure water	997.1	4179	0.613	0.0010003
CuO	6500	540	18	–

The equations of k, ε are

$$\frac{\partial}{\partial x_j}\left(\left(\frac{\mu_t}{\sigma_k}+\mu_{nf}\right)\frac{\partial k}{\partial x_j}\right)-\rho_{nf}\varepsilon+G_k=\frac{\partial}{\partial x_i}(u_i\rho_{nf}k),\quad G_k=-\rho_{nf}\overline{u_j'u_i'}\frac{\partial u_j}{\partial x_i} \tag{8.6}$$

$$\frac{\partial}{\partial x_i}(u_i\rho_{nf}\varepsilon)=\frac{\varepsilon}{k}G_kC_{1\varepsilon}-\rho_{nf}\frac{\varepsilon^2}{k}C_{2\varepsilon}+\frac{\partial}{\partial x_j}\left(\left(\frac{\mu_t}{\sigma_\varepsilon}+\mu_{nf}\right)\frac{\partial \varepsilon}{\partial x_j}\right) \tag{8.7}$$

$$C_{1\varepsilon}=1.42,\quad C_\mu=0.0845,\quad C_{2\varepsilon}=1.68, Pr_t=0.85, \sigma_k=1, \sigma_\varepsilon=1.3 \tag{8.8}$$

$\rho_{nf}, (\rho C_p)_{nf}, k_{nf}$, and μ_{nf} are

$$\rho_{nf}=\rho_f(1-\phi)+\rho_s\phi \tag{8.9}$$

$$(\rho C_p)_{nf}=-(\phi-1)(\rho C_p)_f+(\rho C_p)_s\phi \tag{8.10}$$

$$\frac{k_{nf}}{k_f}=1+3\frac{\phi(k_p/k_f-1)}{-\phi(k_p/k_f-1)+(2+k_p/k_f)}+5\times 10^4 g\prime(\phi,T,d_p)\phi\rho_f c_{p,f}\sqrt{\frac{\kappa_b T}{\rho_p d_p}}$$

$$g\prime(\phi,T,d_p)=\left(a_2Ln(d_p)+a_1+a_5Ln(d_p)^2+a_4Ln(d_p)Ln(\phi)+a_3Ln(\phi)\right)Ln(T)$$

$$+\left(a_{10}Ln(d_p)^2+a_6+a_8Ln(\phi)+a_7Ln(d_p)+a_9Ln(d_p)Ln(\phi)\right),$$

$$\frac{4\times 10^{-8}}{d_p}+1/k_p=1/k_{p,eff}$$

$$\tag{8.11}$$

$$\frac{\mu_{nf}}{\mu_f}=\frac{k_{Brownian}}{k_f\mathrm{Pr}}+(1-\phi)^{-2.5} \tag{8.12}$$

z = 0 and z = L have the following conditions:

$$v_i=0, w_i=cte, u_i=0, I=0.16(\mathrm{Re})^{\frac{-1}{8}}, T_i=cte \tag{8.13}$$

$$\frac{\partial \mathrm{v}}{\partial \mathrm{z}}=\frac{\partial \mathrm{w}}{\partial \mathrm{z}}=\frac{\partial \mathrm{u}}{\partial \mathrm{z}}=0, \frac{\partial \mathrm{T}}{\partial \mathrm{z}}=0 \tag{8.14}$$

In the present work, the definitions of Nu and f are

$$Nu=\frac{hD_h}{k_{nf}} \tag{8.15}$$

$$f=\frac{\Delta p}{\frac{\rho v_m^2}{2}\frac{L}{D_h}} \tag{8.16}$$

8.2 Effects of active parameters

Nanofluid turbulent forced convection and entropy analysis through a heat exchanger which is used for heat recovery is simulated via FVM. Hot air and nanofluid are considered as working fluids of inner and outer pipes. Properties of CuO-water are estimated considering Brownian motion effect. Impacts of Re and N on temperature, velocity, thermal and viscous entropy generation contours are illustrated in Figs. 8.2 and 8.3. As inlet velocity enhances, pressure drop and temperature gradient increase. Increasing N leads to greater temperature gradient with the price of greater Darcy factor. So, stronger vortexes appear with enhancement of N. Therefore convective flow improves with increase of N. Pressure drop enhances with rise of N and Re. Entropy contours indicate that viscous entropy generation enhances with rise of pressure loss, but opposite behavior is seen $S_{gen,th}$. Fig. 8.3

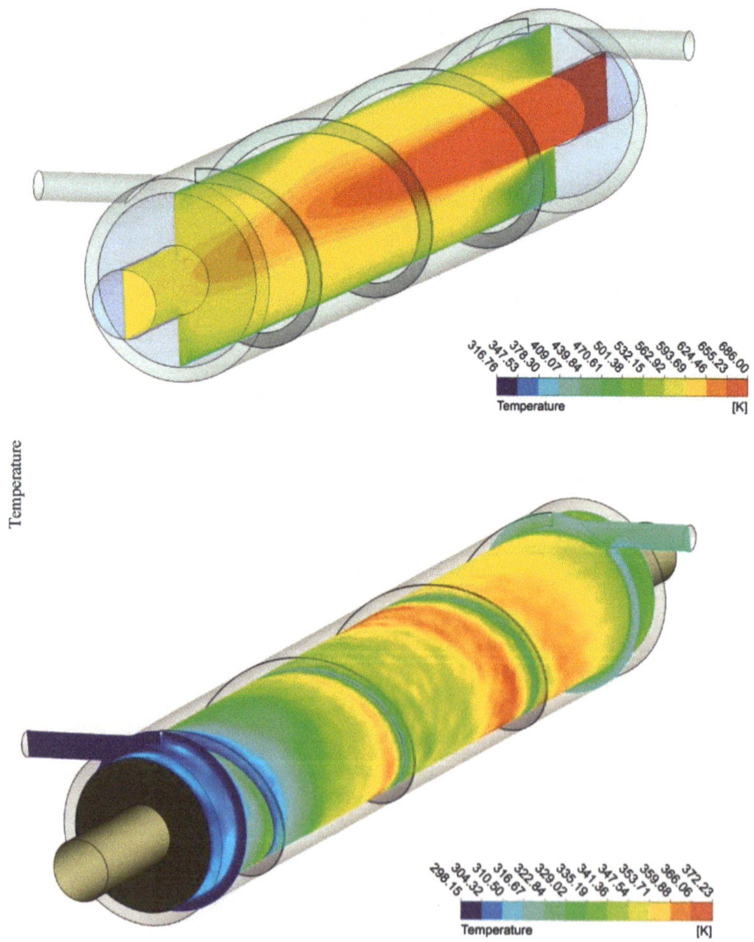

FIGURE 8.2

Temperature, entropy generation contours at Re = 2000, $N = 3$.

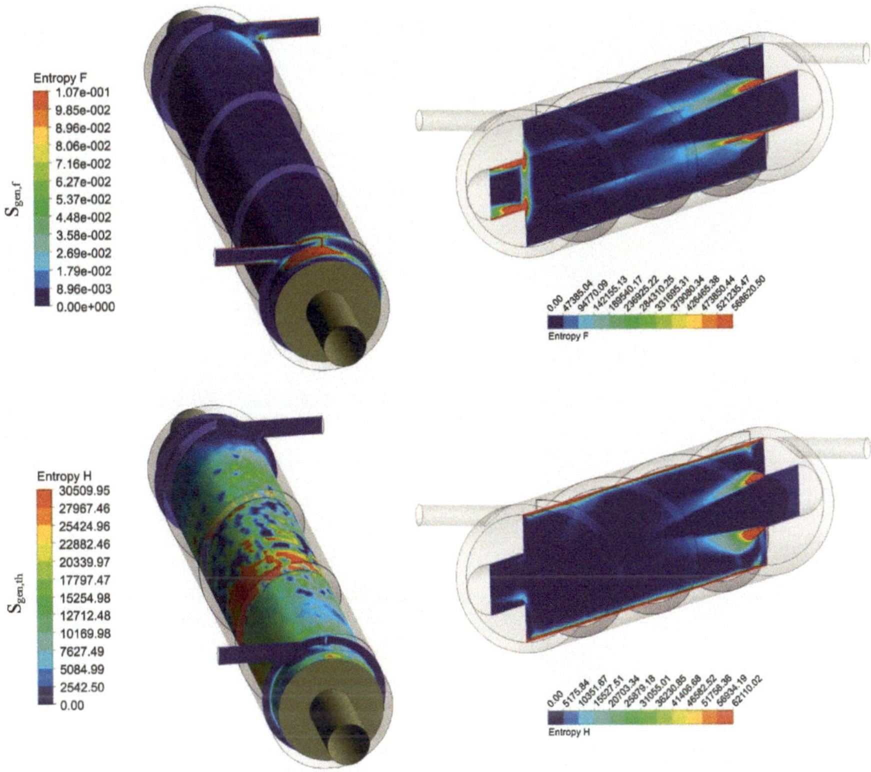

FIGURE 8.2 cont'd.

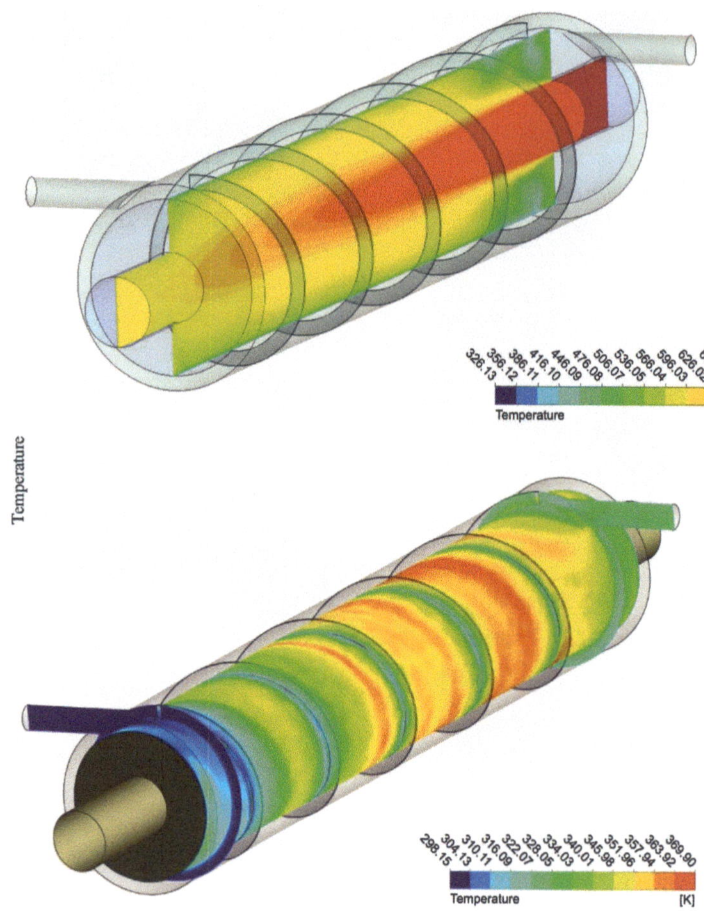

FIGURE 8.3

Temperature, entropy generation contours at Re = 2000, $N = 5$.

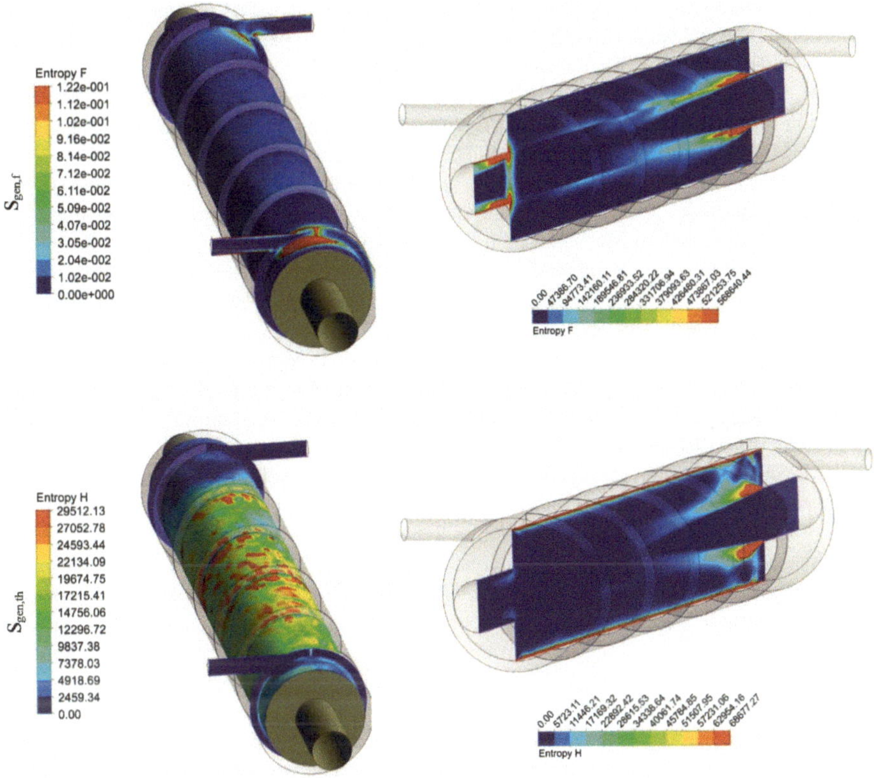

FIGURE 8.3 cont'd.

depicts the impacts of N and Re on entropy components. $S_{gen,th}$ is a decreasing function of Re and N. Roles of significant parameters on $S_{gen,f}$ are differing of $S_{gen,th}$. So, $S_{gen,th}$ enhances with increase of Re and N.

References

[1] T. Wang, Y. Zhang, J. Zhang, Z. Peng, G. Shu, Comparisons of system benefits and thermo-economics for exhaust energy recovery applied on a heavy-duty diesel engine and a light-duty vehicle gasoline engine, Energy Conversion and Management 84 (2014) 97–107.

[2] Y. Qin, Y. Zhao, X. Chen, L. Wang, F. Li, T. Bao, Moist curing increases the solar reflectance of concrete, Construction and Building Materials 215 (2019) 114–118.

[3] M. Sheikholeslami, Effect of uniform suction on nanofluid flow and heat transfer over a cylinder, Journal of the Brazilian Society of Mechanical Sciences and Engineering 37 (2015) 1623–1633.

[4] M. Sheikholeslami, Numerical modeling of nano enhanced PCM solidification in an enclosure with metallic fin, Journal of Molecular Liquids 259 (2018) 424–438.

[5] M. Sheikholeslami, Solidification of NEPCM under the effect of magnetic field in a porous thermal energy storage enclosure using CuO nanoparticles, Journal of Molecular Liquids 263 (2018) 303–315.

[6] M. Sheikholeslami, Influence of magnetic field on Al_2O_3-H_2O nanofluid forced convection heat transfer in a porous lid driven cavity with hot sphere obstacle by means of LBM, Journal of Molecular Liquids 263 (2018) 472–488.

[7] M. Sheikholeslami, New computational approach for exergy and entropy analysis of nanofluid under the impact of Lorentz force through a porous media, Computer Methods in Applied Mechanics and Engineering 344 (2019) 319–333.

[8] M. Sheikholeslami, Numerical approach for MHD Al_2O_3-water nanofluid transportation inside a permeable medium using innovative computer method, Computer Methods in Applied Mechanics and Engineering 344 (2019) 306–318.

[9] X. Ma, M. Sheikholeslami, M. Jafaryar, A. Shafee, T. Nguyen-Thoi, Z. Li, Solidification inside a clean energy storage unit utilizing phase change material with copper oxide nanoparticles, Journal of Cleaner Production (2019), https://doi.org/10.1016/j.jclepro.2019.118888.

[10] M. Sheikholeslami, O. Mahian, Enhancement of PCM solidification using inorganic nanoparticles and an external magnetic field with application in energy storage systems, Journal of Cleaner Production 215 (2019) 963–977.

[11] M. Sheikholeslami, B. Rezaeianjouybari, M. Darzi, A. Shafee, Z. Li, T.K. Nguyen, Application of nano-refrigerant for boiling heat transfer enhancement employing an experimental study, International Journal of Heat and Mass Transfer 141 (2019) 974–980.

[12] M. Sheikholeslami, Lattice Boltzmann Method simulation of MHD non-Darcy nanofluid free convection, Physica B 516 (2017) 55–71.

[13] M. Sheikholeslami, Magnetic field influence on CuO -H_2O nanofluid convective flow in a permeable cavity considering various shapes for nanoparticles, International Journal of Hydrogen Energy 42 (2017) 19611–19621.

[14] M. Sheikholeslami, Numerical simulation for solidification in a LHTESS by means of nano-enhanced PCM, Journal of the Taiwan Institute of Chemical Engineers 86 (2018) 25–41.

[15] M. Sheikholeslami, M. Jafaryar, M. Hedayat, A. Shafee, Z. Li, T.K. Nguyen, M. Bakouri, Heat transfer and turbulent simulation of nanomaterial due to compound turbulator including irreversibility analysis, International Journal of Heat and Mass Transfer 137 (2019) 1290–1300.

[16] M. Sheikholeslami, Application of Darcy law for nanofluid flow in a porous cavity under the impact of Lorentz forces, Journal of Molecular Liquids 266 (2018) 495–503.

[17] M. Sheikholeslami, Finite element method for PCM solidification in existence of CuO nanoparticles, Journal of Molecular Liquids 265 (2018) 347–355.

[18] M. Sheikholeslami, M. Jafaryar, A. Shafee, Z. Li, R.-ul Haq, Heat transfer of nanoparticles employing innovative turbulator considering entropy generation, International Journal of Heat and Mass Transfer 136 (2019) 1233–1240.

[19] M. Sheikholeslami, R.-ul Haq, A. Shafee, Z. Li, Y.G. Elaraki, I. Tlili, Heat transfer simulation of heat storage unit with nanoparticles and fins through a heat exchanger, International Journal of Heat and Mass Transfer 135 (2019) 470–478.

[20] M. Sheikholeslami, R.-ul Haq, A. Shafee, Z. Li, Heat transfer behavior of Nanoparticle enhanced PCM solidification through an enclosure with V shaped fins, International Journal of Heat and Mass Transfer 130 (2019) 1322–1342.

[21] M. Sheikholeslami, S.A. Shehzad, Z. Li, A. Shafee, Numerical modeling for alumina nanofluid magnetohydrodynamic convective heat transfer in a permeable medium using Darcy law, International Journal of Heat and Mass Transfer 127 (2018) 614–622.
[22] M. Sheikholeslami, Z. Li, A. Shafee, Lorentz forces effect on NEPCM heat transfer during solidification in a porous energy storage system, International Journal of Heat and Mass Transfer 127 (2018) 665–674.
[23] M. Sheikholeslami, M. Jafaryar, S. Saleem, Z. Li, A. Shafee, Y. Jiang, Nanofluid heat transfer augmentation and exergy loss inside a pipe equipped with innovative turbulators, International Journal of Heat and Mass Transfer 126 (2018) 156–163.
[24] M. Sheikholeslami, A. Ghasemi, Z. Li, A. Shafee, S. Saleem, Influence of CuO nanoparticles on heat transfer behavior of PCM in solidification process considering radiative source term, International Journal of Heat and Mass Transfer 126 (2018) 1252–1264.
[25] M. Sheikholeslami, M. Darzi, Z. Li, Experimental investigation for entropy generation and exergy loss of nano-refrigerant condensation process, International Journal of Heat and Mass Transfer 125 (2018) 1087–1095.
[26] M. Sheikholeslami, S.A. Shehzad, Z. Li, Water based nanofluid free convection heat transfer in a three dimensional porous cavity with hot sphere obstacle in existence of Lorenz forces, International Journal of Heat and Mass Transfer 125 (2018) 375–386.
[27] M. Sheikholeslami, M. Jafaryar, Z. Li, Nanofluid turbulent convective flow in a circular duct with helical turbulators considering CuO nanoparticles, International Journal of Heat and Mass Transfer 124 (2018) 980–989.
[28] M. Sheikholeslami, A. Ghasemi, Solidification heat transfer of nanofluid in existence of thermal radiation by means of FEM, International Journal of Heat and Mass Transfer 123 (2018) 418–431.
[29] M. Sheikholeslami, S.A. Shehzad, CVFEM simulation for nanofluid migration in a porous medium using Darcy model, International Journal of Heat and Mass Transfer 122 (2018) 1264–1271.
[30] M. Sheikholeslami, M. Darzi, M.K. Sadoughi, Heat transfer improvement and pressure drop during condensation of refrigerant-based nanofluid; an experimental procedure, International Journal of Heat and Mass Transfer 122 (2018) 643–650.
[31] M. Sheikholeslami, S.A. Shehzad, Simulation of water based nanofluid convective flow inside a porous enclosure via Non-equilibrium model, International Journal of Heat and Mass Transfer 120 (2018) 1200–1212.
[32] M. Sheikholeslami, M. Seyednezhad, Simulation of nanofluid flow and natural convection in a porous media under the influence of electric field using CVFEM, International Journal of Heat and Mass Transfer 120 (2018) 772–781.
[33] M. Sheikholeslami, H.B. Rokni, Numerical simulation for impact of Coulomb force on nanofluid heat transfer in a porous enclosure in presence of thermal radiation, International Journal of Heat and Mass Transfer 118 (2018) 823–831.
[34] M. Sheikholeslami, S.A. Shehzad, Numerical analysis of Fe_3O_4 $-H_2O$ nanofluid flow in permeable media under the effect of external magnetic source, International Journal of Heat and Mass Transfer 118 (2018) 182–192.
[35] M. Sheikholeslami, M.K. Sadoughi, Simulation of CuO- water nanofluid heat transfer enhancement in presence of melting surface, International Journal of Heat and Mass Transfer 116 (2018) 909–919.
[36] M. Sheikholeslami, H.B. Rokni, Simulation of nanofluid heat transfer in presence of magnetic field: a review, International Journal of Heat and Mass Transfer 115 (2017) 1203–1233.

[37] M. Sheikholeslami, T. Hayat, A. Alsaedi, On simulation of nanofluid radiation and natural convection in an enclosure with elliptical cylinders, International Journal of Heat and Mass Transfer 115 (2017) 981–991.
[38] M. Sheikholeslami, T. Hayat, A. Alsaedi, S. Abelman, Numerical analysis of EHD nanofluid force convective heat transfer considering electric field dependent viscosity, International Journal of Heat and Mass Transfer 108 (2017) 2558–2565.
[39] M. Sheikholeslami, T. Hayat, A. Alsaedi, Numerical study for external magnetic source influence on water based nanofluid convective heat transfer, International Journal of Heat and Mass Transfer 106 (2017) 745–755.
[40] Y. Qin, H. He, X. Ou, T. Bao, Experimental study on darkening water-rich mud tailings for accelerating desiccation, Journal of Cleaner Production (2019) 118235, https://doi.org/10.1016/j.jclepro.2019.118235.
[41] Y. Qin, J.E. Hiller, D. Meng, Linearity between pavement thermophysical properties and surface temperatures, Journal of Materials in Civil Engineering (2019), https://doi.org/10.1061/(ASCE)MT.1943-5533.0002890.
[42] Y. Qin, A review on the development of cool pavements to mitigate urban heat island effect, Renewable and Sustainable Energy Reviews 52 (2015) 445–459.
[43] Y. Qin, Y. He, J.E. Hiller, G. Mei, A new water-retaining paver block for reducing runoff and cooling pavement, Journal of Cleaner Production 199 (2018) 948–956.
[44] M. Sheikholeslami, S.A. Shehzad, CVFEM for influence of external magnetic source on $Fe_3O_4 - H_2O$ nanofluid behavior in a permeable cavity considering shape effect, International Journal of Heat and Mass Transfer 115 (2017) 180–191.
[45] M. Sheikholeslami, M. Seyednezhad, Nanofluid heat transfer in a permeable enclosure in presence of variable magnetic field by means of CVFEM, International Journal of Heat and Mass Transfer 114 (2017) 1169–1180.
[46] M. Sheikholeslami, H.B. Rokni, Melting heat transfer influence on nanofluid flow inside a cavity in existence of magnetic field, International Journal of Heat and Mass Transfer 114 (2017) 517–526.
[47] M. Sheikholeslami, M. Sadoughi, Mesoscopic method for MHD nanofluid flow inside a porous cavity considering various shapes of nanoparticles, International Journal of Heat and Mass Transfer 113 (2017) 106–114.
[48] M. Sheikholeslami, M.M. Bhatti, Forced convection of nanofluid in presence of constant magnetic field considering shape effects of nanoparticles, International Journal of Heat and Mass Transfer 111 (2017) 1039–1049.
[49] M. Sheikholeslami, M.M. Bhatti, Active method for nanofluid heat transfer enhancement by means of EHD, International Journal of Heat and Mass Transfer 109 (2017) 115–122.
[50] M. Sheikholeslami, Numerical investigation for $CuO-H_2O$ nanofluid flow in a porous channel with magnetic field using mesoscopic method, Journal of Molecular Liquids 249 (2018) 739–746.
[51] M. Sheikholeslami, H.B. Rokni, Numerical modeling of nanofluid natural convection in a semi annulus in existence of Lorentz force, Computer Methods in Applied Mechanics and Engineering 317 (2017) 419–430.
[52] M. Sheikholeslami, A. Zeeshan, Analysis of flow and heat transfer in water based nanofluid due to magnetic field in a porous enclosure with constant heat flux using CVFEM, Computer Methods in Applied Mechanics and Engineering 320 (2017) 68–81.

[53] M. Sheikholeslami, S.A. Shehzad, Thermal radiation of ferrofluid in existence of Lorentz forces considering variable viscosity, International Journal of Heat and Mass Transfer 109 (2017) 82–92.

[54] M. Sheikholeslami, H.B. Rokni, Nanofluid two phase model analysis in existence of induced magnetic field, International Journal of Heat and Mass Transfer 107 (2017) 288–299.

[55] M. Sheikholeslami, S.A. Shehzad, Magnetohydrodynamic nanofluid convection in a porous enclosure considering heat flux boundary condition, International Journal of Heat and Mass Transfer 106 (2017) 1261–1269.

[56] M. Sheikholeslami, R. Ellahi, Three dimensional mesoscopic simulation of magnetic field effect on natural convection of nanofluid, International Journal of Heat and Mass Transfer 89 (2015) 799–808.

[57] M. Sheikholeslami, Numerical investigation of nanofluid free convection under the influence of electric field in a porous enclosure, Journal of Molecular Liquids 249 (2018) 1212–1221.

[58] M. Sheikholeslami, CuO-water nanofluid flow due to magnetic field inside a porous media considering Brownian motion, Journal of Molecular Liquids 249 (2018) 921–929.

[59] M. Sheikholeslami, M.A. Sheremet, A. Shafee, I. Tlili, Simulation of nanoliquid thermogravitational convection within a porous chamber imposing magnetic and radiation impacts, Physica A: Statistical Mechanics and its Applications (2020), https://doi.org/10.1016/j.physa.2019.124058.

[60] M. Sheikholeslami, A.N. Keshteli, H. Babazadeh, Nanoparticles favorable effects on performance of thermal storage units, Journal of Molecular Liquids 300 (2020) 112329.

[61] A. Shafee, M. Sheikholeslami, M. Jafaryar, Houman Babazadeh, Utilizing copper oxide nanoparticles for expedition of solidification within a storage system, Journal of Molecular Liquids 302 (2020) 112371, https://doi.org/10.1016/j.molliq.2019.112371.

[62] Y. Qin, J. Luo, Z. Chen, G. Mei, L.-E. Yan, Measuring the albedo of limited-extent targets without the aid of known-albedo masks, Solar Energy 171 (2018) 971–976.

[63] F. Li, M. Sheikholeslami, Rebwar Nasir Dara, M. Jafaryar, A. Shafee, T. Nguyen-Thoi, Z. Li, Numerical study for nanofluid behavior inside a storage finned enclosure involving melting process, Journal of Molecular Liquids 297 (2020) 111939.

CHAPTER

Ferrohydrodynamic nanofluid flow in heat exchangers

9

9.1 Definition of the problem

Fig. 9.1 demonstrates the schematic of the T-junction in the presence of three wires in normal direction of surface as source of magnetic field [1]. In this study, the radius of the inlet flow is 0.05 m. The applied boundary condition is also presented in the figure. As an electric current goes through wire, the magnetic field is generated in the domain (Fig. 9.1B).

Fig. 9.2 illustrates the grid of a chosen geometry. In this study, all sizes become nondimensionless as shown in Fig. 9.1A. In order to perform grid study, three different grids with 6000, 12,000, and 18,000 elements are chosen as coarse, fine, and very fine grids for the grid independency analysis, respectively. After a comparison with experimental data, a model with 12,000 grids shows a good agreement and chosen as a final grid for further studies.

In order to obtain flow feature and heat transfer inside the T-junction, Navier–Stokes equations along with energy equation are solved simultaneously. To apply the magnetic field on the domain, the components of the magnetic field should be considered in the momentum equations. Other scientists considered different ways for enhancing heat transfer [2–67]. In our previous study, comprehensive details of our study were presented. Therefore, the final equations for our problem are as follows:

Continuity equation:

$$\frac{\partial u}{\partial x}+\frac{\partial v}{\partial y}=0 \tag{9.1}$$

Momentum equation:

$$\rho_m\left(u\frac{\partial u}{\partial x}+v\frac{\partial u}{\partial y}\right)=-\frac{\partial p}{\partial x}+\mu_m\left(\frac{\partial^2 u}{\partial x^2}+\frac{\partial^2 u}{\partial y^2}\right)+F_K(x) \tag{9.2}$$

$$\rho_m\left(u\frac{\partial v}{\partial x}+v\frac{\partial v}{\partial y}\right)=-\frac{\partial p}{\partial y}+\mu_m\left(\frac{\partial^2 v}{\partial x^2}+\frac{\partial^2 v}{\partial y^2}\right)+F_K(y) \tag{9.3}$$

Nanofluid in Heat Exchangers for Mechanical Systems. https://doi.org/10.1016/B978-0-12-821923-2.00009-9

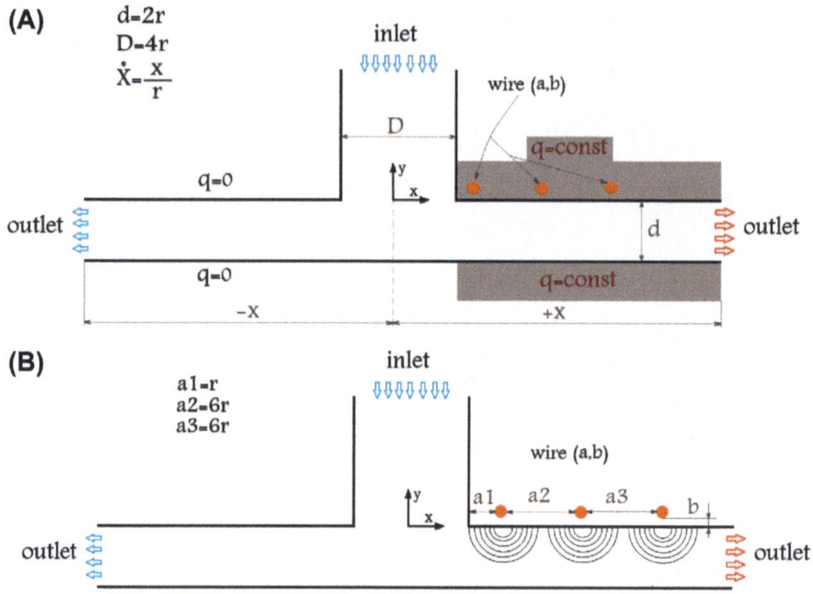

FIGURE 9.1

(A) Schematic of T-junction in presence of magnetic field (wire). (B) The schematic distribution of magnetic field in the domain.

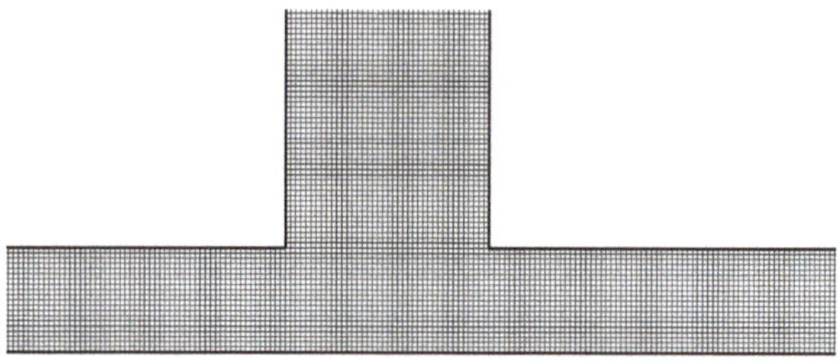

FIGURE 9.2

Two-dimensional generated grid.

Energy equation:

$$(\rho_m C_p)_m \left(u \frac{\partial T}{\partial x} + v \frac{\partial T}{\partial y} \right) = k_m \left(\frac{\partial^2 T}{\partial x^2} + \frac{\partial^2 T}{\partial y^2} \right) \tag{9.4}$$

The terms $F_K(x)$ and $F_K(y)$ correspond to the presence of the magnetic gradient and are known as the Kelvin force. H_x and H_y, the components of the magnetic field in the x and y directions, are determined as follows [36]:

$$H_x(x, y) = \frac{I}{2\pi} \frac{(x-a)}{(x-a)^2 + (y-b)^2} \tag{9.5}$$

$$H_y(x, y) = \frac{I}{2\pi} \frac{(y-b)}{(x-a)^2 + (y-b)^2} \tag{9.6}$$

Moreover, the magnitude of the magnetic field intensity is given by

$$H(x, y, z) = \frac{I}{2\pi} \frac{1}{\sqrt{(x-a)^2 + (y-b)^2}} \tag{9.7}$$

It is worthy to note that the term $\mu_0 M \frac{\partial H}{\partial z}$ should be considered to the momentum equation in the z direction. The magnetization is known by M and is defined as [1].

$$M = \frac{6m_p}{\pi d_p^3} \left[coth(\xi) - \frac{1}{\xi} \right] \tag{9.8}$$

In addition, the particle magnetic moment for the magnetite particles is calculated as

$$m_p = \frac{4\mu_B \pi d_p^3}{6 \times 91.25 \times 10^{-30}} \tag{9.9}$$

Also ξ is the Langevin parameter and is defined as [38].

$$\xi = \frac{\mu_0 m_p H}{k_B T} \tag{9.10}$$

As mentioned in the previous section, Magnetic number (Mn) is reliant on the magnetic field intensity. This shows that Mn increases with a rise in the intensity of magnetic field [38].

$$Mn = \frac{\mu_0 \chi H_r^2 h^2}{\rho_m \alpha_m^2} \tag{9.11}$$

where χ is the magnetic susceptibility of nanofluid. As stated earlier, the magnetic susceptibility of a ferrofluid containing 4 Vol% with a mean diameter of 10 nm is approximately $\chi = 0.348586$ [39]. H_r is the characteristic of magnetic field strength and is calculated by

$$H_r = H(a, 0) = \frac{I}{2\pi b} \tag{9.12}$$

Table 9.1 The value of current equivalent to the chosen magnetic intensity.

Mn	*I*
0	0
3.26×10^{-5}	3
9.06×10^{-5}	5
17.76×10^{-5}	7
52.2×10^{-5}	12

Thus, the value of the current for each magnetic field is as follows (Table 9.1): The mixture physical properties in the above equations are calculated as below. Mixture density [40]:

$$\rho_m = \varphi.\rho_p + (1-\varphi).\rho_f \tag{9.13}$$

Mixture dynamic viscosity:

$$\mu_m = \frac{1}{(1-\varphi)^{2.5}}\mu_f \tag{9.14}$$

For spherical nanoparticles, the mixture thermal conductivity is obtained by the following equation [41]:

$$k_m = \left[\frac{k_p + 2k_f - 2\varphi(k_f - k_p)}{k_p + 2k_f + \varphi(k_f - k_p)}\right]k_f \tag{9.15}$$

Mixture specific heat:

$$cp_m = \frac{\varphi(\rho cp)_p + (1-\varphi)(\rho cp)_f}{\rho_m} \tag{9.16}$$

Reynolds number:

$$\mathrm{Re}_m = \frac{\rho_m v_m d_i}{\mu_m} \tag{9.17}$$

Nusselt number:

$$Nu_{\mathrm{hot}} = \frac{q'' d_i}{k_m\,(T_w - T_b)} \tag{9.18}$$

where d_i is the internal diameter of the small microchannel. k_m is the thermal conductivity of the nanofluid.

The inlet velocity of ferrofluid is corresponding to $Re_m = 50 - 200$. In order to calculate the heat transfer, outcomes of Nusselt number have been computed for water-based ferrofluid containing 4 Vol% Fe_3O_4. The temperature of inlet flow is constant ($T_{hot,in} = T_0$), and a constant heat flux (200 $\frac{W}{m^2K}$) is applied on the left wall of the T-junction as shown in Fig. 9.1A.

9.2 Effects of active parameters

Fig. 9.3 illustrates the streamline patterns inside the T-junction in various inlet velocities in presence of magnetic field. Fig. 9.3A depicts the streamline for

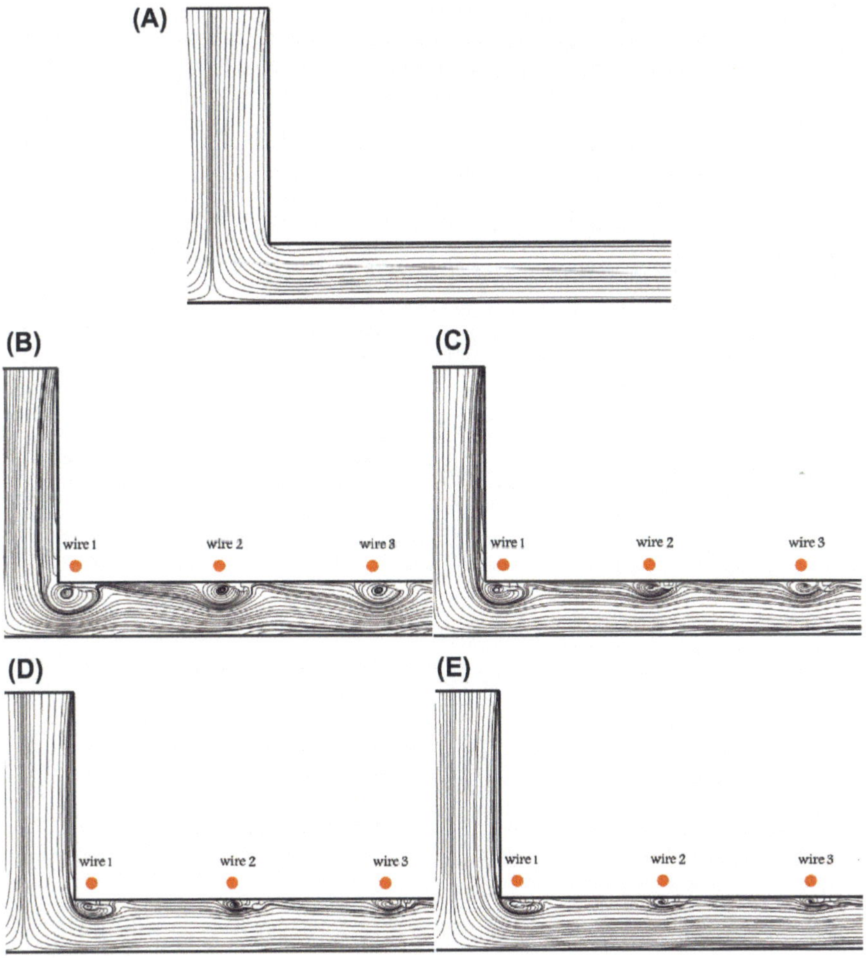

FIGURE 9.3

Streamlines for (A) Re = 50, $Mn = 0$; (B) Re = 50, $Mn = 52.2 \times 10^5$; (C) Re = 100, $Mn = 52.2 \times 10^5$; (D) Re = 150, $Mn = 52.2 \times 10^5$; (E) Re = 200, $Mn = 52.2 \times 10^5$.

Re = 50 without magnetic field. Fig. 9.3B–E shows the effect of nonuniform magnetic field ($Mn = 52.2 \times 10^5$) on the streamlines when Reynolds number is varied from 50 to 200.

As shown in the figure, the occurrence of the magnetic field significantly changes the streamlines in the T-junction. In low Reynolds number (Re = 50), large eddies are formed in the vicinity of the magnetic source close to the wall. As the velocity of the inlet flow is increased, the size of these eddies declines. It is worthy to note that nonuniform cross-magnetic field like nonuniform axial magnetic field is small in magnetohydrodynamic (MHD). Thus, the related term is not considered in the momentum equations.

Fig. 9.4 plots the variation of Nusselt number inside the T-junction for different inlet velocities when magnetic field is presented in the domain. The plots show the

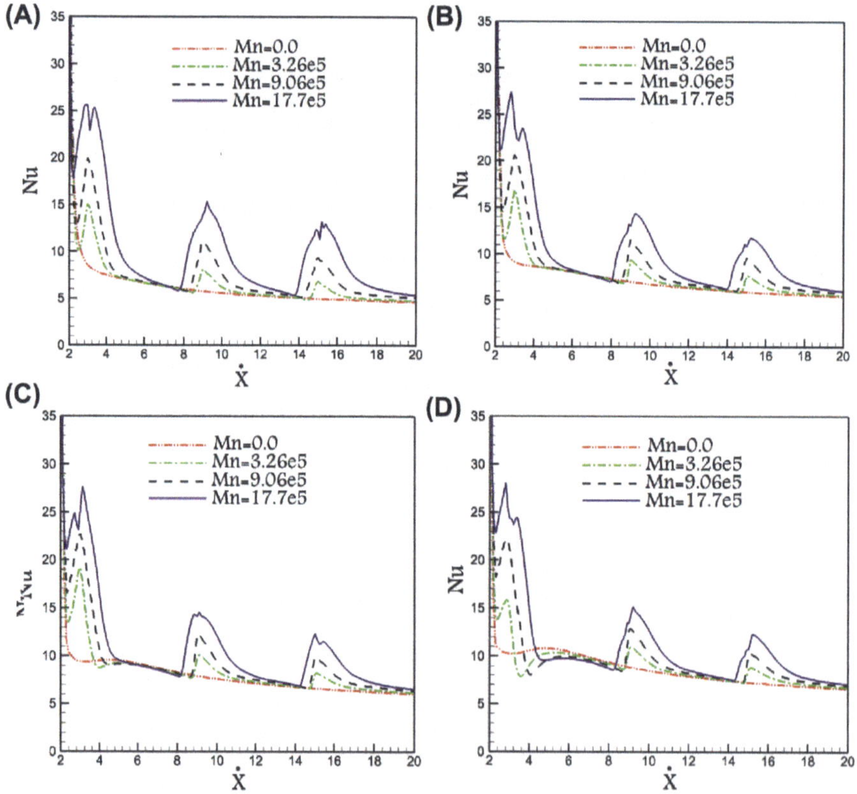

FIGURE 9.4

Variation of Nusselt number along the x direction inside the T-junction in Various Reynolds Number: (A) Re = 50, (B) Re = 100, (C) Re = 150, (D) Re = 200.

effect of the magnetic field on heat transfer is limited when the inlet velocity is increased. Indeed, the range of high Nusselt number is restricted in high Reynolds number. The variation of Nusselt number shows that the value of maximum heat transfer remains approximately similar in different inlet velocities. It is worthy to note that the role of increasing the inlet velocity is not considerable in heat transfer as shown in Fig. 9.5. The maximum difference occurs in the initial distance close to the inlet region due to the flow effect in the T-junction.

One of the main hydrodynamic characteristics of the flow inside the tube is skin friction. Fig. 9.6 illustrates the effect of the magnetic field on the value of skin friction along the x direction for various Reynolds numbers. The presence of the magnetic field significantly intensifies the skin friction in vicinity of the magnetic source. The results clearly show that the effect of magnetic field on skin friction highly declines as the inlet velocity increases.

In order to reveal all effects of the magnetic field, the variation of the skin friction without magnetic source along the x direction is plotted in Fig. 9.7. As expected, raising the inlet velocity decreases the value of skin friction. According to flow feature, the main reason for the increasing of the skin friction is the formation of the vortices inside the domain. Indeed, the vortices occur due to high skin friction inside the domain.

In order to notice the effects of the magnetic field on the thermal characteristics of the domain, the contour of the temperature distribution inside the T-junction is depicted in Fig. 9.8. The comparison of the various models shows that temperature

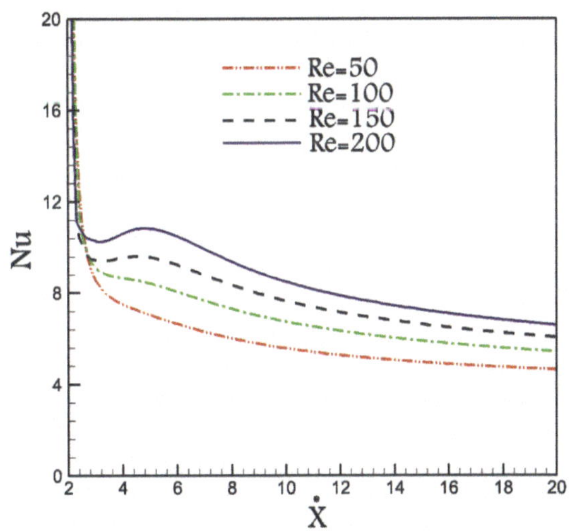

FIGURE 9.5

Comparison of Nusselt number along the x direction for different inlet velocities.

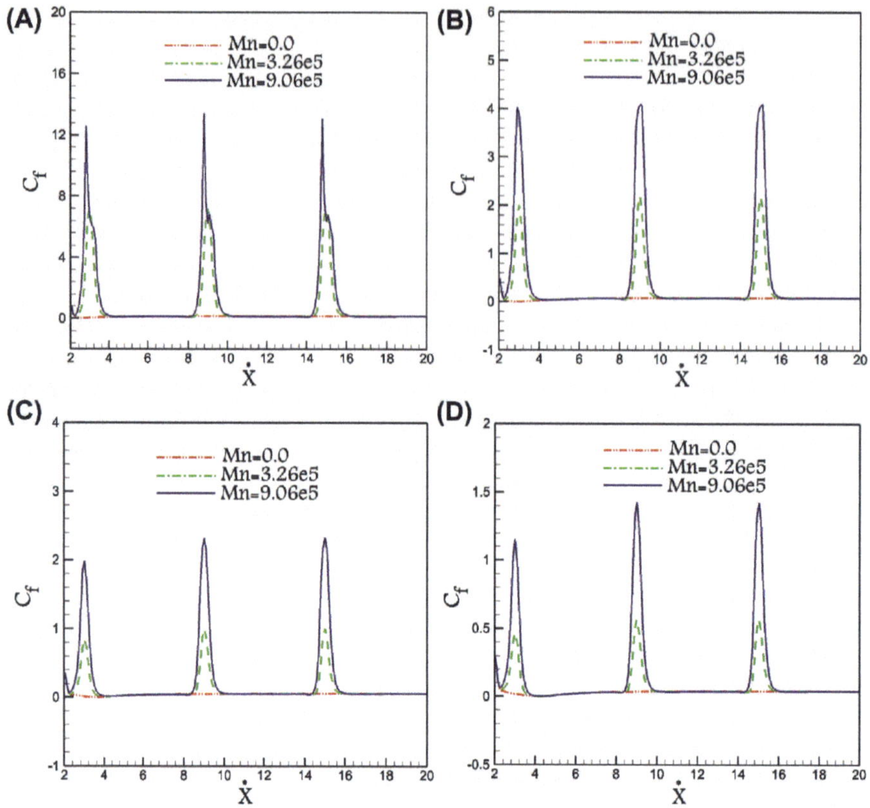

FIGURE 9.6

Distribution of skin friction along the x direction inside the T-junction in Various Reynolds Number: (A) Re = 50, (B) Re = 100, (C) Re = 150, (D) Re = 200.

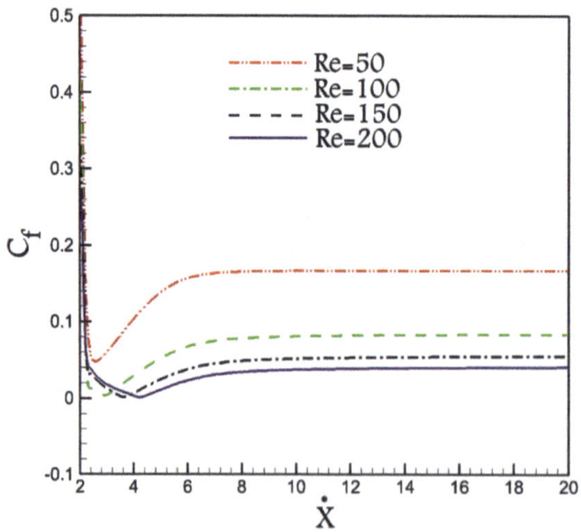

FIGURE 9.7

Comparison of Nusselt number along the x direction for different inlet velocities without magnetic field.

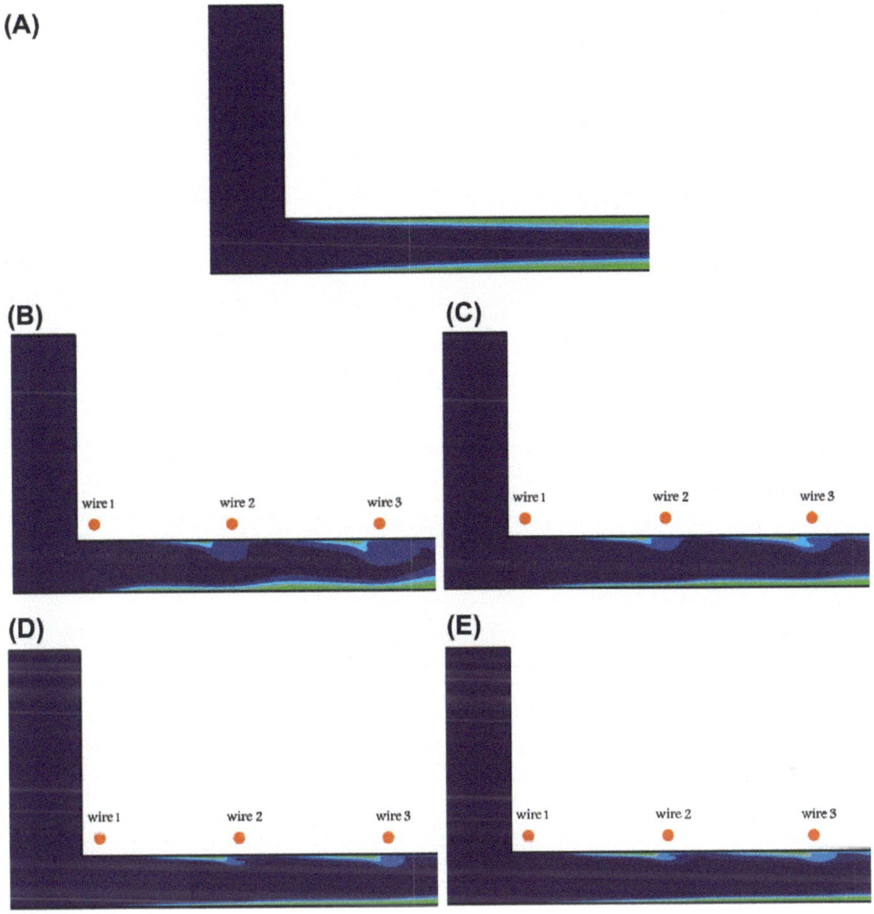

FIGURE 9.8

Comparison of temperature distribution inside the domain in presence of magnetic field ($Mn = 52.2 \times 10^5$) in various inlet velocities: (A) Re = 50, (B) Re = 100, (C) Re = 150, (D) Re = 200.

locally disturbs in the vicinity of the magnetic source. Since the magnetic field induces vortices, these small circulation regions increase the heat transfer and the temperature declines in comparison with no magnetic field condition (Fig. 9.8A). In addition, it is obviously noticed that the temperature decreases in the vicinity of the second and third wire.

Fig. 9.9 plots the variation of the average Nusselt number for various inlet velocities without magnetic field. It is found that the heat transfer intensifies approximately 40% as the Reynolds number is increased in the T-junction. The effect of the three nonuniform magnetic sources on the average Nusselt number

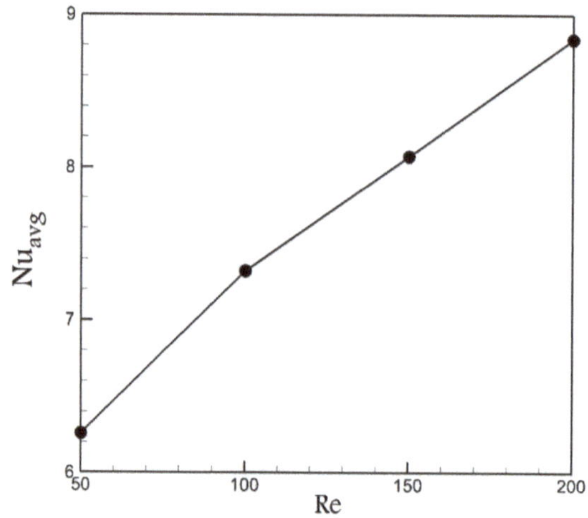

FIGURE 9.9

Variation of the average Nusselt number along the channel for various Reynolds number ($Mn = 0$).

is presented in Fig. 9.10 when the inlet velocity is constant (Re $= 50$). The results clearly indicate that Nusselt number increases 64% as the magnetic field with a strength of $Mn = 52.2 \times 10^5$ is applied in the domain. According to the obtained results, it is found that magnetic field induces vortices which significantly intensifies the heat transfer in the domain.

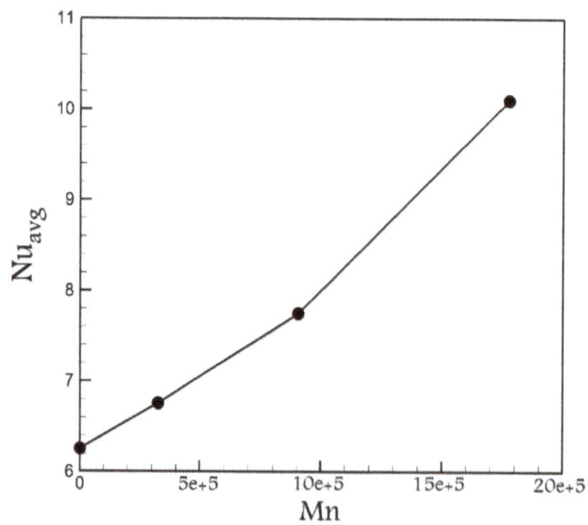

FIGURE 9.10

Effect of strength of magnetic field on the average Nusselt number along the channel ($Re_m = 50$).

References

[1] M. Barzegar Gerdroodbary, M. Sheikholeslami, S. Valiallah Mousavi, A. Anazadehsayed, R. Moradi, The influence of non-uniform magnetic field on heat transfer intensification of ferrofluid inside a T-junction, Chemical Engineering and Processing: Process Intensification 123 (2018) 58–66.

[2] M. Sheikholeslami, M. Barzegar Gerdroodbary, R. Moradi, S. Ahmad, Z. Li, Application of neural network for estimation of heat transfer treatment of Al_2O_3- H_2O nanofluid through a channel, Computer Methods in Applied Mechanics and Engineering 344 (2019) 1–12.

[3] Y. Qin, J.E. Hiller, D. Meng, Linearity between pavement thermophysical properties and surface temperatures, Journal of Materials in Civil Engineering (2019), https://doi.org/10.1061/(ASCE)MT.1943-5533.0002890.

[4] M. Sheikholeslami, Effect of uniform suction on nanofluid flow and heat transfer over a cylinder, Journal of the Brazilian Society of Mechanical Sciences and Engineering 37 (2015) 1623–1633.

[5] M. Sheikholeslami, Lattice Boltzmann method simulation of MHD non-Darcy nanofluid free convection, Physica B 516 (2017) 55–71.

[6] M. Sheikholeslami, Magnetic field influence on CuO -H_2O nanofluid convective flow in a permeable cavity considering various shapes for nanoparticles, International Journal of Hydrogen Energy 42 (2017) 19611–19621.

[7] M. Sheikholeslami, Numerical simulation for solidification in a LHTESS by means of nano-enhanced PCM, Journal of the Taiwan Institute of Chemical Engineers 86 (2018) 25–41.

[8] Y. Qin, H. He, X. Ou, T. Bao, Experimental study on darkening water-rich mud tailings for accelerating desiccation, Journal of Cleaner Production (2019) 118235, https://doi.org/10.1016/j.jclepro.2019.118235.

[9] M. Sheikholeslami, K. Vajravelu, M.M. Rashidi, Forced convection heat transfer in a semi annulus under the influence of a variable magnetic field, International Journal of Heat and Mass Transfer 92 (2016) 339–348.

[10] M. Sheikholeslami, T. Hayat, A. Alsaedi, Numerical study for external magnetic source influence on water based nanofluid convective heat transfer, International Journal of Heat and Mass Transfer 106 (2017) 745–755.

[11] M. Sheikholeslami, T. Hayat, A. Alsaedi, S. Abelman, Numerical analysis of EHD nanofluid force convective heat transfer considering electric field dependent viscosity, International Journal of Heat and Mass Transfer 108 (2017) 2558–2565.

[12] M. Sheikholeslami, Numerical modeling of nano enhanced PCM solidification in an enclosure with metallic fin, Journal of Molecular Liquids 259 (2018) 424–438.

[13] M. Sheikholeslami, Solidification of NEPCM under the effect of magnetic field in a porous thermal energy storage enclosure using CuO nanoparticles, Journal of Molecular Liquids 263 (2018) 303–315.

[14] M. Sheikholeslami, Influence of magnetic field on Al_2O_3-H_2O nanofluid forced convection heat transfer in a porous lid driven cavity with hot sphere obstacle by means of LBM, Journal of Molecular Liquids 263 (2018) 472–488.

[15] M. Sheikholeslami, New computational approach for exergy and entropy analysis of nanofluid under the impact of Lorentz force through a porous media, Computer Methods in Applied Mechanics and Engineering 344 (2019) 319–333.

[16] M. Sheikholeslami, Numerical approach for MHD Al_2O_3-water nanofluid transportation inside a permeable medium using innovative computer method, Computer Methods in Applied Mechanics and Engineering 344 (2019) 306–318.
[17] X. Ma, M. Sheikholeslami, M. Jafaryar, S. Ahmad, T. Nguyen-Thoi, Z. Li, Solidification inside a clean energy storage unit utilizing phase change material with copper oxide nanoparticles, Journal of Cleaner Production (2019), https://doi.org/10.1016/j.jclepro.2019.118888.
[18] M. Sheikholeslami, O. Mahian, Enhancement of PCM solidification using inorganic nanoparticles and an external magnetic field with application in energy storage systems, Journal of Cleaner Production 215 (2019) 963–977.
[19] M. Sheikholeslami, B. Rezaeianjouybari, M. Darzi, S. Ahmad, Z. Li, T.K. Nguyen, Application of nano-refrigerant for boiling heat transfer enhancement employing an experimental study, International Journal of Heat and Mass Transfer 141 (2019) 974–980.
[20] Y. Qin, M. Zhang, J.E. Hiller, Theoretical and experimental studies on the daily accumulative heat gain from cool roofs, Energy 129 (2017) 138–147.
[21] M. Sheikholeslami, M. Jafaryar, M. Hedayat, S. Ahmad, Z. Li, T.K. Nguyen, M. Bakouri, Heat transfer and turbulent simulation of nanomaterial due to compound turbulator including irreversibility analysis, International Journal of Heat and Mass Transfer 137 (2019) 1290–1300.
[22] M. Sheikholeslami, Application of Darcy law for nanofluid flow in a porous cavity under the impact of Lorentz forces, Journal of Molecular Liquids 266 (2018) 495–503.
[23] M. Sheikholeslami, Finite element method for PCM solidification in existence of CuO nanoparticles, Journal of Molecular Liquids 265 (2018) 347–355.
[24] Y. Qin, H. He, A new simplified method for measuring the albedo of limited extent targets, Solar Energy 157 (Suppl. C) (2017) 1047–1055.
[25] M. Sheikholeslami, M. Jafaryar, S. Ahmad, Z. Li, R-ul Haq, Heat transfer of nanoparticles employing innovative turbulator considering entropy generation, International Journal of Heat and Mass Transfer 136 (2019) 1233–1240.
[26] M. Sheikholeslami, R.-ul Haq, S. Ahmad, Z. Li, Y.G. Elaraki, I. Tlili, Heat transfer simulation of heat storage unit with nanoparticles and fins through a heat exchanger, International Journal of Heat and Mass Transfer 135 (2019) 470–478.
[27] M. Sheikholeslami, R.-ul Haq, S. Ahmad, Z. Li, Heat transfer behavior of nanoparticle enhanced PCM solidification through an enclosure with V shaped fins, International Journal of Heat and Mass Transfer 130 (2019) 1322–1342.
[28] M. Sheikholeslami, S.A. Shehzad, Z. Li, S. Ahmad, Numerical modeling for alumina nanofluid magnetohydrodynamic convective heat transfer in a permeable medium using Darcy law, International Journal of Heat and Mass Transfer 127 (2018) 614–622.
[29] M. Sheikholeslami, Z. Li, S. Ahmad, Lorentz forces effect on NEPCM heat transfer during solidification in a porous energy storage system, International Journal of Heat and Mass Transfer 127 (2018) 665–674.
[30] M. Sheikholeslami, M. Jafaryar, S. Saleem, Z. Li, S. Ahmad, Y. Jiang, Nanofluid heat transfer augmentation and exergy loss inside a pipe equipped with innovative turbulators, International Journal of Heat and Mass Transfer 126 (2018) 156–163.
[31] Y. Qin, J. Luo, Z. Chen, G. Mei, L.-E. Yan, Measuring the albedo of limited-extent targets without the aid of known-albedo masks, Solar Energy 171 (2018) 971–976.

[32] M. Sheikholeslami, A. Ghasemi, Z. Li, S. Ahmad, S. Saleem, Influence of CuO nanoparticles on heat transfer behavior of PCM in solidification process considering radiative source term, International Journal of Heat and Mass Transfer 126 (2018) 1252–1264.

[33] M. Sheikholeslami, M. Darzi, Z. Li, Experimental investigation for entropy generation and exergy loss of nano-refrigerant condensation process, International Journal of Heat and Mass Transfer 125 (2018) 1087–1095.

[34] M. Sheikholeslami, S.A. Shehzad, Z. Li, Water based nanofluid free convection heat transfer in a three dimensional porous cavity with hot sphere obstacle in existence of Lorenz forces, International Journal of Heat and Mass Transfer 125 (2018) 375–386.

[35] Y. Qin, Y. Zhao, X. Chen, L. Wang, F. Li, T. Bao, Moist curing increases the solar reflectance of concrete, Construction and Building Materials 215 (2019) 114–118.

[36] M. Sheikholeslami, M. Jafaryar, Z. Li, Nanofluid turbulent convective flow in a circular duct with helical turbulators considering CuO nanoparticles, International Journal of Heat and Mass Transfer 124 (2018) 980–989.

[37] M. Sheikholeslami, A. Ghasemi, Solidification heat transfer of nanofluid in existence of thermal radiation by means of FEM, International Journal of Heat and Mass Transfer 123 (2018) 418–431.

[38] Y. Qin, Y. He, J.E. Hiller, G. Mei, A new water-retaining paver block for reducing runoff and cooling pavement, Journal of Cleaner Production 199 (2018) 948–956.

[39] M. Sheikholeslami, S.A. Shehzad, CVFEM simulation for nanofluid migration in a porous medium using Darcy model, International Journal of Heat and Mass Transfer 122 (2018) 1264–1271.

[40] M. Sheikholeslami, M. Darzi, M.K. Sadoughi, Heat transfer improvement and pressure drop during condensation of refrigerant-based nanofluid; an experimental procedure, International Journal of Heat and Mass Transfer 122 (2018) 643–650.

[41] M. Sheikholeslami, S.A. Shehzad, Simulation of water based nanofluid convective flow inside a porous enclosure via Non-equilibrium model, International Journal of Heat and Mass Transfer 120 (2018) 1200–1212.

[42] M. Sheikholeslami, M. Seyednezhad, Simulation of nanofluid flow and natural convection in a porous media under the influence of electric field using CVFEM, International Journal of Heat and Mass Transfer 120 (2018) 772–781.

[43] Y. Qin, A review on the development of cool pavements to mitigate urban heat island effect, Renewable and Sustainable Energy Reviews 52 (2015) 445–459.

[44] M. Sheikholeslami, H.B. Rokni, Numerical simulation for impact of Coulomb force on nanofluid heat transfer in a porous enclosure in presence of thermal radiation, International Journal of Heat and Mass Transfer 118 (2018) 823–831.

[45] M. Sheikholeslami, S.A. Shehzad, Numerical analysis of Fe_3O_4 $-H_2O$ nanofluid flow in permeable media under the effect of external magnetic source, International Journal of Heat and Mass Transfer 118 (2018) 182–192.

[46] M. Sheikholeslami, M.K. Sadoughi, Simulation of CuO- water nanofluid heat transfer enhancement in presence of melting surface, International Journal of Heat and Mass Transfer 116 (2018) 909–919.

[47] M. Sheikholeslami, H.B. Rokni, Simulation of nanofluid heat transfer in presence of magnetic field: a review, International Journal of Heat and Mass Transfer 115 (2017) 1203–1233.

[48] M. Sheikholeslami, T. Hayat, A. Alsaedi, On simulation of nanofluid radiation and natural convection in an enclosure with elliptical cylinders, International Journal of Heat and Mass Transfer 115 (2017) 981–991.
[49] M. Sheikholeslami, S.A. Shehzad, CVFEM for influence of external magnetic source on Fe3O4 – H2O nanofluid behavior in a permeable cavity considering shape effect, International Journal of Heat and Mass Transfer 115 (2017) 180–191.
[50] M. Sheikholeslami, M. Seyednezhad, Nanofluid heat transfer in a permeable enclosure in presence of variable magnetic field by means of CVFEM, International Journal of Heat and Mass Transfer 114 (2017) 1169–1180.
[51] M. Sheikholeslami, B.H. Rokni, Melting heat transfer influence on nanofluid flow inside a cavity in existence of magnetic field, International Journal of Heat and Mass Transfer 114 (2017) 517–526.
[52] M. Sheikholeslami, M. Sadoughi, Mesoscopic method for MHD nanofluid flow inside a porous cavity considering various shapes of nanoparticles, International Journal of Heat and Mass Transfer 113 (2017) 106–114.
[53] M. Sheikholeslami, M.M. Bhatti, Forced convection of nanofluid in presence of constant magnetic field considering shape effects of nanoparticles, International Journal of Heat and Mass Transfer 111 (2017) 1039–1049.
[54] M. Sheikholeslami, M.M. Bhatti, Active method for nanofluid heat transfer enhancement by means of EHD, International Journal of Heat and Mass Transfer 109 (2017) 115–122.
[55] M. Sheikholeslami, S.A. Shehzad, Thermal radiation of ferrofluid in existence of Lorentz forces considering variable viscosity, International Journal of Heat and Mass Transfer 109 (2017) 82–92.
[56] M. Sheikholeslami, M.A. Sheremet, A. Shafee, I. Tlili, Simulation of nanoliquid thermogravitational convection within a porous chamber imposing magnetic and radiation impacts, Physica A: Statistical Mechanics and its Applications (2020), https://doi.org/10.1016/j.physa.2019.124058.
[57] M. Sheikholeslami, H.B. Rokni, Nanofluid two phase model analysis in existence of induced magnetic field, International Journal of Heat and Mass Transfer 107 (2017) 288–299.
[58] M. Sheikholeslami, A.N. Keshteli, H. Babazadeh, Nanoparticles favorable effects on performance of thermal storage units, Journal of Molecular Liquids 300 (2020) 112329.
[59] M. Sheikholeslami, S.A. Shehzad, Magnetohydrodynamic nanofluid convection in a porous enclosure considering heat flux boundary condition, International Journal of Heat and Mass Transfer 106 (2017) 1261–1269.
[60] M. Sheikholeslami, R. Ellahi, Three dimensional mesoscopic simulation of magnetic field effect on natural convection of nanofluid, International Journal of Heat and Mass Transfer 89 (2015) 799–808.
[61] M. Sheikholeslami, Numerical investigation of nanofluid free convection under the influence of electric field in a porous enclosure, Journal of Molecular Liquids 249 (2018) 1212–1221.
[62] M. Sheikholeslami, CuO-water nanofluid flow due to magnetic field inside a porous media considering Brownian motion, Journal of Molecular Liquids 249 (2018) 921–929.
[63] M. Sheikholeslami, Numerical investigation for CuO-H_2O nanofluid flow in a porous channel with magnetic field using mesoscopic method, Journal of Molecular Liquids 249 (2018) 739–746.

[64] A. Shafee, M. Sheikholeslami, M. Jafaryar, H. Babazadeh, Utilizing copper oxide nanoparticles for expedition of solidification within a storage system, Journal of Molecular Liquids 302 (2020) 112371, https://doi.org/10.1016/j.molliq.2019.112371.

[65] M. Sheikholeslami, H.B. Rokni, Numerical modeling of nanofluid natural convection in a semi annulus in existence of Lorentz force, Computer Methods in Applied Mechanics and Engineering 317 (2017) 419–430.

[66] M. Sheikholeslami, A. Zeeshan, Analysis of flow and heat transfer in water based nanofluid due to magnetic field in a porous enclosure with constant heat flux using CVFEM, Computer Methods in Applied Mechanics and Engineering 320 (2017) 68–81.

[67] F. Li, M. Sheikholeslami, R.N. Dara, M. Jafaryar, A. Shafee, T. Nguyen-Thoi, Z. Li, Numerical study for nanofluid behavior inside a storage finned enclosure involving melting process, Journal of Molecular Liquids 297 (2020) 111939.

CHAPTER

Solar heat exchanger with twist tape

10

10.1 Definition of the problem

In this chapter, a circular pipe fitted with helical tape has been considered as solar heat exchanger. Three-dimensional (3D) view of pipe with solar absorber has been exhibited in Fig. 10.1. Carrier fluid is Al_2O_3-water nanofluid [1]. In addition, Table 10.1 exhibits the formulation properties. Solar heat flux condition was assumed for upper side of pipe and plates. For outlet and inlet sections, we assumed pressure outlet and velocity inlet conditions. Steady 3D flow in turbulent regime, neglecting buoyancy effect and considering realizable k−ε model, leads to final equations which are solved via ANSYS FLUENT 18.1. Nondimensional parameters have been listed in Table 10.2. Settings of such software were classified in previous publication [1]. According to Refs. [1–4], the governing PDEs are summarized as follows:

$$\frac{\partial}{\partial x_i}(\rho u_i) = 0 \tag{10.1}$$

$$\begin{aligned}
-\overline{u_i{}' u_j{}'} &= \mu_t\left(\frac{\partial u_i}{\partial x_j} + \frac{\partial u_j}{\partial x_i}\right) - \frac{2}{3}\left(\rho K + \mu_t\frac{\partial u_k}{\partial x_k}\right)\delta_{ij}, \\
\mu_t &= \frac{\rho_{nf} C_\mu K^2}{\varepsilon}, \\
\frac{\partial}{\partial x_j}(\rho_{nf} u_i u_j) &= -\frac{\partial p}{\partial x_i} + \frac{\partial}{\partial x_j}\left[\mu\left(\frac{\partial u_i}{\partial x_j} - \rho\overline{u_i{}' u_j{}'}\right)\right],
\end{aligned} \tag{10.2}$$

$$\frac{\partial}{\partial x_i}(\rho u_i T) = \frac{\partial}{\partial x_j}\left((\Gamma + \Gamma_t)\frac{\partial T}{\partial x_j}\right), \quad E = h - \frac{p}{\rho} + \frac{u^2}{2} \tag{10.3}$$

$$\frac{\partial}{\partial x_j}(\rho_{nf} K u_j) = \frac{\partial}{\partial x_j}\left[\left(\mu_{nf} + \frac{\mu_t}{\sigma_k}\right)\frac{\partial K}{\partial x_j}\right] + G_k - \rho_{nf}\varepsilon \tag{10.4}$$

Nanofluid in Heat Exchangers for Mechanical Systems. https://doi.org/10.1016/B978-0-12-821923-2.00010-5

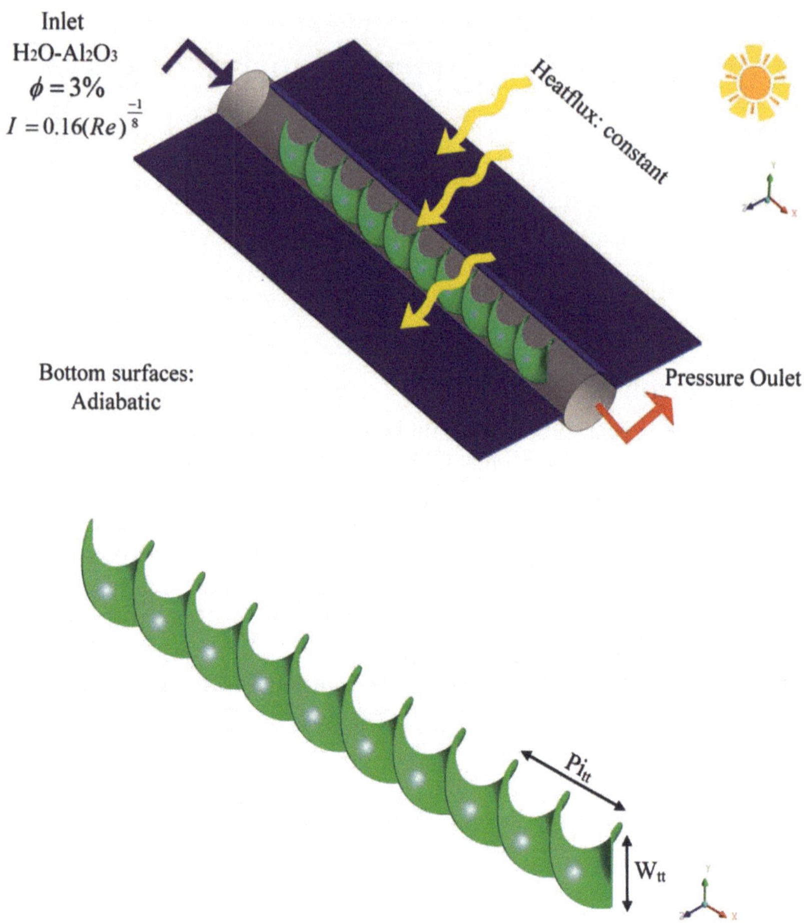

FIGURE 10.1

Shape of turbulator, mesh, and summarized boundary condition.

$$\frac{\partial}{\partial x_j}(\rho_{nf}\varepsilon u_j) = \frac{\partial}{\partial x_j}\left[\left(\mu_{nf} + \frac{\mu_t}{\sigma_\varepsilon}\right)\frac{\partial \varepsilon}{\partial x_j}\right] + \rho_{nf}C_1 S\varepsilon - C_2\rho_{nf}\frac{\varepsilon^2}{K + \sqrt{\nu\varepsilon}} \tag{10.5}$$

The bottom surface is isolated. Inlet and outlet follow the following conditions:

$$w_i = 0, v_i = 0, I = 0.16(Re)^{\frac{-1}{8}}, T_i = cte, u_i = cte \tag{10.6}$$

$$\frac{\partial \mathrm{v}}{\partial \mathrm{x}} = \frac{\partial \mathrm{w}}{\partial \mathrm{x}} = \frac{\partial \mathrm{u}}{\partial \mathrm{x}} = 0, \frac{\partial \mathrm{T}}{\partial \mathrm{x}} = 0 \tag{10.7}$$

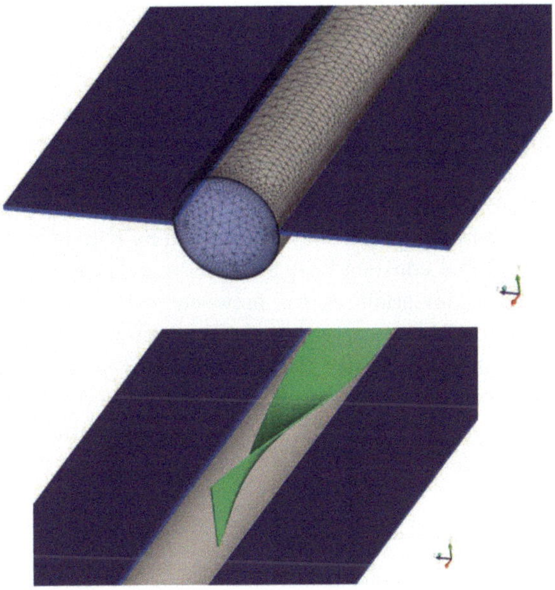

FIGURE 10.1 cont'd.

Table 10.1 Applied models for nanofluid properties.

Density	$\rho_{nf} = \rho_f(1 - \phi) + \rho_s\phi$
Heat capacity	$(\rho C_p)_{nf} = (1 - \phi)(C_p)_f + \rho_s\phi$
Viscosity	$\mu_{nf} = \dfrac{\mu_f}{(1 - \phi)^{2.5}}$
Thermal conductivity	$k_{nf} = k_f\left[\dfrac{k_p - 2\phi(k_f - k_p) + 2k_f}{k_p + \phi(k_f - k_p) + 2k_f}\right]$

Table 10.2 Simulation parameter variables.

Reynolds number	$\mathrm{Re} = \frac{\rho u D}{\mu}$
Friction factor	$f = \frac{2D\Delta p}{\rho L_{pipe} {u_m}^2}$
Number of revolution	$N = \frac{L_{tt}}{Pi_{tt}}$
Diameter ratio	$D^* = \frac{w_{tt}}{L_{tt}} \times 10$
Nusselt number	$Nu = \frac{hD}{k_{nf}}, h = \frac{q''}{(T_w - T_b)}$
Thermal efficiency	$\eta = \frac{Nu/Nu_P}{(f/f_P)^{1/3}}$

10.2 Effects of active parameters

This chapter describes a macroscopic method for modeling the turbulent flow of nanoparticle inside a duct with twisted-tape (TT) turbulator. Upper side of the system gets uniform flux from the sun. Assumption of constant physical properties for Al_2O_3-water nanomaterial and the involving k−ε model lead to obtain a set of equations which were solved via FVM. For validation, the outputs from current work have been compared with the outputs reported by Kim et al. [2] for the same kind of working fluid and constant heat flux condition. From Fig. 10.2, it can be concluded that small deviation exists between two results. So, reasonable agreement can be offered with current code. This outputs obtained by involving k−ε model and y^+ ranges guarantee the accuracy. In Fig. 10.1, sample grid has been shown and one can see that we utilized hybrid method and mesh was denser near the TT. With respect to Fig. 10.3, selecting grid with 6,953,150 element results in correct outputs. Using various external devices was examined in recent publications [5−69].

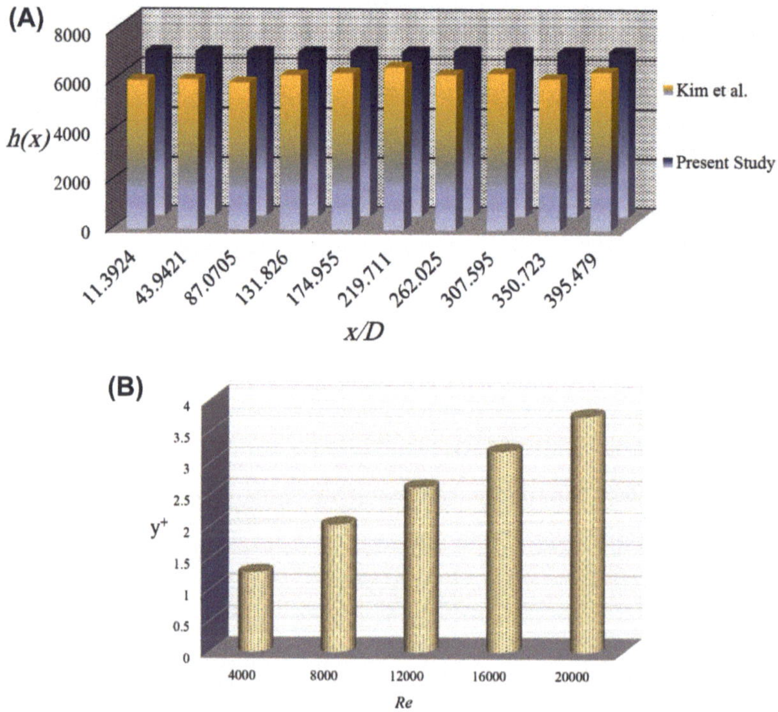

FIGURE 10.2

(A) Validation of numerical approach by reproducing outputs of Ref. [2]; (B) Ranges of y^+ when $D^* = 0.14$, $N = 4$.

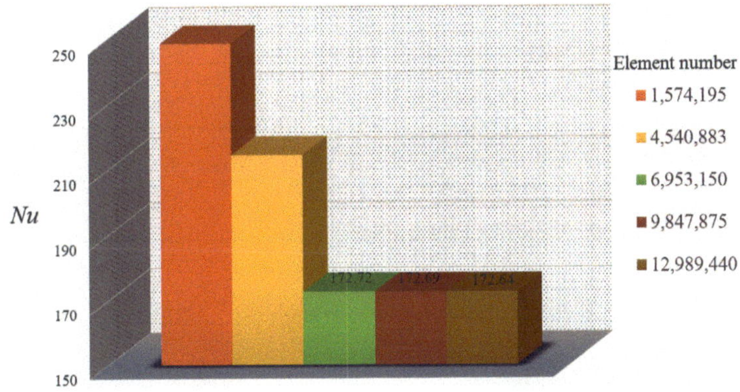

FIGURE 10.3

Mesh analysis for case with $D^* = 0.14$, $N = 7$, and Re $= 20{,}000$.

Figs. 10.4–10.6 were confirmed to investigate the best design with respect to variable parameters. Utilizing greater diameter ratio generates a rotating secondary vortex resulting in a thinner thermal or hydrodynamic boundary layer. Better efficiency has been reported for higher number of revolutions because of induces in turbulent intensity. Augmenting inlet velocity, the temperature near the wall is reduced because of stronger secondary flows. Intensification of turbulence intensity has been occurred for greater diameter of TT. Also, pressure loss augments with rise of all variables. More effective mixing occurs within the pipe with involving more revolution which is described by a stronger swirl flow. To reduce the thickness of thermal boundary layer and reaching greater turbulent intensity, turbulator with greater revolution number can be selected. Stronger turbulence intensity can be reported as a result of augmenting D^*. Figs. 10.7–10.9 are exhibited to capture the differences of various cases. Augmenting both geometric variables results in not only increment of convective flow but also augmentation of pressure drop. So, selecting best design needs to analyzing thermal performance. Selecting greater values of D^* and N are commendable for maximum thermal performance. Performance tends to decline with augment of inlet velocity. Greater turbulence intensity is found in the cases with greater N, which is the reason why Nusselt number has direct relationship with N. Influence of D^* on rate of heat transfer is greater than those of on pressure drop. Consequently, using greater D^* gives higher performance. Diameter ratio is more contributive than N to the benefit of efficiency. Only in lowest Re, we can find the greater η than unity. The maximum is about 1.08 which was reported for cases with Re $= 4000$, $N = 7$, and $D^* = 0.14$.

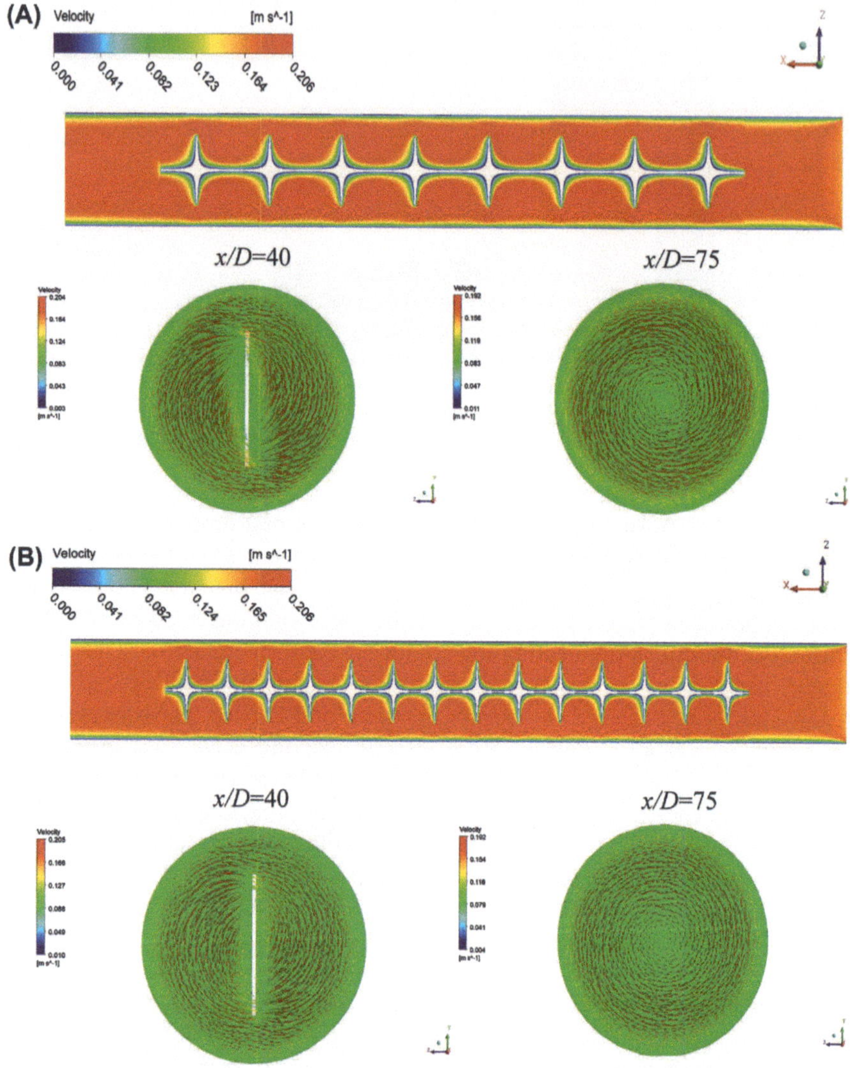

FIGURE 10.4

Contour and vector plot of velocity in different planes. (A) $N = 4$, $D^* = 0.1$, and Re = 4000; (B) $N = 7$, $D^* = 0.1$, and Re = 4000; (C) $N = 7$, $D^* = 0.1$, and Re = 20,000; (D) $N = 7$, $D^* = 0.14$, and Re = 4000.

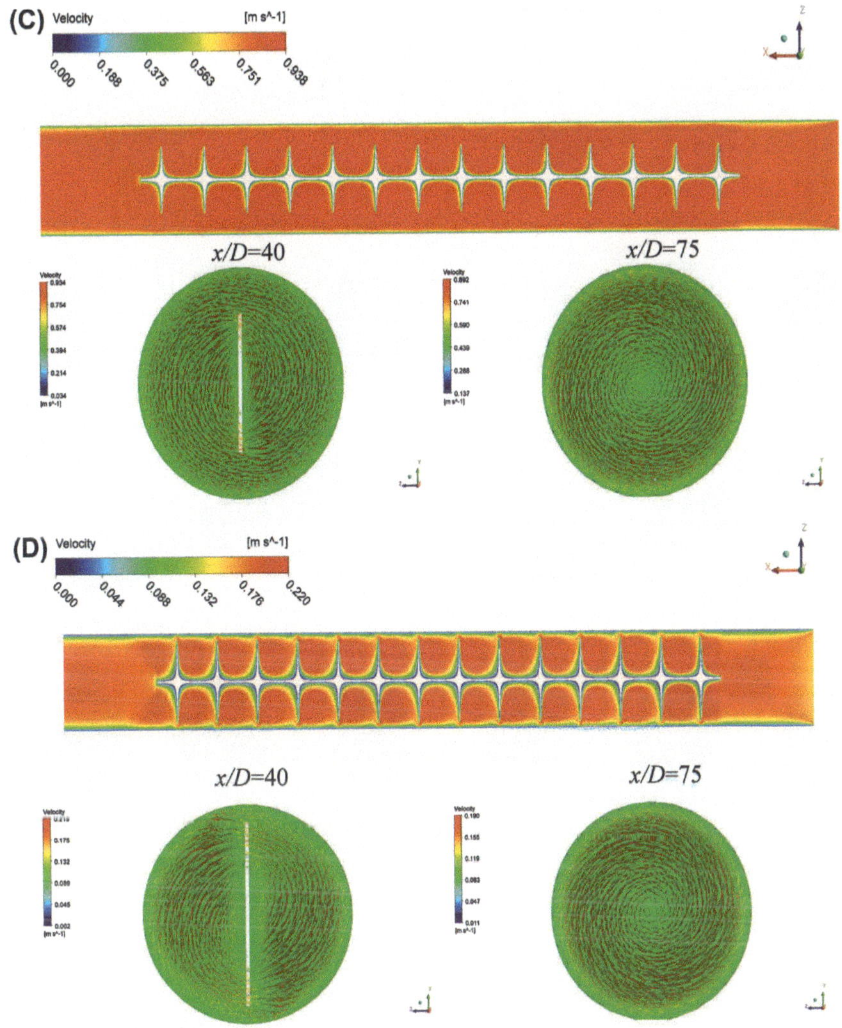

FIGURE 10.4 cont'd.

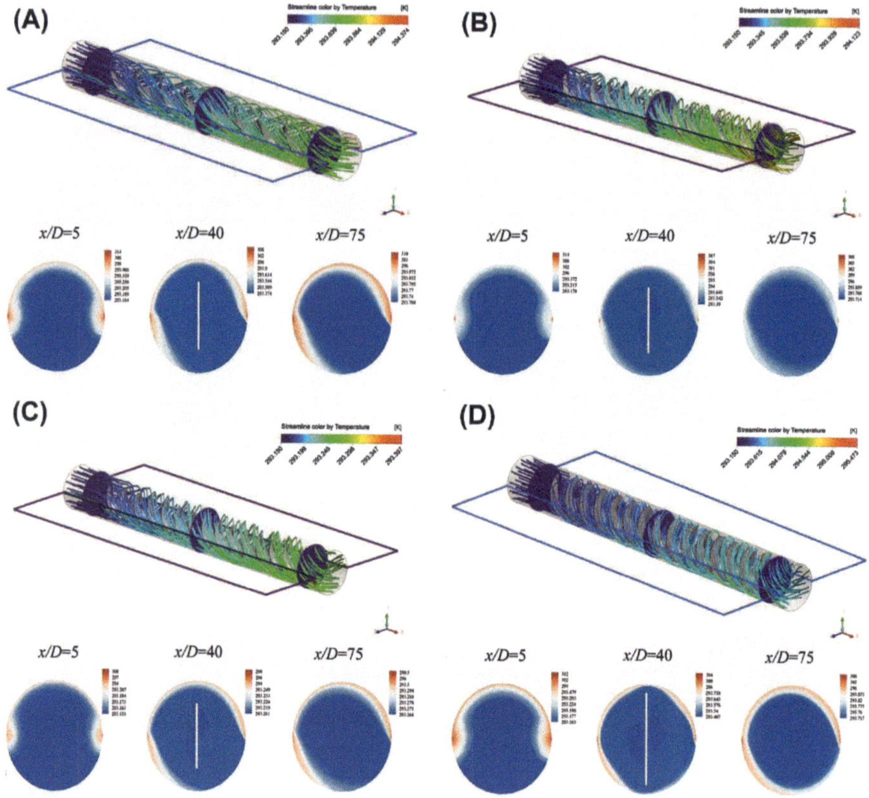

FIGURE 10.5

Contour plots of temperature and streamline. (A) $N = 4$, $D^* = 0.1$, and Re = 4000; (B) $N = 7$, $D^* = 0.1$, and Re = 4000; (C) $N = 7$, $D^* = 0.1$, and Re = 20,000; (D) $N = 7$, $D^* = 0.14$, and Re = 4000.

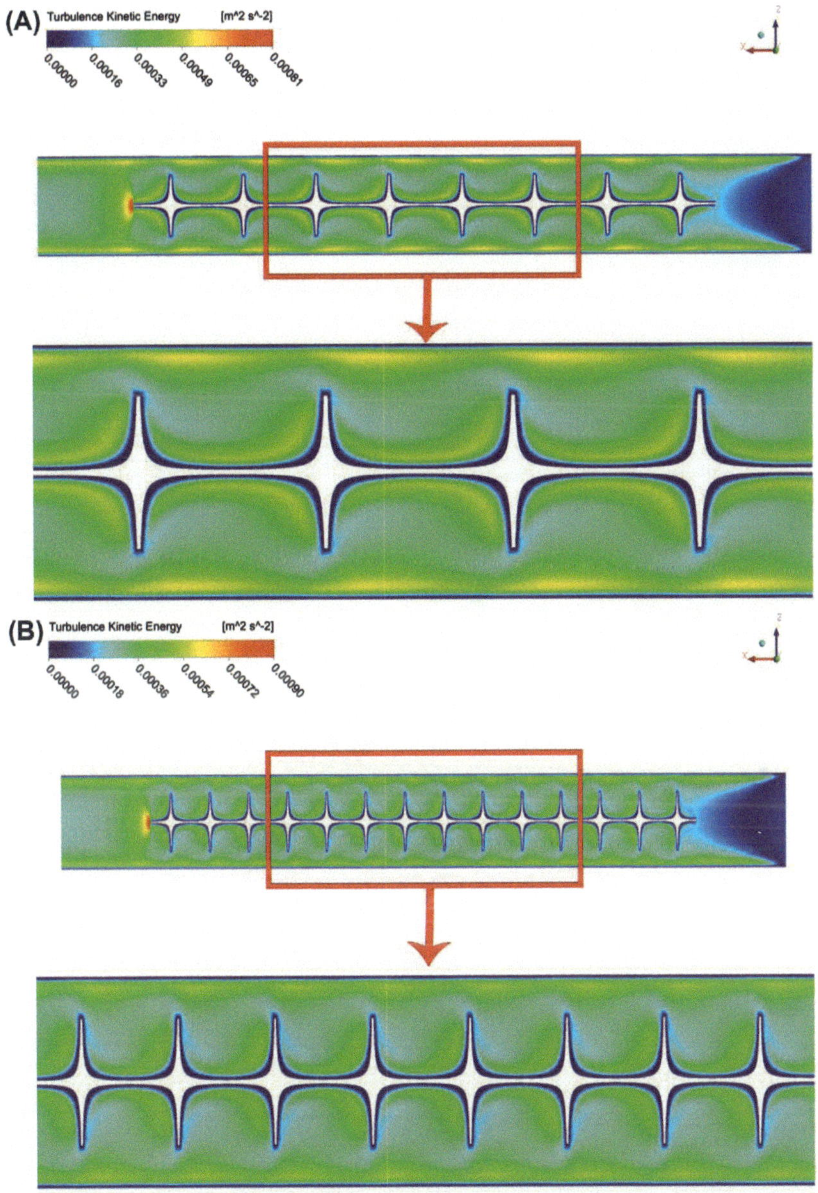

FIGURE 10.6

Contour plots of Turbulence kinetic energy. (A) $N = 4$, $D^* = 0.1$, and Re = 4000; (B) $N = 7$, $D^* = 0.1$, and Re = 4000; (C) $N = 7$, $D^* = 0.1$, and Re = 20,000; (D) $N = 7$, $D^* = 0.14$, and Re = 4000.

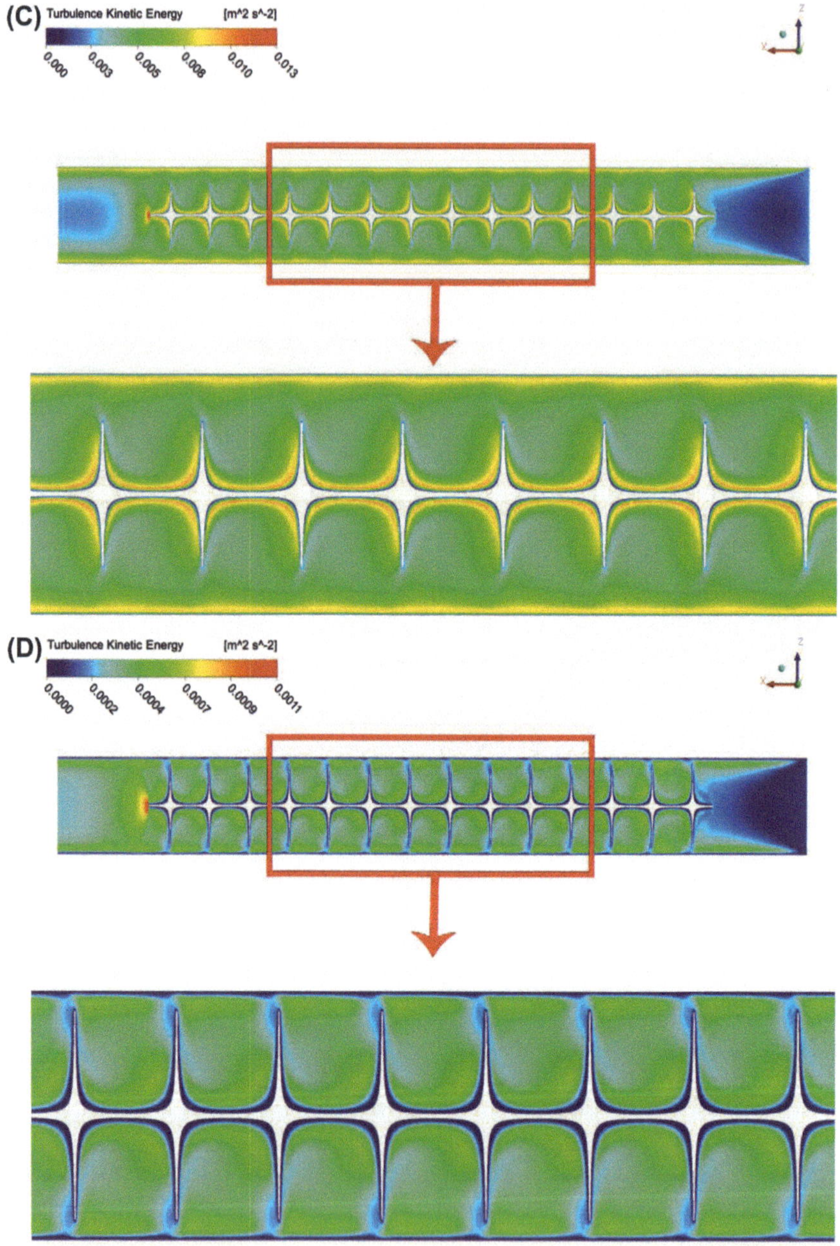

FIGURE 10.6 cont'd.

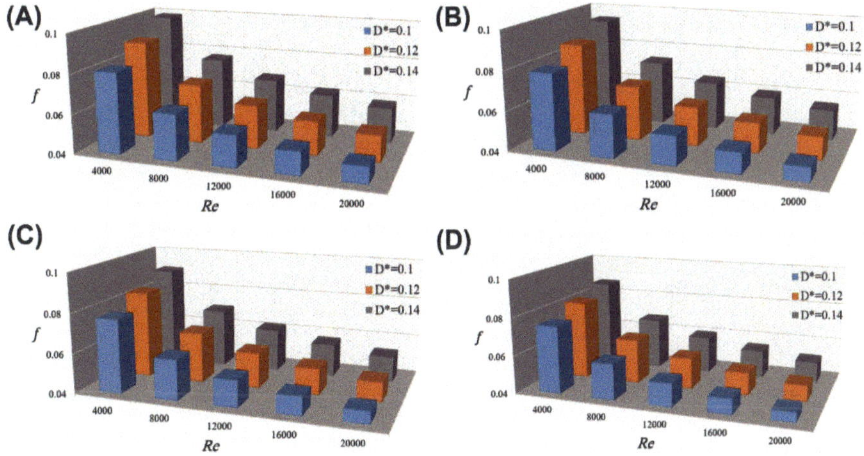

FIGURE 10.7

Influence of N and D^* on f. $N=7$ (A), $N=6$ (B), $N=5$ (C), $N=4$ (D).

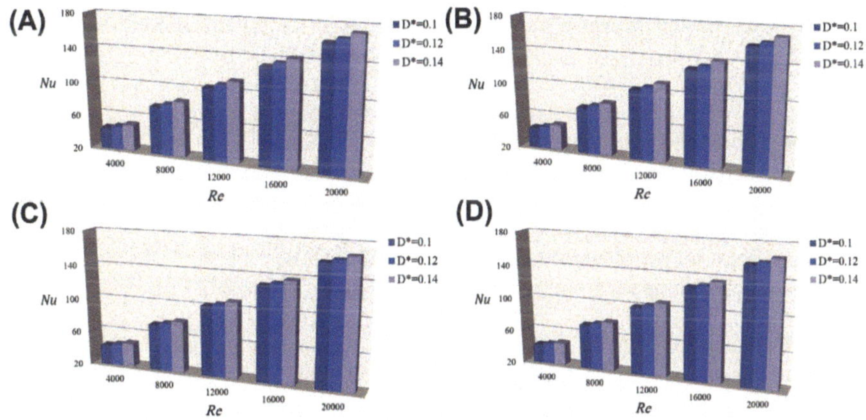

FIGURE 10.8

Influence of N and D^* on Nu. $N=7$ (A), $N=6$ (B), $N=5$ (C), $N=4$ (D).

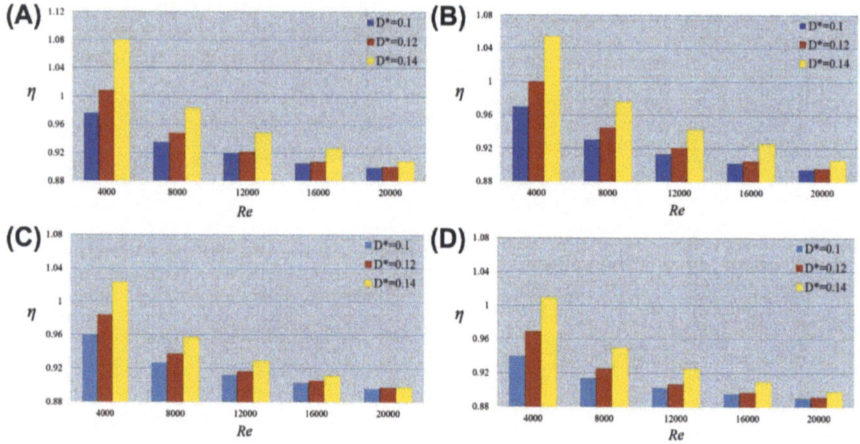

FIGURE 10.9

Influence of N and D^* on η. $N = 7$ (A), $N = 6$ (B), $N = 5$ (C), $N = 4$ (D).

References

[1] S. Ali Farshad, M. Sheikholeslami, S.H. Hosseini, A. Shafee, Z. Li, Nanofluid turbulent forced convection through a solar flat plate collector with Al_2O_3 nanoparticles, Microsystem Technologies (2019), https://doi.org/10.1007/s00542-019-04430-2.

[2] D. Kim, Y. Kwon, Y. Cho, C. Li, S. Cheong, Y. Hwang, J. Lee, D. Hong, S. Moona, Convective heat transfer characteristics of nanofluids under laminar and turbulent flow conditions, Current Applied Physics 9 (2009) 119–123.

[3] S. Ali Farshad, M. Sheikholeslami, FVM modeling of nanofluid forced convection through a solar unit involving MCTT, International Journal of Mechanical Sciences 159 (2019) 126–139.

[4] S. Ali Farshad, M. Sheikholeslami, Nanofluid flow inside a solar collector utilizing twisted tape considering exergy and entropy analysis, Renewable Energy 141 (2019) 246–258.

[5] S. Ali Farshad, M. Sheikholeslami, Simulation of nanoparticles second law treatment inside a solar collector considering turbulent flow, Physica A: Statistical Mechanics and Its Applications 525 (2019) 1–12.

[6] S. Ali Farshad, M. Sheikholeslami, Turbulent nanofluid flow through a solar collector influenced by multi-channel twisted tape considering entropy generation, The European Physical Journal Plus 134 (2019) 149, https://doi.org/10.1140/epjp/i2019-12606-2.

[7] S. Ali Farshad, M. Sheikholeslami, Simulation of exergy loss of nanomaterial through a solar heat exchanger with insertion of multi-channel twisted tape, Journal of Thermal Analysis and Calorimetry 138 (2019) 795–804.

[8] M. Sheikholeslami, T. Hayat, A. Alsaedi, S. Abelman, Numerical analysis of EHD nanofluid force convective heat transfer considering electric field dependent viscosity, International Journal of Heat and Mass Transfer 108 (2017) 2558–2565.

[9] L. Yang, W. Ji, J.N. Huang, G. Xu, An updated review on the influential parameters on thermal conductivity of nano-fluids, Journal of Molecular Liquids 296 (2019) 111780.

[10] M. Sheikholeslami, M. Sadoughi, Mesoscopic method for MHD nanofluid flow inside a porous cavity considering various shapes of nanoparticles, International Journal of Heat and Mass Transfer 113 (2017) 106−114.

[11] M. Sheikholeslami, M.M. Bhatti, Forced convection of nanofluid in presence of constant magnetic field considering shape effects of nanoparticles, International Journal of Heat and Mass Transfer 111 (2017) 1039−1049.

[12] M. Sheikholeslami, M.M. Bhatti, Active method for nanofluid heat transfer enhancement by means of EHD, International Journal of Heat and Mass Transfer 109 (2017) 115−122.

[13] M. Sheikholeslami, S.A. Shehzad, Thermal radiation of ferrofluid in existence of Lorentz forces considering variable viscosity, International Journal of Heat and Mass Transfer 109 (2017) 82−92.

[14] M. Sheikholeslami, H.B. Rokni, Nanofluid two phase model analysis in existence of induced magnetic field, International Journal of Heat and Mass Transfer 107 (2017) 288−299.

[15] M. Sheikholeslami, S.A. Shehzad, Magnetohydrodynamic nanofluid convection in a porous enclosure considering heat flux boundary condition, International Journal of Heat and Mass Transfer 106 (2017) 1261−1269.

[16] M. Sheikholeslami, R. Ellahi, Three dimensional mesoscopic simulation of magnetic field effect on natural convection of nanofluid, International Journal of Heat and Mass Transfer 89 (2015) 799−808.

[17] M. Sheikholeslami, Numerical investigation of nanofluid free convection under the influence of electric field in a porous enclosure, Journal of Molecular Liquids 249 (2018) 1212−1221.

[18] M. Sheikholeslami, CuO-water nanofluid flow due to magnetic field inside a porous media considering Brownian motion, Journal of Molecular Liquids 249 (2018) 921−929.

[19] M. Sheikholeslami, Numerical investigation for CuO-H_2O nanofluid flow in a porous channel with magnetic field using mesoscopic method, Journal of Molecular Liquids 249 (2018) 739−746.

[20] M. Sheikholeslami, H.B. Rokni, Numerical modeling of nanofluid natural convection in a semi annulus in existence of Lorentz force, Computer Methods in Applied Mechanics and Engineering 317 (2017) 419−430.

[21] M. Sheikholeslami, A. Zeeshan, Analysis of flow and heat transfer in water based nanofluid due to magnetic field in a porous enclosure with constant heat flux using CVFEM, Computer Methods in Applied Mechanics and Engineering 320 (2017) 68−81.

[22] L. Yang, W. Ji, Z. Zhang, X. Jin, Thermal conductivity enhancement of water by adding graphene Nano-sheets: consideration of particle loading and temperature effects, International Communications in Heat and Mass Transfer 109 (2019) 104353.

[23] L. Yang, J.N. Huang, W. Ji, M. Mao, Investigations of a new combined application of nanofluids in heat recovery and air purification, Powder Technology 360 (2020), https://doi.org/10.1016/j.powtec.2019.10.053.

[24] M. Sheikholeslami, H.B. Rokni, Numerical simulation for impact of Coulomb force on nanofluid heat transfer in a porous enclosure in presence of thermal radiation, International Journal of Heat and Mass Transfer 118 (2018) 823–831.

[25] M. Sheikholeslami, S.A. Shehzad, Numerical analysis of $Fe_3O_4-H_2O$ nanofluid flow in permeable media under the effect of external magnetic source, International Journal of Heat and Mass Transfer 118 (2018) 182–192.

[26] M. Sheikholeslami, M.K. Sadoughi, Simulation of CuO- water nanofluid heat transfer enhancement in presence of melting surface, International Journal of Heat and Mass Transfer 116 (2018) 909–919.

[27] M. Sheikholeslami, H.B. Rokni, Simulation of nanofluid heat transfer in presence of magnetic field: a review, International Journal of Heat and Mass Transfer 115 (2017) 1203–1233.

[28] M. Sheikholeslami, T. Hayat, A. Alsaedi, On simulation of nanofluid radiation and natural convection in an enclosure with elliptical cylinders, International Journal of Heat and Mass Transfer 115 (2017) 981–991.

[29] M. Sheikholeslami, S.A. Shehzad, CVFEM for influence of external magnetic source on $Fe_3O_4 - H_2O$ nanofluid behavior in a permeable cavity considering shape effect, International Journal of Heat and Mass Transfer 115 (2017) 180–191.

[30] M. Sheikholeslami, M. Seyednezhad, Nanofluid heat transfer in a permeable enclosure in presence of variable magnetic field by means of CVFEM, International Journal of Heat and Mass Transfer 114 (2017) 1169–1180.

[31] M. Sheikholeslami, H.B. Rokni, Melting heat transfer influence on nanofluid flow inside a cavity in existence of magnetic field, International Journal of Heat and Mass Transfer 114 (2017) 517–526.

[32] M. Sheikholeslami, M. Barzegar Gerdroodbary, R. Moradi, A. Shafee, Z. Li, Application of Neural Network for estimation of heat transfer treatment of Al_2O_3- H_2O nanofluid through a channel, Computer Methods in Applied Mechanics and Engineering 344 (2019) 1–12.

[33] M. Sheikholeslami, T. Hayat, A. Alsaedi, Numerical study for external magnetic source influence on water based nanofluid convective heat transfer, International Journal of Heat and Mass Transfer 106 (2017) 745–755.

[34] Y. Qin, J.E. Hiller, D. Meng, Linearity between pavement thermophysical properties and surface temperatures, Journal of Materials in Civil Engineering (2019), https://doi.org/10.1061/(ASCE)MT.1943-5533.0002890.

[35] Y. Qin, H. He, X. Ou, T. Bao, Experimental study on darkening water-rich mud tailings for accelerating desiccation, Journal of Cleaner Production (2019) 118235, https://doi.org/10.1016/j.jclepro.2019.118235.

[36] Y. Qin, Y. Zhao, X. Chen, L. Wang, F. Li, T. Bao, Moist curing increases the solar reflectance of concrete, Construction and Building Materials 215 (2019) 114–118.

[37] Y. Qin, Y. He, J.E. Hiller, G. Mei, A new water-retaining paver block for reducing runoff and cooling pavement, Journal of Cleaner Production 199 (2018) 948–956.

[38] M. Sheikholeslami, Effect of uniform suction on nanofluid flow and heat transfer over a cylinder, Journal of the Brazilian Society of Mechanical Sciences and Engineering 37 (2015) 1623–1633.

[39] M. Sheikholeslami, Lattice Boltzmann Method simulation of MHD non-Darcy nanofluid free convection, Physica B 516 (2017) 55–71.

[40] M. Sheikholeslami, Magnetic field influence on CuO -H_2O nanofluid convective flow in a permeable cavity considering various shapes for nanoparticles, International Journal of Hydrogen Energy 42 (2017) 19611−19621.

[41] M. Sheikholeslami, Numerical simulation for solidification in a LHTESS by means of nano-enhanced PCM, Journal of the Taiwan Institute of Chemical Engineers 86 (2018) 25−41.

[42] M. Sheikholeslami, Numerical modeling of Nano enhanced PCM solidification in an enclosure with metallic fin, Journal of Molecular Liquids 259 (2018) 424−438.

[43] M. Sheikholeslami, Solidification of NEPCM under the effect of magnetic field in a porous thermal energy storage enclosure using CuO nanoparticles, Journal of Molecular Liquids 263 (2018) 303−315.

[44] M. Sheikholeslami, Influence of magnetic field on Al_2O_3-H_2O nanofluid forced convection heat transfer in a porous lid driven cavity with hot sphere obstacle by means of LBM, Journal of Molecular Liquids 263 (2018) 472−488.

[45] M. Sheikholeslami, New computational approach for exergy and entropy analysis of nanofluid under the impact of Lorentz force through a porous media, Computer Methods in Applied Mechanics and Engineering 344 (2019) 319−333.

[46] M. Sheikholeslami, Numerical approach for MHD Al_2O_3-water nanofluid transportation inside a permeable medium using innovative computer method, Computer Methods in Applied Mechanics and Engineering 344 (2019) 306−318.

[47] X. Ma, M. Sheikholeslami, M. Jafaryar, A. Shafee, T. Nguyen-Thoi, Z. Li, Solidification inside a clean energy storage unit utilizing phase change material with copper oxide nanoparticles, Journal of Cleaner Production (2019), https://doi.org/10.1016/j.jclepro.2019.118888.

[48] M. Sheikholeslami, O. Mahian, Enhancement of PCM solidification using inorganic nanoparticles and an external magnetic field with application in energy storage systems, Journal of Cleaner Production 215 (2019) 963−977.

[49] M. Sheikholeslami, B. Rezaeianjouybari, M. Darzi, A. Shafee, Z. Li, T.K. Nguyen, Application of nano refrigerant for boiling heat transfer enhancement employing an experimental study, International Journal of Heat and Mass Transfer 141 (2019) 974−980.

[50] M. Sheikholeslami, M. Jafaryar, M. Hedayat, A. Shafee, Z. Li, T.K. Nguyen, M. Bakouri, Heat transfer and turbulent simulation of nanomaterial due to compound turbulator including irreversibility analysis, International Journal of Heat and Mass Transfer 137 (2019) 1290−1300.

[51] M. Sheikholeslami, Application of Darcy law for nanofluid flow in a porous cavity under the impact of Lorentz forces, Journal of Molecular Liquids 266 (2018) 495−503.

[52] M. Sheikholeslami, Finite element method for PCM solidification in existence of CuO nanoparticles, Journal of Molecular Liquids 265 (2018) 347−355.

[53] M. Sheikholeslami, M. Jafaryar, A. Shafee, Z. Li, R.-ul Haq, Heat transfer of nanoparticles employing innovative turbulator considering entropy generation, International Journal of Heat and Mass Transfer 136 (2019) 1233−1240.

[54] M. Sheikholeslami, R.-ul Haq, A. Shafee, Z. Li, Y.G. Elaraki, I. Tlili, Heat transfer simulation of heat storage unit with nanoparticles and fins through a heat exchanger, International Journal of Heat and Mass Transfer 135 (2019) 470−478.

[55] M. Sheikholeslami, R.-ul Haq, A. Shafee, Z. Li, Heat transfer behavior of Nanoparticle enhanced PCM solidification through an enclosure with V shaped fins, International Journal of Heat and Mass Transfer 130 (2019) 1322–1342.

[56] M. Sheikholeslami, S.A. Shehzad, Z. Li, A. Shafee, Numerical modeling for alumina nanofluid magnetohydrodynamic convective heat transfer in a permeable medium using Darcy law, International Journal of Heat and Mass Transfer 127 (2018) 614–622.

[57] M. Sheikholeslami, Z. Li, A. Shafee, Lorentz forces effect on NEPCM heat transfer during solidification in a porous energy storage system, International Journal of Heat and Mass Transfer 127 (2018) 665–674.

[58] M. Sheikholeslami, M. Jafaryar, S. Saleem, Z. Li, A. Shafee, Y. Jiang, Nanofluid heat transfer augmentation and exergy loss inside a pipe equipped with innovative turbulators, International Journal of Heat and Mass Transfer 126 (2018) 156–163.

[59] M. Sheikholeslami, A. Ghasemi, Z. Li, A. Shafee, S. Saleem, Influence of CuO nanoparticles on heat transfer behavior of PCM in solidification process considering radiative source term, International Journal of Heat and Mass Transfer 126 (2018) 1252–1264.

[60] M. Sheikholeslami, M. Darzi, Z. Li, Experimental investigation for entropy generation and exergy loss of nano-refrigerant condensation process, International Journal of Heat and Mass Transfer 125 (2018) 1087–1095.

[61] M. Sheikholeslami, S.A. Shehzad, Z. Li, Water based nanofluid free convection heat transfer in a three dimensional porous cavity with hot sphere obstacle in existence of Lorenz forces, International Journal of Heat and Mass Transfer 125 (2018) 375–386.

[62] M. Sheikholeslami, M. Jafaryar, Z. Li, Nanofluid turbulent convective flow in a circular duct with helical turbulators considering CuO nanoparticles, International Journal of Heat and Mass Transfer 124 (2018) 980–989.

[63] Y. Qin, J. Luo, Z. Chen, G. Mei, L.-E. Yan, Measuring the albedo of limited-extent targets without the aid of known-albedo masks, Solar Energy 171 (2018) 971–976.

[64] Y. Qin, A review on the development of cool pavements to mitigate urban heat island effect, Renewable and Sustainable Energy Reviews 52 (2015) 445–459.

[65] M. Sheikholeslami, A. Ghasemi, Solidification heat transfer of nanofluid in existence of thermal radiation by means of FEM, International Journal of Heat and Mass Transfer 123 (2018) 418–431.

[66] M. Sheikholeslami, S.A. Shehzad, CVFEM simulation for nanofluid migration in a porous medium using Darcy model, International Journal of Heat and Mass Transfer 122 (2018) 1264–1271.

[67] M. Sheikholeslami, M. Darzi, M.K. Sadoughi, Heat transfer improvement and pressure drop during condensation of refrigerant-based nanofluid; an experimental procedure, International Journal of Heat and Mass Transfer 122 (2018) 643–650.

[68] M. Sheikholeslami, S.A. Shehzad, Simulation of water based nanofluid convective flow inside a porous enclosure via Non-equilibrium model, International Journal of Heat and Mass Transfer 120 (2018) 1200–1212.

[69] M. Sheikholeslami, M. Seyednezhad, Simulation of nanofluid flow and natural convection in a porous media under the influence of electric field using CVFEM, International Journal of Heat and Mass Transfer 120 (2018) 772–781.

Index

'*Note:* Page numbers followed by "f" indicate figures and "t" indicate tables.'

I

K

L

M

N

T

U

V

W

Kolibri 360 GmbH
Ferdinand-Jühlke-Straße 7
99095 Erfurt, Deutschland
produktsicherheit@kolibri360.de